T4-AKH-272

ADVANCES IN GENETICS

VOLUME 30

Contributors to This Volume

Nora J. Besansky

Harald Biessmann

Michel Caboche

Vicki L. Chandler

Frank H. Collins

Victoria G. Finnerty

Edith Butler Gralla

Kristine J. Hardeman

Daniel J. Kosman

James M. Mason

Frédérique Pelsy

Virginia M. Peschke

Ronald L. Phillips

Richard E. Tashian

ADVANCES IN GENETICS

Edited by

JOHN G. SCANDALIOS

*Department of Genetics
North Carolina State University
Raleigh, North Carolina*

THEODORE R. F. WRIGHT

*Department of Biology
University of Virginia
Charlottesville, Virginia*

VOLUME 30

WITHDRAWN
FAIRFIELD UNIVERSITY
LIBRARY

ACADEMIC PRESS, INC.
Harcourt Brace Jovanovich, Publishers

San Diego   New York   Boston
London   Sydney   Tokyo   Toronto

This book is printed on acid-free paper. ∞

Copyright © 1992 by ACADEMIC PRESS, INC.
All Rights Reserved.
No part of this publication may be reproduced or transmitted in any form or by any means, electronic or mechanical, including photocopy, recording, or any information storage and retrieval system, without permission in writing from the publisher.

Academic Press, Inc.
1250 Sixth Avenue, San Diego, California 92101-4311

United Kingdom Edition published by
Academic Press Limited
24–28 Oval Road, London NW1 7DX

Library of Congress Catalog Number: 47-30313

International Standard Book Number: 0-12-017630-0

PRINTED IN THE UNITED STATES OF AMERICA
92 93 94 95 96 97   BC   9 8 7 6 5 4 3 2 1

CONTENTS

Molecular Genetics of Nitrate Reductase in Higher Plants

FRÉDÉRIQUE PELSY AND MICHEL CABOCHE

Genetic Implications of Somaclonal Variation in Plants

VIRGINIA M. PESCHKE AND RONALD L. PHILLIPS

268503

The *Mu* Elements of *Zea mays*

VICKI L. CHANDLER AND KRISTINE J. HARDEMAN

Molecular Perspectives on the Genetics of Mosquitoes

NORA J. BESANSKY, VICTORIA G. FINNERTY, AND FRANK H. COLLINS

Genetics and Molecular Biology of Telomeres

HARALD BIESSMANN AND JAMES M. MASON

Molecular Genetics of Superoxide Dismutases in Yeasts and Related Fungi

EDITH BUTLER GRALLA AND DANIEL J. KOSMAN

Genetics of the Mammalian Carbonic Anhydrases

RICHARD E. TASHIAN

CONTRIBUTORS TO VOLUME 30

Numbers in parentheses indicate the pages on which the authors' contributions begin.

NORA J. BESANSKY (123), *Malaria Branch, Division of Parasitic Diseases, Center for Infectious Diseases, Centers for Disease Control, Atlanta, Georgia 30333, and Department of Biology, Emory University, Atlanta, Georgia 30322*

HARALD BIESSMANN (185), *Developmental Biology Center, University of California, Irvine, Irvine, California 92717*

MICHEL CABOCHE (1), *Laboratoire de Biologie Cellulaire, INRA-Versailles, 78026 Versailles, France*

VICKI L. CHANDLER (77), *Institute of Molecular Biology, University of Oregon, Eugene, Oregon 97403*

FRANK H. COLLINS (123), *Malaria Branch, Division of Parasitic Diseases, Center for Infectious Diseases, Centers for Disease Control, Atlanta, Georgia 30333, and Department of Biology, Emory University, Atlanta, Georgia 30322*

VICTORIA G. FINNERTY (123), *Department of Biology, Emory University, Atlanta, Georgia 30322*

EDITH BUTLER GRALLA (251), *Department of Chemistry and Biochemistry, University of California, Los Angeles, Los Angeles, California 90024*

KRISTINE J. HARDEMAN (77), *Institute of Molecular Biology, University of Oregon, Eugene, Oregon 97403*

DANIEL J. KOSMAN (251), *Department of Biochemistry, The State University of New York at Buffalo, Buffalo, New York 14214*

JAMES M. MASON (185), *Experimental Carcinogenesis and Mutagenesis Branch, National Institute of Environmental Health Sciences, Research Triangle Park, North Carolina 27709*

FRÉDÉRIQUE PELSY (1), *Laboratoire de Biologie Cellulaire, INRA-Versailles, 78026 Versailles, France*

VIRGINIA M. PESCHKE (41), *Department of Biology, Washington University, St. Louis, Missouri 63130*

RONALD L. PHILLIPS (41), *Department of Agronomy and Plant Genetics and the Plant Molecular Genetics Institute, University of Minnesota, St. Paul, Minnesota 55108*

RICHARD E. TASHIAN (321), *Department of Human Genetics, University of Michigan Medical School, Ann Arbor, Michigan 48109*

PREFACE

Advances in Genetics was the first serial publication devoted solely to the burgeoning field of genetics. The serial was founded in 1946 by Dr. Milislav Demerec, then director of the Genetics Department at the Carnegie Institution of Washington in Cold Spring Harbor, New York. The stated purpose for the series was "that critical summaries of outstanding genetic problems, written by prominent geneticists in such form that they will be useful as reference material for geneticists and also as a source of information to nongeneticists, may appear in a single publication." Over the years, the goals set forth initially have been more than fulfilled, and a lasting tradition of excellence has been established.

In more recent years, our field has experienced some revolutionary developments emanating from the enormous technological advances that have occurred. Recombinant DNA and related molecular technologies now make possible the intricate manipulation of genetic information, in virtually every cell type and organism, that could not even have been imagined at the time when *Advances in Genetics* was initiated. These developments have led to an unparalleled information explosion.

Because of the diversity of genetics as a science, *Advances in Genetics* has adhered to the policy of publishing a series of outstanding but largely unrelated articles in each volume, and this policy will be maintained. However, the editors on occasion will depart from this format and review a central topic in a special "topical" or "thematic" volume, as this is essential in view of the extremely rapid developments in genetics. Four such volumes (Volumes 22, 24, 27, and 28) have been published to date and have been well received by the scientific community; others are in preparation.

Our purpose is not merely to inform but also to stimulate the reader— whether a beginning or an advanced scholar—to explore, question, and, whenever possible, test various hypotheses advanced herein. We intend that each volume covers some material of lasting value, in view of the very rapid developments in this field.

JOHN G. SCANDALIOS
THEODORE R. F. WRIGHT

MOLECULAR GENETICS OF NITRATE REDUCTASE
IN HIGHER PLANTS

Frédérique Pelsy and Michel Caboche

Laboratoire de Biologie Cellulaire,
INRA-Versailles 78026, Versailles, France

I. Introduction

Nitrogen is involved in the building of molecules essential for the viability of living organisms. During the last century it was found that nitrogen fertilizers increase biomass production dramatically; nitrogen availability is therefore considered as a limiting factor for

1

ADVANCES IN GENETICS, Vol. 30

Copyright © 1992 by Academic Press, Inc.
All rights of reproduction in any form reserved.

most plant crops. Various types of nitrogen-containing molecules are found in the environment. Microorganisms break down the nitrogen-containing molecules from soil organic matter to produce ammonium ions, which in turn can be oxidized to nitrate. Nitrate may be converted by soil bacteria to gaseous forms, such as dinitrogen gas, which is released to the atmosphere. Nitrate is the main nitrogen source assimilated by plants, but ammonium or other nitrogen-containing organic molecules can also be used. Plants have also developed alternative mechanisms of nitrogen supply, such as fixation of atmospheric nitrogen in symbiosis with prokaryotes or associations with mycorrhizal fungi.

Nitrate uptake by plant roots is the first step of nitrate assimilation: treatments that interfere with metabolic processes (e.g., metabolic inhibitors, low temperatures, anaerobic conditions, and darkness) cause substantial reduction of net nitrate uptake by roots (Clarkson, 1986; Lycklama, 1963). The absorption of nitrate by plant roots is commonly associated with alkalinization of external media, but it is not clear whether this is due to a nitrate/cation symport (Ullrich-Eberius *et al.*, 1981) or a nitrate/anion antiport (Thibaud and Grignon, 1981). Apart from being a nitrogen source, nitrate may also play a role as an osmoticum when accumulated in plant vacuoles.

In higher plants nitrate is reduced in two steps by nitrate reductase and nitrite reductase, which successively catalyze the reduction of nitrate to nitrite and then to the ammonium ion in an eight-electron reduction process. Incorporation of ammonium into amino acids is carried out by the glutamine synthetase–glutamate synthase pathway (Suzuki and Gadal, 1984; Oaks and Hirel, 1985).

Plant physiologists have been interested for many years in the biochemical, physiological, and genetic characterization of nitrate reductase because of its pivotal role in nitrate assimilation, and different reviews have been published on the topic (Wray, 1986, 1988; Kleinhofs *et al.*, 1989; Solomonson and Barber, 1990; Caboche and Rouzé, 1990). Also, a conference on nitrate assimilation was held (Crawford and Campbell, 1990; Becker and Caboche, 1990).

In this article, we will consider the biochemical characteristics of the nitrate reductase of higher plants and then focus on the structure and regulation of the apoenzyme genes, including recent data, such as the identification of the loci involved in nitrate reduction in *Arabidopsis thaliana* and *Nicotiana plumbaginifolia*. The larger part of the review will deal with genetic aspects of nitrate reduction by considering plant mutants deficient for this activity.

II. Function of Nitrate Reduction in Living Organisms

Nitrate reductases from fungi, algae, or higher plants are typical examples of multicenter redox enzymes that catalyze the two-electron reduction of nitrate to nitrite using pyridine dinucleotides as electron donors. Much progress has been made in the characterization of the nitrate assimilation pathway, primarily in fungi, wherein it is genetically characterized perhaps better than in any eukaryotic organism (Cove, 1979). In higher plants nitrate reductase has been widely studied because it was considered to be the limiting step of the nitrate assimilation pathway.

Unlike in eukaryotic organisms, bacterial nitrate reduction can serve two main purposes. Nitrate may be reduced to ammonia and used as a nitrogen source for the synthesis of cellular components. Alternatively, nitrate can be reduced to nitrite, ammonia, nitrous oxide, or dinitrogen gas for anaerobic respiration, which generates energy for cell growth. The derived metabolites are then generally released to the environment of the anaerobically growing cells. The enzymes that catalyze the initial step of these respective pathways have been classified historically as assimilatory and dissimilatory nitrate reductases (Cole, 1989).

In general, assimilatory nitrate reductases are expressed under both aerobic and anaerobic conditions and are regulated by N metabolites, whereas the dissimilatory nitrate reductases are produced only under anaerobic conditions. The dissimilatory enzymes are repressed by oxygen and are induced during anaerobic growth by their respective oxidized substrates. Transcription of their structural genes is tightly regulated by a transcriptional activator, the FNR protein (Cole, 1988). In organisms in which both pathways exist, as in *Salmonella typhymurium*, the two types of nitrate reductase are encoded by separate genes that are regulated differentially by the above physiological parameters (Barrett and Riggs, 1982).

The diversity and heterogeneity of bacterial nitrate reductases, which can be found in all cellular compartments, reflect the wide diversity of the physiological role for reduction. That is, free-living bacteria and the bacteroids of *Rhizobium japonicum* contain both membrane-bound and soluble dissimilatory activity (Kennedy *et al.*, 1975), but this activity is membrane bound only in *Escherichia coli* (Boxer, 1989). *Rhodopseudomonas capsulata* respiratory nitrate reductase is located in the periplasmic space, although the assimilatory reducing pathway is located on the cytoplasmic side of the bacterial membrane (McEwan *et al.*, 1984).

III. Biochemistry of Nitrate Reductases

A. The Apoenzyme

Nitrate reductases isolated from eukaryotic sources have been shown to be homomultimeric proteins, each subunit containing three prosthetic groups—flavin, heme, and a molybdenum cofactor (MoCo)—housed in different domains (for review, see Solomonson and Barber, 1990). They catalyze the reduction of nitrate to nitrite using NAD(P)H as electron donor.

Nitrate reductase protein structure has been widely studied in *Chlorella*, an organism that contains a homotetrameric enzyme that is stable and easy to isolate. Radiation inactivation analysis, a technique used to determine the minimal functional size of enzymes, gave a target size of 100 kDa for *Chlorella* NADH:nitrate reductase activity, which corresponds to the subunit size and confirms that each subunit contains a full set of prosthetic groups (Solomonson and McCreery, 1986).

In higher plant species, nitrate reductase is a nuclear-encoded, cytosolic enzyme that is present in small amounts and is highly labile, and consequently difficult to purify. Through successful purification it has been established that nitrate reductase is a homodimer with a subunit molecular mass of 100–120 kDa, depending on the species, each subunit containing one molecule of each prosthetic group (Fig. 1).

Immunoaffinity-purified corn and squash nitrate reductase were shown to have an interchain disulfide bond as well as a reactive thiol group. Because the interchain disulfide bond can be reduced without loss of nitrate reductase activity, it probably stabilizes the dimeric structure of the enzyme without any direct role in its catalytic functions (Hyde *et al.*, 1989).

B. The Cofactors

1. NADH or NADPH as Electron Donor

There are three closely related forms of nitrate reductases (NRs)—NADH:NR (EC 1.6.6.1), NAD(P)H:NR (EC 1.6.6.2), and NADPH:NR (EC 1.6.6.3). NADH:NR is the most common form in algae and higher plants; some species also contain NAD(P)H:NR, whereas NADPH:NR occurs in fungi. In *Nicotiana tabacum* and *N. plumbaginifolia*, only NADH NR isoforms have been isolated. In barley, NADH and NAD(P)H NRs have been isolated. The NADH NR enzyme predominates in leaf tissues under most environmental conditions. The NAD(P)H-specific

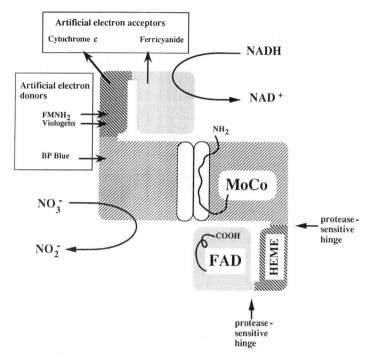

F𝐼𝐺. 1. Hypothetical model for higher plant nitrate reductase. Electron donor and acceptor sites for NADH:NR and partial enzymatic activities are shown on one subunit (left). The main structural features are shown on the other subunit (right): the three catalytic domains containing their respective prosthetic groups, the amino-terminal domain (possibly involved in the interaction between subunits) and the amino and carboxy termini of the polypeptide chain.

nitrate reductase is, on the other hand, detected in all plant tissues, in particular in the root system. The situation is different in rice, in which both the NADH and NAD(P)H NRs occur in the leaves of wild-type seedlings (Shen *et al.*, 1976). In nitrogen-fixing plants, peculiar situations have been described: a so-called constitutive nitrate reductase (expressed in the absence of nitrate) was identified in several leguminous species (Andrews *et al.*, 1990) and in alder (Benamar *et al.*, 1989). In soybean, three isoforms have been isolated: one inducible NADH:NR and two constitutive nitrate reductases, one NADH dependent, and one bispecific (Streit *et al.*, 1987).

2. The Prosthetic Groups

Three prosthetic groups are involved in the catalytic functions of nitrate reductases: FAD, heme, and MoCo. Garrett and Nason (1967) have shown that visible spectra of purified nitrate reductases are indicative of a cytochrome and that they contain an Fe-heme. The heme is housed in a cytochrome *b*-like section of the apoenzyme, and Campbell (1988) assumes that it enables nitrate reductase to reduce cytochrome *c* as natural function.

The MoCo is a complex of the metal Mo with an organic moiety, molybdopterin, which is a common cofactor of Mo-containing enzymes except nitrogenase. The structure of the MoCo has recently been elucidated (Kramer *et al.*, 1987) and appears to be slightly different in prokaryotes and eukaryotes (Johnson *et al.*, 1990).

Thus nitrate reductase functions as an electron transport system with internal electron transfer from the reduced FAD to the cytochrome *b*, then to the molybdopterin.

3. *The Artificial Electron Donors and Acceptors*

Apart from the reduction of nitrate by NADH, nitrate reductase also catalyzes a variety of partial activities that use alternative artificial electron donors or acceptors involving one or more of the prosthetic groups. These partial activities have been divided into two classes, referred to as either "diaphorase" or "nitrate-reducing" activities. Diaphorase activities use NAD(P)H as the electron donor and include NADH:ferricyanide reductase (NADH:FR) and NADH:cytochrome *c* reductase (NADH:CR) activities. Nitrate-reducing activities include reduced flavin:nitrate reductase (FH:NR), reduced methyl viologen: nitrate reductase (MV:NR), and reduced bromophenol blue:nitrate reductase (BB:NR) activities. Radiation inactivation analyses have provided apparent target sizes for NADH:CR, NADH:FR, and MV:NR that were in each case different and significantly smaller than the minimum polypeptide molecular mass, suggesting that these activities reside on functionally independent domains (Solomonson and McCreery, 1986). Limited proteolysis studies have established that NADH:FR activity requires FAD whereas NADH:CR activity requires both FAD and heme prosthetic groups (Solomonson and McCreery, 1986; Kubo *et al.*, 1988, Notton *et al.*, 1989). Reduced flavin:NR and MV:NR activities require heme and MoCo whereas BB:NR activity requires only MoCo (Chérel *et al.*, 1990; Meyer *et al.*, 1987).

C. Structure of Nitrate Reductase Subunits Deduced from cDNA Analysis

The availability of antibodies prepared against nitrate reductase has enabled the cloning of nitrate reductase cDNAs from several plant species—barley (Cheng *et al.*, 1986), squash (Crawford *et al.*, 1986), and tobacco (Calza *et al.*, 1987)—by screening cDNA expression libraries. Because nitrate reductase antigenic sites are conserved in plants (Chérel *et al.*, 1986), it was expected that nitrate reductase protein sequences would also be conserved.

The sequence data derived from the analyses of these different nitrate reductase cDNAs were compared with other protein sequences and suggested organization in three functional domains of this enzyme. Three catalytic domains are well defined in the primary sequence of these NRs by homology with other redox proteins: cytochrome b_5 reductase, cytochrome b_5, and sulfite oxidase. Being homologous to cytochrome b_5 reductase (Calza *et al.*, 1987), the FAD-containing C-terminal domain of nitrate reductase is confirmed to transfer electrons from NADH to the heme domain, just as the cytochrome b_5 reductase does to cytochrome b_5. Surprisingly, the MoCo-binding domain has no clear homology with other molybdoproteins, even with the *E. coli* nitrate reductase amino acid sequence (Kinghorn and Campbell, 1989), except with sulfite oxidase (Crawford *et al.*, 1988); it includes the N-terminal moiety of the nitrate reductase sequence, except for the first 80 amino acids. The absence of homology of nitrate reductase with xanthine oxidase, another molybdenum cofactor-containing enzyme, might be explained by a different catalytic function of the cofactor in the two enzymes, i.e., oxidation of the substrate in one enzyme and reduction in the other. The comparison of plant nitrate reductase coding sequences also reveals that catalytic domains are separated by sequences that appear to diverge rapidly through evolution (Daniel-Vedele *et al.*, 1989). Limited proteolysis experiments (Kubo *et al.*, 1988) combined with protein sequencing (Nakagawa *et al.*, 1990) have shown that these hinge regions between the different domains of nitrate reductase are preferential sites for proteolytic cleavage.

D. ANALYSIS OF STRUCTURE–ACTIVITY RELATIONSHIPS IN PLANT NITRATE REDUCTASES BY EXPRESSION IN HETEROLOGOUS HOSTS

Site-directed mutagenesis can be used to study the function of the different catalytic domains, the binding of the prosthetic groups, or to devise nitrate reductase proteins that are less susceptible to proteolytic cleavage. Such an approach would be simplified if expression of higher plant nitrate reductases could be obtained by the use of appropriate expression vectors in heterologous bacterial or eukaryotic cells. However, up to now, no fully functional enzyme has been expressed from cDNA expression vectors in heterologous systems. In bacterial systems, Hyde and Campbell (1990) have been able to express the C-terminal part of nitrate reductase involved in the ferricyanide reductase activity of the enzyme. Expression of the full-length cDNA of tobacco nitrate reductase was detected in *Saccharomyces cerevisiae* cells and led to the production of a polypeptide of the expected size able to organize in dimeric structure. This production was correlated with the detection of cytochrome *c* reductase activity, involving the functionality of the heme and FAD domains of the enzyme. Neither a full NADH NR activity nor any of the partial activities involving a functional MoCo domain could be detected in cellular extracts (Truong *et al.*, 1991). This can be explained by observing that in yeast no molybdenum cofactor is produced (Rajagopalan, personal communication).

E. INTRACELLULAR LOCALIZATION OF NITRATE REDUCTASE

It is now well established that nitrate reduction can occur in both roots and shoots of higher plants (Beevers and Hageman, 1983), but it is proposed that a greater proportion of nitrate assimilation occurs in the shoot when the rate of nitrate uptake increases without accompanying induction of nitrate reductase level in the roots. Thus a major part of the nitrate taken up is translocated by xylem sap and reduced in the leaves (Andrews, 1986).

Surprisingly, intracellular localization of higher plant nitrate reductase is not yet firmly established, although different features converge to assume that it is a cytoplasmic enzyme but located very close to the chloroplast: (1) the majority of the plant nitrate reductase activity has been found in the soluble fraction; (2) the predominantly present form of the physiological electron donor is NADH, mainly present in the cytosol, but is not NADPH, which is mainly in the chloroplast; (3) the produced nitrite, which is toxic, necessitates an efficient transfer into the chloroplast to nitrite reductase, which is clearly a chloroplastic enzyme. Other evidence suggests an apparent coupling or close association between nitrate uptake and nitrate reduction (Ingemarsson, 1987). Recent results by Tischner *et al.* (1989) suggest that a plasma membrane-bound form of nitrate reductase is involved in nitrate uptake in *Chlorella*, although a normal uptake of nitrate is measured in nitrate-deficient mutants (Warner and Huffaker, 1989).

The availability of monospecific antibodies has not enabled the final localization of the nitrate reductase by immunoelectron microscopy—this technique yielded conflicting results, with reports on a cytosolic location (Vaughn and Campbell, 1988), on a chloroplastic location (Kamachi *et al.*, 1987), or both locations (Roldan *et al.*, 1987).

Moreover, Ekes (1981) localized ferricyanide reduction in the chloroplastic envelope and suggested that nitrate reductase was present in this compartment.

IV. Structure and Expression of Nitrate Reductase Genes

A. NITRATE REDUCTASE STRUCTURAL *Nia* GENES

Nia genes have been cloned and sequenced in tobacco (Vaucheret *et al.*, 1989), tomato (Daniel-Vedele *et al.*, 1989), rice (Choi *et al.*, 1989), bean (Hoff *et al.*, 1991), *A. thaliana* (Cheng *et al.*, 1988), and the green algae *Chlamydomonas reinhardtii* (Fernandez *et al.*, 1989). Southern hybridization studies with nitrate reductase cDNA probes have been performed to study the number of nitrate reductase structural genes. In solanaceous species, one *Nia* sequence was detected per haploid genome, as expected from genetic data (Müller, 1983). In *A. thaliana*, two distinct sequences have been identified as *Nia* loci; the genetic analysis of nitrate reductase expression is in agreement with these molecular data. It would seem that until now no *Nia* pseudogenes have been observed in the genome of dicotyledonous plants. A more complex situation exists in rice and barley, in which reconstruction experiments and dot–blot analyses suggest the presence of one to three copies per haploid genome for the major NADH NR gene of barley, and six to eight copies per haploid genome in rice, depending on the subspecies (Kleinhofs *et al.*, 1988).

A comparison of *Nia* gene sequences shows that their overall structure is well conserved among plants, suggesting that one structural model can describe every eukaryotic assimilatory nitrate reductase. The general organization shows a coding sequence interrupted by three introns (two in *A. thaliana*) of variable size but located at strictly conserved positions (Daniel-Vedele *et al.*, 1989; Hoff *et al.*, 1991). The sequence analysis of the *C. reinhardtii Nit1* gene, however, indicates that the intron positions differ from those found in higher plants (Fernandez *et al.*, 1989).

B. EFFECTORS AFFECTING EXPRESSION OF *Nia* GENES

Nitrate induction of nitrate assimilatory enzymes was probably the first "substrate induction" phenomenon recognized in higher plants. Nitrate reductase and nitrite reductase have been found to be regulated by a number of environmental factors (Campbell, 1988), such as nitrate, nitrogen metabolites, light, and circadian rhythms. Plant hormones have also been shown to influence the level of nitrate reductase in plant tissues.

1. Induction by Nitrate

The induction of nitrate reductase activity by nitrate was recognized over 30 years ago and was finally demonstrated to occur in barley and maize leaves when antibodies specific for nitrate reductase were used to detect *de novo* synthesis of the nitrate reductase protein simultaneously with a measure of nitrate reductase activity (Somers *et al.*, 1983; Remmler and Campbell, 1986). More recently, the molecular cloning of *Nia* cDNAs has provided molecular probes to study the influence of nitrate on *Nia* gene expression in conjunction with the analysis of nitrate reductase protein and activity. *Nia* mRNA accumulates in plants grown on nitrate (Cheng *et al.*, 1986; Crawford *et al.*, 1988). Galangau *et al.* (1988) have shown that nitrogen-starved tobacco plants rapidly accumulated *Nia* mRNA in their leaves during the minutes following nitrate replenishment whereas nitrate reductase protein accumulation proceeds more slowly after a lag of several hours. In barley, as with maize, the increases in nitrate reductase mRNA level induced by nitrate were due, at least in part, to an increased transcription of the gene (Meltzer *et al.*, 1989; Lu *et al.*, 1990). When reporter sequences were transcriptionally fused to a 3-kb *Nia* promoter fragment and introduced back into the nuclear genome of *N. plumbaginifolia*, the inducibility of transcription of these sequences by nitrate was observed (Marion-Poll and Vaucheret, personal communication). The molecular mechanism underlying the nitrate induction is still unknown, although it is assumed that the nitrate ion is not likely to interact directly with the nitrate reductase promoter, but rather a putative regulatory protein could mediate this process.

2. Light Regulation

Although nitrate reductase activity can be induced by nitrate in darkness, the levels of activity are generally low compared to those induced in light (Duke and Duke, 1984; Oaks *et al.*, 1988). Cytokinins and light have a synergistic effect on nitrate reductase mRNA induction in etiolated barley leaves (Lu *et al.*, 1990). A plastic derived factor has been postulated to be involved in the regulation of nitrate reductase expression in mustard cotyledons (Schuster *et al.*, 1989). Preliminary experiments performed on squash have suggested that this light regulation of *Nia* mRNA level is under phytochrome control (Rajasekhar *et al.*, 1988). In tobacco leaves, the kinetics of light-induced accumulation of mRNA coding for nitrate reductase protein or for the well-studied light-regulated small subunit of ribulose1,5-bisphosphate carboxylase are very similar (Deng et al., 1990).

In addition, diurnal variations in nitrate reductase activity have been found in various species (Lillo and Henriksen, 1984). This rhythmic variation has been studied at the nitrate reductase mRNA level in tobacco leaves of nitrate-fed tobacco plants, over a 24-hour light–dark cycle. In these conditions, the mRNA level increased throughout the night with a maximum at the end of the night period, then it decreased to a minimum at the end of the day period. This circadian rhythm remains for 32 hours at the mRNA level when plants are transferred to continuous light or dark conditions. A rhythmic variation of nitrate reductase protein level and activity remains in continuous light but disappears in darkness (Deng et al., 1990).

3. Regulation by Nitrogen Metabolites

If the regulation of nitrate reductase by nitrogen metabolites is well documented in fungi, the situation is still unclear for higher plants. A number of physiological studies have suggested that plant nitrate reductase is not regulated by nitrogen metabolites. For instance, ammonium does not significantly affect the accumulation of leaf *Nia* transcripts in *A. thaliana* plantlets grown on nitrate (Crawford et al., 1988).

On the contrary, other evidence would suggest that nitrogen metabolites can negatively regulate nitrate reductase expression inplant tissues, depending on the ratio of nitrate/nitrogen. Martino and Smarrelli (1989) have reported that an exogenous glutamine supply reduced significantly the level of nitrate reductase transcript in squash cotyledons. In tobacco plants grown on nitrate, *Nia* transcript accumulation in roots was strongly reduced by ammonium succinate or glutamine addition to the nutrient solution (Deng et al., 1991). These results suggest that glutamine and/or other nitrogen metabolites may exert a negative control on nitrate reductase expression.

This aspect of regulation needs further characterization in order to distinguish between these two conflicting hypotheses, i.e., differences that could be explained by differences in translocation to the leaves or the different intracellular compartmentation of nitrogen metabolites.

C. COUPLED REGULATION OF NITRATE AND NITRITE REDUCTASES

The nitrate and nitrite reductase genes are controlled primarily by the substrate nitrate and by a number of environmental stimuli that modulate their expression. As for nitrate reductase, Back et al. (1991) have shown that nitrate inducibility of the nitrite reductase gene is under transcriptional control; the transcriptional fusion of the promoter of spinach nitrite reductase with *GUS* as a reporter gene, introduced in tobacco, was found to be nitrate inducible. Because both genes are regulated by nitrate, it may be expected that they share a similar regulatory element within their respective promoter regions that is responsible for nitrate induction. A computer comparison of the spinach nitrite reductase gene and the tobacco nitrate reductase gene has not revealed any striking sequence homology (Meyer, personal communication).

The expression of these two genes has also been analyzed under constant nitrate conditions, varying only the light regime. In maize shoots, in which a diurnal variability has been shown for nitrate reductase mRNA and activity, both mRNA and activity decrease rapidly in the absence of light. For nitrite reductase, the corresponding mRNA shows a pattern similar to that of nitrate reductase but the enzyme activity stays fairly constant over the day. This may be due to the fact that this protein is more stable than nitrate reductase (Bowsher *et al.*, 1991). In *N. plumbaginifolia* wild-type plants, nitrate and nitrite reductase mRNA levels fluctuate according to a circadian rhythm, with similar timing of maximal and minimal transcript accumulation (Faure *et al.*, 1991).

In addition, simultaneous accumulation of nitrite reductase and nitrate reductase transcripts in *N. plumbaginifolia* leaves is observed in mutants defective for nitrate reductase activity. When a low level of nitrate reductase activity is restored, a normal level of nitrite reductase transcript is observed. These data suggested that there is a coregulation of the nitrate and nitrite reductase expression (Faure *et al.*, 1991).

V. Genetics of Nitrate Reductase

The process of generation, recovery, and analysis of mutants is a powerful tool to study a metabolic pathway. In microorganisms, where it is relatively simple to screen large populations for a specific variant, rapid advances have been made. As an example, nitrate reductase mutants have been extensively characterized in prokaryotes, algae, and fungi. In plants, this task is much more tedious, partly due to the diploid—or multiploid—nature of their nuclear genome and the long generation time.

A. Cell Mutagenesis versus M2 Seed Mutagenesis

Different mutagenic agents are available to obtain plant mutants, and the mutagenesis can be applied either at the plant cell level, prior to selection in cell culture, or directly at the seed stage to recover mutants screened in the second generation of mutagen-treated seeds (M2).

Mutagenesis at the cell level has preferentially been performed on protoplasts derived from haploid tissues, to enable the detection of recessive mutations. Testing of clones can then be carried out directly on the calli derived from these mutagenized cells. To perform progeny analysis, the recovery of fertile diploid plants is necessary; this procedure is thus restricted to those species that can be regenerated into whole plants. The diploidization of haploid calli into diploid fertile plants occurs spontaneously, as has been shown for *N. plumbaginifolia* (Gabard *et al.*, 1987). Surprisingly enough, the selection of mutants at the cell level can also be performed in diploid cell cultures. It has been shown that most of the *nia* mutants obtained from diploid material were found to carry two mutant alleles with different characteristics, detectable by progeny segregation analysis, thus suggesting that the mutagenic treatment had been efficient enough to induce the simultaneous mutation of the two alleles of the *Nia* genes carried by a diploid cell (Annie Marion-Poll, personal communication).

Nitrate reductase-deficient mutants have been obtained after *N*-ethylnitrosourea treatment of tobacco cells (Müller and Grafe, 1978), by *N*-methyl-*N'*-nitro-*N*-nitrosoguanidine (MNNG) treatment of mesophyll protoplasts of haploid *Hyoscyamus muticus* (Strauss *et al.*, 1981), and γ or UV irradiation of *N. plumbaginifolia* haploid protoplasts (Grafe *et al.*, 1986). Moreover, a number of nitrate reductase-deficient mutants have arisen spontaneously in plant cell cultures of *Datura innoxia* (King and Khanna, 1980) and in haploid protoplast preparations of *N. plumbaginifolia* (Gabard *et al.*, 1988). These mutants are generated by so-called somaclonal variation events (Evans, 1989) resulting in a variety of mutation types, such as point mutations, frameshift mutations, and transposon insertions (Meyer *et al.*, 1991 and Meyer, personal communication).

Another way to isolate nitrate reductase-deficient mutants is by the screening of M2 seedlings of self-pollinating M1 plants that have been mutagenized. This procedure directly leads to the isolation of homozygous recessive mutants, which are usually fertile, and overcomes the problems that sometimes occur in the regeneration of plants from mutant cell lines. Different mutagens can be used—nitrate reductase-deficient mutants have been isolated by MNNG (Oostinder-Braaksma and Feenstra, 1973), γ irradiation (Wilkinson and Crawford, 1991) of *A. thaliana* seeds, sodium azide treatment of barley and pea seeds (Kleinhofs *et al.*, 1978), ethyl methane sulfonate (EMS) treatment of *Pisum sativum* seeds (Feenstra and Jacobsen, 1980) and of *N. plum-*

baginifolia seeds (Pelsy *et al.*, 1991) as well as tomato seeds (Schoen-makers *et al.*, 1991).

B. SELECTION FOR NITRATE REDUCTASE MUTANTS

Three methods have been successfully used to select for nitrate re-ductase-deficient mutants: chlorate resistance, nitrate reductase ac-tivity assay, and nitrate utilization auxotrophy. Other methods of screening mutants affected in nitrate reductase expression will also be briefly mentioned.

1. Chlorate Resistance

The first higher plant mutants completely lacking nitrate reductase activity were isolated by the screening of allodihaploid *N. tabacum* cell cultures for chlorate resistance (Müller and Grafe, 1978; Müller, 1983). The selection method was based on the observation that the toxicity of chlorate to plant cells depends on the nitrate reductase activity, pre-sumably because it catalyzes the reduction of chlorate to the highly toxic chlorite (Åberg, 1947). The same procedure has enabled the se-lection of two chlorate-resistant cell lines of *D. innoxia* (King and Khanna, 1980), 211 fully deficient mutants of *N. plumbaginifolia* (Ga-bard *et al.*, 1987), and five mutants of tomato (Schoenmarkers *et al.*, 1992).

Chlorate-resistant mutants have been isolated directly from M2 seedlings of *A. thaliana* (Oostinder-Braaksma and Feenstra, 1973; Braaskma and Feenstra, 1982), *P. sativum* (Feenstra and Jacobsen, 1980), soybean (Nelson *et al.*, 1983), and *N. plumbaginifolia* (Pelsy *et al.*, 1991). Numerous mutants, expressing low levels or almost lacking nitrate reductase, as well as mutants exhibiting reduced uptake of chlorate and nitrate, have also been recovered.

2. Lack of Nitrate Reductase Activity

Rapid assay procedures on individual seedlings for a low nitrate re-ductase activity have been used to select nine barley nitrate reductase-deficient mutants from 6000 M2 seedlings tested (Warner *et al.*, 1977). Although rather laborious, this technique enabled the selection of de-fined genetic variants at the seedling or whole plant level, depending on the development stage at which the screening procedures could be applied. This technique can keep the plants viable and nitrate reduc-tase-deficient seedlings have been transplanted to soil and fertilized with ammonium sulfate. Moreover, the use of an efficient mutagenesis procedure such as sodium azide treatment has significantly reduced the size of the population to be screened.

3. Nitrate Utilization Auxotrophy

Nitrate utilization deficiency can also be selected as a conditional auxotrophy marker. A mutant cell line of *H. muticus* has been isolated in cell culture by replica plating on the basis of its nitrate auxotrophy and shown to be devoid of *in vivo* nitrate reductase activity under inducing conditions (Strauss *et al.*, 1981). Many *nia* mutants display a typical phenotype of bleaching at the two-cotyledon stage of development when germinated on a medium containing nitrate as sole nitrogen source (Fig. 3). We have shown recently that nitrate reductase-deficient mutants of *N. plumbaginifolia* can be identified by M2 seed screening on the basis of their inability to grow with nitrate as the sole nitrogen source (Pelsy *et al.*, 1991).

4. Chlorate Hypersensitivity

Hypersensitivity to chlorate has been developed as a screening procedure to obtain novel mutants affected in nitrate assimilation in *A. thaliana*. EMS-treated M2 plants showing early susceptibility symptoms have been isolated in this way. One of these mutants was shown to have increased nitrate reductase activity but nearly normal nitrate uptake characteristics. Another mutant displayed a higher rate of uptake of nitrate; both mutations have been characterized as monogenic and recessive (Wang *et al.*, 1986, 1988).

5. Resistance to Elevated Nitrate Concentrations

Mutants resistant to elevated nitrate concentration have been obtained by M2 seed screening of *A. thaliana*. The selection procedure was based on the toxicity of high potassium nitrate concentrations (higher than 150 mM) for *A. thaliana* seedlings. Several M3 lines showing resistance to nitrate were isolated and most of them showed improved resistance, which was also transmissible to the progeny (Debeys *et al.*, 1990). The molecular basis of this tolerance is not known.

C. PHENOTYPE AND CLASSIFICATION INTO COMPLEMENTATION GROUPS

Fully deficient mutants are not normally viable under standard greenhouse growth conditions, due to their nitrate utilization deficiency. They can, however, be propagated permanently *in vitro* when provided with a reduced nitrogen source, such as the ammonium ion. In the greenhouse they can be grated onto wild-type stocks to obtain

FIG. 2. A nitrate reductase-deficient grafted mutant (left) compared to the wild-type *N. plumbaginifolia* (right). Grafted mutants are characterized by chlorotic, poorly pigmented, crinkled leaves and reduced photosynthesis.

viable and flowering plants. The analysis of a number of these mutants revealed that nitrate reductase deficiency was systematically correlated with a chlorotic phenotype and with a severe defect in the production of organic acids in the leaves (Saux *et al.*, 1987) (see Fig. 2).

Genetic classification of mutants into complementation groups gives an indication of the minimal number of loci implicated in the first step of the nitrate assimilation pathway in higher plants. In *A. nidulans*, for example, nitrate assimilation is controlled by at least 16 genes (Cove, 1979). This classification can be carried out either by somatic hybridization or by sexual crosses between two deficient mutants to analyze their ability to recover nitrate reductase-expressing heterozygotes by complementation. All nitrate reductase-deficient mutations so far characterized have been found to be recessive.

The nitrate reductase-deficient mutants isolated from different plant species have been biochemically classified into two main groups, depending on the presence or absence of a xanthine dehydrogenase (XDH) activity. Those showing xanthine dehydrogenase activity were tentatively classified as apoenzyme-deficient mutants. In contrast, those lacking both nitrate reductase and xanthine dehydrogenase activities were classified as cofactor-deficient mutants affected in the biosynthesis of the MoCo.

D. The Cofactor Mutants

The mutants that show a pleiotropic lack of nitrate reductase and xanthine dehydrogenase activities were assumed to be defective in the biosynthesis or processing of the molybdenum cofactor. Several loci assumed to be involved in MoCo synthesis have been identified. Five loci have been characterized in barley (*Nar2–Nar6*) (Kleinhofs *et al.*, 1989) and three in *A. thaliana* B25, B73, and G1 (Braaskma and Feenstra, 1982; Scholten *et al.*, 1985)); B25 and B73 loci have been located on chromosomes 1 and 5, respectively. In tomato, four nonallelic cofactor mutants have been isolated to date (Schoenmakers *et al.*, 1991).

In the genus *Nicotiana*, these mutants are designated *cnx* (cofactor of *n*itrate reductase and *x*anthine dehydrogenase). So far, two loci have been described for *N. tabacum* by complementation studies using somatic cell fusion (Xuan *et al.*, 1983) and six *Cnx* loci (*CnxA–CnxF*) have been described for *N. plumbaginifolia* both by somatic hybridization and by genetic crosses of regenerated plants (Dirk *et al.*, 1985; Gabard *et al.*, 1988). Interspecific somatic hybridizations between *N. tabacum* and *N. plumbaginifolia* mutants were performed by Xuan *et al.* (1983) in order to classify functionally the *cnx* mutants of both species in the same group. The same type of experiment has been carried out by Làzàr *et al.* (1983) to establish functional relationships between *H. muticus* and *N. tabacum cnx* mutants.

In *N. tabacum*, as in *N. plumbaginifolia*, the distribution of the *cnx* mutants in the different complementation groups is very heterogeneous; half of them have been characterized as *cnxA* mutants (Gabard *et*

TABLE 1
Genetic Data on XDH-Deficient Mutants in Various Plant Species

Species	Gene[a]	Localization	Number of tested mutants (name)	NR activity (wt %)	Mutant growth on nitrate medium	Chlorate resistance	Molybdenum restoration	Ref.[b]
Arabidopsis thaliana	Cnx	Chromosome 5	1 (B73)	15	Poor	Resistance	Yes	1, 15
	Rgn	Chromosome 1	1 (B25)	3	Poor	Resistance	No	1, 15
	nd	nd	1 (G1)	0	No	Resistance	Yes	2
Hordeum vulgare	Nar2	Chromosome 7	3 (Az 34, R9401, R9201)	0–5	No/poor	Resistance	No	3, 4
	Nar3	nd	3 (Az 71; Xno 18, 19)	2–5	Poor	Resistance	nd	3
	Nar4	nd	1 (Az 72)	0	No		nd	3
	Nar5	nd	3 (Az 62, 66, 68)	5–6	Poor		nd	3
	Nar6	nd	1 (Az 69)	2	No		nd	3
Hyoscyamus muticus	CnxA	nd	2 (Ma 2, D12)	0	No	Resistance	Yes	5
	nd	nd	2 (IV C2, XIV E9)	0	No	Resistance	No	5
Lycopersicon esculentum	nd	nd	1 (A29)	62	Poor	Low resistance	No	6
	nd	nd	1 (A57)	0	No	High resistance	No	6
	nd	nd	1 (C3)	0	No	High resistance	No	6
	nd	nd	1 (C42)	0	No	High resistance	No	6
Nicotiana plumbaginifolia	CnxA	nd	32 alleles	0	No	Resistance	Yes	8, 9, 10, 11
	CnxB	nd	4 alleles	0	No	Resistance	No	8, 9, 10
	CnxC	nd	5 alleles	0	No	Resistance	No	8, 9, 10
	CnxD	nd	12 alleles	0	No	Resistance	No	8, 9, 10
	CnxE	nd	4 alleles	0	No	Resistance	No	9, 10
	CnxF	nd	9 alleles	0	No	Resistance	No	9, 10
Nicotiana tabacum	CnxA	nd	4 alleles	0	No	Resistance	Yes	12, 13
	CnxB	nd	4 alleles	0	No	Resistance	No	12, 13
Pisum sativum	Nar2	nd	1 (A 300)	20	Yes	nd	nd	14
	nd	nd	1 (E1)	4	Poor	Resistance	nd	15

[a]nd, Not determined.

[b]References cited: (1) Braaskma and Feenstra (1982); (2) Scholten et al. (1985); (3) Kleinhofs et al. (1989); (4) Melzer et al. (1989); (5) Frankhauser et al. (1984); (6) Schoenmakers et al. (1992); (8) Dirk et al. (1985); (9) Gabard et al. (1988); (10) Marion-Poll et al. (1991); (11) Pelsy et al. (1991); (12) Grafe and Müller (1983); (13) Xuan et al. (1983); (14) Warner et al. (1982); (15) Jacobsen et al. (1984).

al., 1988) and only two *cnxB* mutants have so far been identified from a collection of 70 mutants isolated by cell screening. The proportion of *cnx* mutants isolated either at the cell level or by the M2 seed-screening procedure are not the same: 33% of the total isolated mutants in the first experiment and 66% in the second one were further classified as *cnx*. This result suggests that there is a negative bias in the isolation of *cnx* mutants at the cellular level (Pelsy *et al.*, 1991).

In *N. plumbaginifolia*, none of the *cnx* mutants tested was able to grow normally with hypoxanthine as the sole nitrogen source. This result was expected, as a functional xanthine dehydrogenase is needed to convert hypoxanthine into xanthine and subsequently to uric acid, two steps of the degradative pathway of purines to urea and ammonium (Gabard *et al.*, 1988).

Further characterization of *cnx* mutants has not been achieved at the biochemical level with respect to the MoCo intermediates involved; however, specific features have been associated with some of the complementation groups. For a summary of XDH-deficient mutant characteristics, see Table 1.

1. Restoration of Nitrate Reductase Activity in cnxA Mutants by Providing Molybdate in Excess

Mendel and Müller (1985) first showed that nitrate reductase-deficient tobacco cell lines, classified in the same *CnxA* complementation group, were repairable by nonphysiological high levels of molybdate (1 mM). A similar *cnxA* mutant phenotype was detected in *H. muticus* and *A. thaliana* mutants (Fankhauser *et al.*, 1984; Braaskma and Feenstra, 1982; Scholten *et al.*, 1985) as well as in *N. plumbaginifolia* mutants (Gabard *et al.*, 1988). Nitrate reductase activity of mutants classified into the other complementation groups cannot be restored by treatment with high levels of molybdate.

These results support the conclusion that the product of the *CnxA* gene would control the insertion of the molybdenum ion in the molybdopterin chelating structure (Mendel and Müller, 1985).

2. Biochemical Analysis

The nitrate reductase-associated NADH cytochrome *c* reductase (CcR) partial activity, which does not involve the MoCo, has been detected in all *cnx* mutants tested so far and is associated with each nitrate reductase subunit (MacDonald *et al.*, 1974; De Vries *et al.*, 1986; Mendel *et al.*, 1986; Steven *et al.*, 1989). Biochemical characterization of the *N. plumbaginifolia cnx* mutants (Marion-Poll *et al.*, 1991) showed that some classes of *N. plumbaginifolia cnx* mutants overexpressed the nitrate reductase-associated CcR activity. This expression was correlated with an increased level of *Nia* transcript. However, in all the extracts prepared from *cnx* mutants, two-site enzyme-linked immunosorbent assay (ELISA) tests performed with a monoclonal antibody and a polyclonal serum led to the detection of cross-reacting material at different levels: low levels in CNX B, C, D, E, and F extracts, whereas CNX A extracts invariably exhibited very high levels of the nitrate reductase apoenzyme. These results could suggest an altered nitrate reductase protein structure in all *cnx* mutants with the exception of *cnx*A, which is probably due to an absence of a functional MoCo.

Tobacco, barley, and *N. plumbaginifolia cnx* mutants have been used to study the structure of nitrate reductase and the involvement of the MoCo in the assembly of the subunits of the enzyme. All *cnxA* mutants express a dimeric inactive structure, but such dimeric forms of nitrate reductase have also been detected in several *Cnx* complementation classes, even for nonleaky mutants (De Vries *et al.,* 1986; Mendel *et al.,* 1986; Marion-Poll *et al.,* 1991). Barley mutants (*nar*), impaired in MoCo synthesis, are either monomeric or dimeric, depending on the mutant tested (Steven *et al.,* 1989). This heterogeneity has been attributed to the leaky characteristics of the *nar* mutations studied. We therefore assume that the MoCo is not required for enzyme dimerization but more probably for the stabilization of the dimer and its protection from proteolytic degradation. The so-called monomeric forms of nitrate reductase would then correspond either to dimeric enzymes degraded by proteolysis to approximately half the size of the dimer, or alternatively, partial proteolysis would destabilize the dimeric structure and release the monomer. Indeed, it was observed that nitrate reductase is susceptible to proteolysis mediated by proteolytic enzymes of higher plants (Hamano *et al.,* 1984; Poulle *et al.,* 1987; Sueyoshi *et al.,* 1989). There is some support for this idea based on the observation in the yeast *S. cerevisiae,* in which no MoCo is synthesized, that the expression of the tobacco nitrate reductase apoenzyme leads to the detection of a dimeric protein as tested by nondenaturing polyacrylamide gel electrophoresis and Western blotting (Truong *et al.,* 1991).

Apart from being impaired in nitrate metabolism, *cnx* mutants are also defective in other functions involving the molybdenum cofactor. Walker-Simmons *et al.* (1989) have described a barley *cnx* mutant with a thermosensitive wilty phenotype. The basis of this phenotype has been attributed to a defect in abscisic acid (ABA) biosynthesis, possibly linked to the absence of MoCo-dependent aldehyde oxidase activity.

E. The Apoenzyme Mutants

Nitrate reductase-deficient mutants showing normal MoCo levels are assumed to be affected in the apoenzyme gene locus and are generally classified as *nia* mutants. A variety of fully deficient mutants have been classified in the *Nia* complementation group, among which some still catalyze various partial activities retained by their defective nitrate reductase (Chérel *et al.*, 1990).

In *A. thaliana* the analysis of chlorate-resistant mutants showed the direct role of two loci, *Chl2* and *Chl3*, in the biosynthesis of the nitrate reductase apoenzyme. The mutant class *Chl3* has a low levels of nitrate reductase activity (10–20% of that in the wild type) as do *Chl2* mutants (5–10%), which can, although poorly, grow with nitrate as the sole nitrogen source (Braaskma and Feenstra, 1982). Based on these data, it was originally proposed that *Chl2* and *Chl3* genes are nitrate reductase structural genes.

In barley, *nar1* mutants have been isolated and characterized. The *Nar1* locus has been identified as the NADH-specific nitrate reductase structural gene (Kleinhofs *et al.*, 1989). None of the *nar1* mutants is completely deficient in nitrate reductase activity. This is a result of a residual nitrate reductase activity due to the expression of a second locus, *Nar7*, which is presumed to be the NAD(P)H NR structural gene (Warner *et al.*, 1987). As opposed to *nar1* and *nar7* mutants, a *nar1–nar7* double mutant is unable to grow on nitrate, suggesting that these two genes represent the major if not the only functional nitrate reductase structural genes in barley.

In contrast to most other species, in *N. plumbaginifolia* fully deficient *nia* mutants are all allelic (Gabard *et al.*, 1987). In *N. tabacum*, which is an amphidiploid species resulting from the hybridization of *Nicotiana sylvestris* and *Nicotiana tomentosiformis*, two functional homeologous *Nia* loci (*Nia1* and *Nia2*) were identified, which correspond to one gene per haploid genome (Müller and Mendel, 1989).

1. Genetic Mapping

In barley, the *Nar1* and *Nar7* loci are not linked and the *Nar1* locus has been mapped on the short arm of chromosome 6 (Kleinhofs *et al.*, 1989).

In *A. thaliana*, Braaskma and Feenstra (1982) showed that the *Chl2* locus is located on chromosome 2 and *Chl3* is on chromosome 1. Restriction fragment length polymorphism (RFLP) analysis has positioned the *Nia2* gene in the middle of chromosome 1 near the *Chl3* locus (Chang *et al.*, 1988; Cheng *et al.*, 1988; Nam *et al.*, 1989). The *Nia1* gene mapped also to chromosome 1 near *Gl-2*, which is not close to any of the known chlorate-resistant loci (Cheng *et al.*, 1988). These results say little about the identity of the *Chl2* gene, which is apparently not a *Nia* locus and must therefore be involved in another aspect of nitrate reductase expression. However, Wilkinson and Crawford (1991) demonstrated that the gene at the *Chl3* locus is the *Nia2* gene. As a deletion mutant in this gene retains only 10% of the *in vitro* nitrate reductase activity in wild-type plants, the *Nia2* gene is responsible for 90% of the nitrate reductase activity in *A. thaliana*. This result explains why one can find chlorate-resistant mutants defective in the *Nia2* gene, even though there is another nitrate reductase gene in the plant, and, conversely, why no chlorate-resistant mutants have been found with lesions in the *Nia1* gene.

2. Biochemical and Immunological Classification

A common feature of *Nia* complementation groups is the presence of mutants defective for some, and sometimes all, of the partial catalytic activity normally carried out by nitrate reductase.

All *nar1* mutants of barley are characterized by greatly reduced but significant NADH and NADPH NR activities in young seedling shoots, although wild-type seedling shoots have high NADH NR activity but no NADPH NR activity. Regarding NR-associated cytochrome *c* reductase activities, they varied from zero to greater than the wild type according to the mutant tested (Kleinhofs *et al.*, 1989).

The biochemical characterization of *N. plumbaginifolia nia* mutants has shown that they belong to four classes, depending on the catalytic domains affected by the mutation (Chérel *et al.*, 1990). The first one lacks both ELISA-detectable protein and all of the partial catalytic activities normally carried out by nitrate reductase. The second and the third classes are characterized by the presence of an ELISA-detectable nitrate reductase protein and nitrate-reducing or diaphorase activities, respectively, and are thought to be affected in the FAD and in the MoCo domain, respectively. Class four is negative in the ELISA test but bromophenol blue nitrate reductase activity is detectable and is presumably affected in the heme domain, where the epitope recognized by the monoclonal antibody ZM 96(9)25, used for the ELISA tests, is located. Molecular analysis of these mutants confirms that reduction of nitrate by the artificial electron donor, bromophenol blue, requires only MoCo as a cofactor (Meyer *et al.*, 1987), whereas reduction by methyl viologen or by reduced flavin has been shown to need functional MoCo and heme domains (Chérel *et al.*, 1990). The expression of these different functional domains, linked, in the heterologous host *S. cerevisiae* (Truong *et al.*, 1991), confirmed the conclusion of Kubo *et al.* (1988) that the cytochrome *c* reductase activity involves the FAD and heme domains.

3. Genetic and Biochemical Evidence of Intragenic Complementation at the Nia Locus

Intragenic complementation at the *Nia* locus was observed in *N. plumbaginifolia* (Pelsy and Gonneau, 1991) as well as in tobacco (Müller and Mendel, 1989) and *C. reinhardtii* (Fernandez *et al.*, 1989) apoenzyme mutants. Systematic crosses between mutants classified in the *Nia* complementation group in *N. plumbaginifolia* (Table 2) led to the observation of a complementation between allelic *nia* mutants, which then enabled growth on nitrate of the heterozygote. Such a complementation was observed only when the homozygous parents were affected on two different functional domains, but never when affected on the same domain. A reconstitution of the nitrate reductase activity has been obtained by mixing inactive enzymatic extracts from two complementing *nia* homozygous mutants, suggesting that nitrate reductase intragenic complementation results either from the formation of heteromeric nitrate reductase or from the interaction between two modified enzymes to reconstitute an electron transfer chain that incorporates redox centers from different subunits (Pelsy and Gonneau, 1991).

4. Molecular Characterization of nia Mutations

Different types of mutational events have been shown to lead to nitrate reductase null mutants by insertion of a mobile element or by single base changes. The isolation of the first tobacco transposable element was possible after its transposition into the *Nia* locus, generating nitrate reductase-deficient mutants (Grandbastien *et al.*, 1989).

Analysis of *N. plumbaginifolia nia* mutants classified in the same biochemical group, lacking both all partial activities involving the heme domain and the epitope for the monoclonal antibody specific for this domain, has been performed by polymerase chain reaction amplification and sequence determination of the corresponding cDNA sequences. Three different mutations have been identified in the sequence corresponding to the cytochrome b_5 domain of nitrate reductase: one conversion led to the substitution into an asparagine of one of the histidine residues implicated in the heme-binding interaction, and two nonsense mutations were a result of a single base change and

TABLE 2
Intragenic Complementation at the *Nia* Locus[a]

	D125	E77	E87	E122	F19	H22	I9	A1	E64	K21	E56	D51	D57	D64	E82	E83	H11	H29	K25	D80	I2	E23
ELISA	+							·			·	+								·		·
Cyt. c red	+							·			·	·								·		·
MV-NR	·							·			+	+								·		·
BPB-NR	·							+			+	+								·		·
Class 3																						
D125	·																					
E77	·	·																				
E87	·	·																				
E122	·	·	·																			
F19	·	·	·	·																		
H22	·	·	·	·	·																	
I9	·	·	·	·	·	·	·															
Class 4																						
A1	· / +/–	+/– / +	·	·	+/–	·	+/– / ·	·	·	·	· ·											
E64	+/–	+/– / +	·	·	+/–	·	· ·		·	·	· ·											
K21	+/– / ·	+/– / +	·	·	+/–	·	+/– / ·	·	· ·	·	· ·											
E56	++	++ / ++	·	·	++	+/– / +	++ / ·	· ·	· ·	· ·	·											

D51	++ ++	+/- +++	·	++ ·	+/- 	++	+/- ·	++ +++	· ·	· ·	·	·	·	·	·	·	·	·			
D57	++ ++	· ++	·	++ ·	·	++	· ·	++ +++	· ·	· ·	·	·	·	·	·	·	·	·			
D64	++ ++	· +	·	++ ·	·	++	· ·	++ +	· ·	· ·	·	·	·	·	·	·	·	·			
E82	++ ++	· ++	·	++ ·	·	++	· ·	++ +++	· ·	· ·	·	·	·	·	·	·	·	·			
E83	++ ++	· ++	·	++ ·	·	++	· ·	++ +	· ·	· ·	·	·	·	·	·	·	·	·			
H11		++		++		++		++	·	·	·	·	·	·							
H29	++ ++	· +++	·	++ ·	·	++	· ·	++ ++	· ·	· ·	·	·	·	·	·	·	·	·			
K25	+/- +/-	· +	·	+/- ·	·	+/-	· ·	+/- ++	· ·	· ·	·	·	·	·	·	·	·	·	·	·	
D80	++ ++	· ++	·	+/- ·	·	++	· ·	++ +	· ·	· ·	·	·	·	·	·	·	·	·	·	·	·
I2	· ·	·	·	·	·	·	· ·	++ ++	· ·	· ·	·	·	·	·	·	·	·	·	·	·	·
Class 1 **E23**	·	·	·	·	·	·	·	·	·	·	·	·	·	·	·	·	·	·		·	·

[a]Biochemical characteristics of each class of *nia* mutants: ELISA, two-site test using nitrate reductase-specific antibodies; Cyt. *c* Red; cytochrome *c* reductase activity; MV-NR, reduced methyl viologen nitrate reductase activity; BPB-NR, reduced bromophenol blue nitrate reductase activity. Symbols within each box have the following meanings: First line: ability to grow with nitrate as sole nitrogen source for heterozygous *nia* mutant plants (++, growth similar to the wild type; +/−, poor growth *in vitro* up to four chlorotic leaves: −, no growth). Second line: reconstituted NADH NR activity after mixing equal amounts of protein extracts of homozygous *nia* mutants (relative reconstituted activity level, +++ to +/−).

a frameshift (Meyer *et al.*, 1992). Other classes of *nia* mutations have been shown to be due to the insertion of mobile elements or to deletion or inversion (C. Meyer, personal communication).

F. Transcription of the *Nia* Gene in Mutants

Pouteau *et al.* (1989) have compared the light and substrate regulation of the *Nia* gene in wild-type and in *N. plumbaginifolia* mutants as classified in their different complementation groups. Mutants that retained an apoenzyme, functional (as for *cnx*) or defective (as for *nia*), overexpressed the corresponding *Nia* mRNA, whereas *nia* mutants devoid of detectable nitrate reductase protein had reduced or undetectable nitrate reductase mRNA levels. Although the diurnal oscillations of the *Nia* mRNA were abolished in all mutants, the observed mutants still retained a significant nitrate-inducible nitrate reductase transcript expression.

The accumulation of *Nia* transcripts in the leaves of nitrate reductase-deficient mutants grown on nitrate can be interpreted as a consequence of nitrogen metabolite regulation, because nitrate reductase deficiency would prevent the accumulation of nitrogen metabolites derived from nitrate and lead to the overexpression of the *Nia* gene. These results could suggest that glutamine and/or other nitrogen metabolites may exert a negative control on nitrate reductase expression.

G. Regulation of Nitrate Assimilation in Fungi as a Model

In fungi, a model for nitrate reductase regulation has been postulated on the basis of genetic and molecular evidence (Cove and Pateman, 1969). Nitrogen metabolite repression ensures preferential utilization of the favored nitrogen sources, ammonium and glutamine. The general positive regulatory factor of the nitrogen circuit, encoded by the gene *AreA* in *A. nidulans* or *Nit2* in *Neurospora crassa,* turns on the transcription of a number of nitrogen-related structural genes, including nitrate reductase, during conditions of nitrogen limitation. Although the mechanism by which these factors act is not yet completely understood, they have been shown to possess a zinc finger DNA-binding domain implicated in the regulatory function (Kudla *et al.*, 1990; Fu and Marzluf, 1990). The absence of this factor leads to an inability to utilize nitrate, nitrite, and other nitrogen sources except ammonium and glutamine.

In addition, another regulatory gene, *NirA* in *A. nidulans* (Scazzoc-chio and Arst, 1989) or *Nit4* in *N. crassa* (Marzluf and Fu, 1989), is specific for the nitrate assimilation pathway. It codes for a positive regulator, which turns on the expression of the nitrate and nitrite re-ductase genes in the presence of nitrate. In the absence of nitrate, ni-trate reductase is thought to prevent the transcription of its structural gene by interfering with the positive regulator. Structural defects af-fecting the structure of nitrate reductase, including the absence of MoCo in the apoenzyme, are associated with a constitutive expression of nitrate-regulated genes. These data led to the postulation of an au-toregulatory role of the nitrate reductase enzyme. It may act as a re-pressor of the expression of the specific activator gene responsible for nitrate inducibility.

The autoregulatory model in fungi implies that, in the absence of nitrate, the direct interaction of the nitrate reductase protein with the positive transcription factor prevents the transcription of the nitrate and nitrite reductase genes. Two lines of evidence suggest that this model does not fit with the data derived from *N. plumbaginifolia*. First, there is the observation of Pouteau *et al.* (1989) that nitrate reductase-deficient mutants, with no detectable nitrate reductase protein, ex-press a nitrate-inducible *Nia* transcript. In this instance, nitrate reduc-tase cannot be postulated as a regulatory element of its own synthesis. Second, a mutant that does not produce any detectable *Nia* transcript expresses a nitrite reductase transcript that is still nitrate inducible. If the autocatalytic model were valid for higher plants, this mutant, lacking the *Nia* transcript, would express a constitutive nitrite reduc-tase that is not observed (Faure *et al.*, 1991).

H. Intra- or Interspecific Functional Complementation and Expression of the *Nia* Gene or Corresponding cDNA in *nia* Mutants

A direct demonstration of the function.of a cloned gene can be ob-tained by complementing the mutant phenotype. These experiments are also useful for further characterization of its regulation. Such ex-periments with *Nia* genes have been successfully carried out on *C. reinhardtii* (Fernandez *et al.*, 1989) and on *Nicotiana* species, due to their ability to be easily transformed. Vaucheret *et al.* (1990) described the functional complementation of tobacco and *N. plumbaginifolia* de-ficient mutants by transformation with the wild-type alleles of the to-bacco structural genes. In *A. thaliana* the *Chl3* locus has been identified as the *Nia2* gene by *Agrobacterium*-mediated introduction of the *Nia2* allele into three *chl3* deletion mutants (Wilkinson and Crawford, 1991). In these experiments, carried out on three plant species, a low number of transformants has been generally obtained, suggesting a weak complementation efficiency. The introduction of a gene into a homologous host may display atypical characteristics, due, for instance, to cosuppression events (Napoli *et al.*, 1990; Van der Krol *et al.*, 1990). When a cloned tomato *Nia* gene has been introduced into a *nia N. plumbaginifolia* mutant, totally defective for both the expression of the *Nia* transcript and the corresponding nitrate reductase protein, an efficient complementation has been obtained that enables the transgenic plant to grow normally on nitrate. In this experiment the level of expression was not higher than 10% of wild type. The transcription of the introduced gene was found to be properly regulated by a circadian rhythm as well as by light and nitrate, as in tomato (Dorbe *et al.*, 1992).

VI. Nitrate Reductase Gene as a Tool for Genetic Engineering

The *Nia* gene appears in many respects to be a unique tool for plant molecular biology studies. First, the *nia* mutation can be selected as a marker either negatively (inability to grow on nitrate) or positively (chlorate resistance) at the whole plant level or in cell culture (see Fig. 3).

Nussaume *et al.* (1992) have developed a counterselection system based on the use of an engineered nitrate reductase gene. Seedlings and cells derived from transgenic *N. plumbaginifolia* plants transformed with the *Nia* cDNA placed under the control of the CaMV promoter express a constitutive nitrate reductase (Vincentz and Caboche, 1991). These transgenic plants are efficiently killed by chlorate on a medium containing ammonium as sole nitrogen source, whereas wild-type plants are not affected by chlorate, as the endogenous nitrate reductase is not induced under these conditions.

The combination of *nia* mutations together with dominant resistant markers in the same genome offers the possibility to use the corresponding cells as universal hybridizers for somatic hybridization experiments: only hybrids between the marker line and nitrate-utilizing cells would simultaneously express the dominant marker and grow with nitrate as sole nitrogen source.

The *Nia* gene can also be used as a transposon trap for the identification of new transposable elements. This possibility has been demonstrated by the cloning of Tnt-1, and could be extended to other

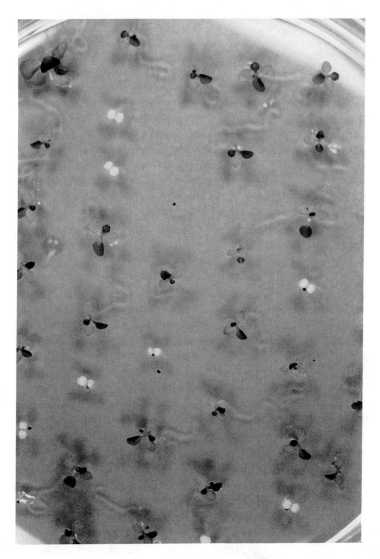

FIG. 3. Mendelian segregation in the progeny of a plant obtained after crossing a homozygous nitrate reductase-deficient mutant with the wild type. Note the bleached phenotype of mutants on the nitrate-containing medium.

plant species. The *Nia* locus can also be envisaged as a target for homologous recombination studies. However, these last two approaches require the presence of only one functional *Nia* gene in the genome of interest, in plant species amenable to cell manipulation *in vitro*.

VII. Conclusions

Nitrate assimilation catalyzed by nitrate and nitrite reductases is an energy-consuming and tightly regulated process in higher plants, and was first through to be a growth-limiting step. However, this hypothesis is not corroborated by recent genetic and molecular data: (1) *Arabidopsis thaliana chl3* deletion mutants with a low level of nitrate reductase activity (10% of the wild type) can grow on saturating or limiting nitrate concentrations as well as the wild-type plants do (Wilkinson and Crawford, 1991). (2) The activities detected in heterozygous complementing *nia* mutant plants represent 10–15% of the wild type (measured under the same conditions) and enable normal growth rates (Pelsy and Gonneau, 1991). (3) Transgenic *N. plumbaginifolia* plants expressing as little as 10% of the wild-type activity (Dorbe *et al.*, 1992) recovered growth rates comparable to that of the wild type. Thus the level of nitrate reductase appears to be in 10-fold excess when fully induced and not to limit the vegetative growth of plants under these experimental conditions. At least in these species, a nitrate reductase overexpression obtained by genetic engineering is probably not a route of choice for the improvement of biomass production.

Nitrate assimilation cannot be considered as an autonomous pathway, as the energy it requires originates in photosynthesis. In support, it has been shown that nitrate reductase activity of spinach leaves underwent rapid changes when rates of net photosynthesis were modulated by varying the CO_2 supply (Kaiser and Brendle-Behnisch, 1991). Cytosolic ATP/AMP levels would be the link between nitrate reductase activity in the cytosol and photosynthesis in the cytoplasm, through regulatory factors (Kaiser and Spill, 1991). Moreover, in CO_2-saturating concentrations, nitrate enhanced the rate of O_2 evolution, the extent of this stimulation being maximal at saturating photon flux densities. This indicates that in plant leaves, nitrate assimilation increases the capacity of the photosynthetic apparatus for noncyclic electron flow, overcoming the limitation imposed by the rate of CO_2 fixation (De la Torre *et al.*, 1992). From these physiological studies it appears that there is a metabolic coupling between photosynthesis and nitrate reduction. Whether this coupling is strictly metabolic, or whether a specific component such as a "plastidic factor" is involved in this coupling, as suggested by Mohr and co-workers (Oelmüller *et al.*, 1988), will be understood by further characterizing photosynthetic mutants specifically affected in electron transfer or in the biogenesis of the chloroplast (Börner *et al.*, 1986). The exploitation of nitrate reductase for crop improvement may be possible when the interactions between nitrate reductase and the photosynthetic apparatus are better understood.

Future studies are likely to concentrate on elucidating the regulation of nitrate reductase expression. As nitrate reductase is a tightly regulated enzyme, one might expect that numerous regulatory genes are involved in its regulation. Unlike the fungi, no such regulatory mutants have been identified from any higher plant species. Constitutive expression of nitrate reductase has been obtained by genetic engineering in *N. plumbaginifolia nia*-deficient mutants complemented by transformation with a chimeric nitrate reductase gene fused to the CaMV-derived 35S promoter. The obtained transgenic plants constitutively expressed from one-fifth to three times the wild-type nitrate reductase activity in their leaves. An analysis of the chimeric nitrate reductase gene expression showed a loss of regulation by light, nitrate, and circadian rhythm of transcript accumulation. However, light was still required for the accumulation of nitrate reductase protein (Vincentz and Caboche, 1991). This points out the possibility of posttranscriptional regulation events. This deregulated expression was without effect on vegetative development and seed production. We therefore presume that a deregulated expression of nitrate reductase expression due to a genetic defect is not per se incompatible with plant viability. Three possibilities can still explain why the search for regulatory mutants has not been so far successful. (1) The corresponding genes are redundant, (2) the regulatory genes are also involved in other aspects of plant physiology and a deficiency is lethal for this reason, and (3) the screening procedures are not adequate for their identification. This latter hypothesis is supported in our laboratory by the fact that putative regulatory mutants have been recently identified in an M2 seed-screening protocol for auxotrophs. A leaky mutant, able to grow with high nitrate concentrations but not with low nitrate concentrations, appears to express nitrate reductase in a deregulated way.

It is now well established that the nitrate reductase gene is a powerful tool for plant cell genetics as a marker that can be selected or counterselected. Moreover, it has enabled the trapping of transposable elements and can also be envisaged as a target for homologous recombination studies. Because nitrate reductase mutants can apparently be isolated in any plant species, the corresponding gene can be the target of these different approaches. Experiments are underway in different laboratories to identify transposable elements in petunia and tomato, by using a nitrate reductase gene as a transposon trap. Conversely, a constitutively expressed nitrate reductase could also be used as a transposon trap in the species in which it would be introduced by transformation.

ACKNOWLEDGMENTS

We wish to thank colleagues in our laboratory for many helpful comments during preparation of this manuscript.

REFERENCES

Åberg, B. (1947). On the mechanism of the toxic action of chlorates and some related substances upon young wheat plants. *Lantbrukshoegsk, Ann.* **15**, 37–107.
Andrews, M. (1986). The partioning of nitrate assimilation between root and shoot of higher plants. *Plant, Cell Environ.* **9**, 511–519.
Andrews, M., De Faria, S. M., McInroy, S. G., and Sprent, J. I. (1990). Constitutive nitrate reductase activity in the leguminosae. *Phytochemistry* **29**, 49–54.
Back, E., Dunne, W., Schneiderbauer, A., de Framont, A., Rastogi, A., and Rothstein, S. J. (1991). Isolation of the spinach nitrite reductase gene promoter which confers nitrate inducibility on GUS gene expression in transgenic tobacco. *Plant Mol. Biol.* **17**, 9–18.
Barrett, E. L., and Riggs, D. L. (1982). Evidence for a second nitrate reductase activity that is distinct from the respiratory enzyme in *Salmonella typhymurium*. *J. Bacteriol.* **150**, 563–571.
Becker, T., and Caboche, M. (1990). Third international symposium on nitrate assimilation. *Plant Mol. Biol. Rep.* **8**, 304–312.
Beevers, L., and Hageman, R. H. (1983). Uptake and reduction of nitrate: Bacteria and higher plants. *In* "Encyclopedia of Plant Physiology" (A. Lauchli and R. L. Bieleski, eds.), Vol. 15A, pp. 351–375. Springer-Verlag, Berlin.
Benamar, S., Pizelle, G., and Thiéry, G. (1989). Activité nitrate réductase non induite par le nitrate dans les feuilles d'*Alnus glutinosa*. *Plant Physiol. Biochem.* **27**, 107–112.
Börner, T., Mendel, R. R., and Schiemann, J. (1986). Nitrate reductase is not accumulated in chloroplast-ribosome-deficient mutants of higher plants. *Planta* **169**, 202–207.
Bowsher, C. G., Long, D. M., Oaks, A., and Rothstein, S. J. (1991). Effect of light–dark cycles on expression of nitrate assimilatory genes in maize shoots and roots. *Plant Physiol.* **95**, 281–287.
Boxer, D. H. (1989). Involvement of the chlorate resistance loci and the molybdenum cofactor in the biosynthesis of the *Escherichia coli* respiratory nitrate reductase. *In* "Molecular and Genetic Aspects of Nitrate Assimilation" (J. Wray and J. Kinghorn, eds.), pp. 27–39. Oxford Science Publications, Oxford.
Braaskma, F. J., and Feenstra, W. J. (1982). Isolation and characterization of nitrate reductase-deficient mutants of *Arabidopsis thaliana*. *Theor. Appl. Genet.* **64**, 83–90.
Caboche, M., and Rouzé, P. (1990). Nitrate reductase: A target for molecular and cellular studies in higher plants. *Trends Genet.* **6**, 187–192.
Calza, R., Huttner, E., Vincentz, M., Rouzé, P., Galangau, F., Vaucheret, H., Chérel, I.,

Meyer, C., Kronenberger, J., and Caboche, M. (1987). Cloning of DNA fragments complementary to tobacco nitrate reductase mRNA and encoding epitopes common to the nitrate reductases from higher plants. *Mol. Gen. Genet.* **209**, 552–562.

Campbell, W. H. (1988). Higher plant nitrate reductase: Arriving at the molecular view. *Curr. Top. Plant Biochem. Physiol.* **7**, 1–15.

Chang, C., Bowman, J. L., DeJohn, A. W., Lander, E. S., and Meyerowitz, E. M. (1988). Restriction fragment length polymorphism linkage map for *Arabidopsis thaliana.* *Proc. Natl. Acad. Sci. U.S.A.* **85**, 6856–6860.

Cheng, C. L., Dewdney, J., Kleinhofs, A., and Goodman, H. M. (1986). Cloning and nitrate induction of nitrate reductase mRNA. *Proc. Natl. Acad. Sci. U.S.A.* **83**, 6825–6828.

Cheng, C. L., Dewdney, J., Nam, H., Den Boer, B. G. W., and Goodman, H. M. (1988). A new locus (*NIA 1*) in *Arabidopsis thaliana* encoding nitrate reductase. *EMBO J.* **7**, 3309–3314.

Chérel, I., Marion-Poll, A., Meyer, C., and Rouzé, P. (1986). Immunological comparisons of nitrate reductases of different plant species using monoclonal antibodies. *Plant Physiol.* **81**, 376–378.

Chérel, I., Gonneau, M., Meyer, C., Pelsy, F., and Caboche, M. (1990). Biochemical and immunological characterization of nitrate reductase-deficient *nia* mutants of *Nicotiana plumbaginifolia. Plant Physiol.* **92**, 659–665.

Choi, H., Kleinhofs, A., and An, G. (1989). Nucleotide sequence of rice nitrate reductase genes. *Plant Mol. Biol.* **13**, 731–733.

Clarkson, D. T. (1986). Regulation of the absorption and release of nitrate by plant cells: A review of current ideas and methodology. *In* "Fundamental, Ecological and Agricultural Aspects of Nitrogen Metabolism in Higher Plants" (H. Lambers, J. J. Neeteson, and I. Stulen, eds.), pp. 3–27. Martinus Nijhoff Publishers, Dordrecht, The Netherlands.

Cole, J. A. (1988). Assimilatory and dissimilatory reduction of nitrate to ammonia. *Symp. Soc. Gen. Microbiol.* **42**, 281–329.

Cole, J. A. (1989). Physiology, biochemistry, and genetics of nitrite reduction by *Escherichia coli. In* "Molecular and Genetic Aspects of Nitrate Assimilation" (J. Wray and J. Kinghorn, eds.), pp. 229–243. Oxford Science Publications, Oxford.

Cove, D. J. (1979). Genetic studies of nitrate assimilation in *Aspergillus nidulans. Biol. Rev. Cambridge Philos. Soc.* **54**, 291–327.

Cove, D. J., and Pateman, J. A. (1969). Autoregulation of the synthesis of nitrate reductase in *Aspergillus nidulans. J. Bacteriol.* **97**, 1374–1378.

Crawford, N., and Campbell, W. H. (1990). Fertile fields. *Plant Cell* **2**, 829–835.

Crawford, N., Campbell, W. H., and Davis, R. W. (1986). Nitrate reductase from squash: cDNA cloning and nitrate regulation. *Proc. Natl. Acad. Sci. U.S.A.* **83**, 8073–8076.

Crawford, N., Smith, M., Bellissimo, D., and Davis, R. W. (1988). Sequence and nitrate regulation of the *Arabidopsis thaliana* mRNA encoding nitrate reductase, a metallo-flavoprotein with three functional domains. *Proc. Natl. Acad. Sci. U.S.A.* **85**, 5006–5010.

Daniel-Vedele, F., Dorbe, M. F., Caboche, M., and Rouzé, P. (1989). Cloning and analysis of the nitrate reductase gene from tomato: A comparison of nitrate reductase protein sequences in higher plants. *Gene* **85**, 371–380.

Debeys, D., Cammaerts, D., and Jacobs, M. (1990). Isolation of mutants resistant to high nitrate concentrations. *In* "Abstracts of Third International Symposium on Nitrate Assimilation, Molecular and Genetic Aspects," p. 118. Bombannes, INRA-Versailles, France.

De la Torre, A., Delgado, B., and Lara, C. (1991). Nitrate-dependent O_2 evolution in intact leaves. *Plant Physiol.* **26**, 898–901.

Deng, M., Moureaux, T., Leydecker, M. T., and Caboche, M. (1990). Nitrate reductase expression is under the control of a circadian rhythm and is light inducible in *Nicotiana tabacum* leaves. *Planta* **180**, 257–261.

Deng, M., Moureaux, T., Chérel, I., Boutin, J. P., and Caboche, M. (1991). Effect of nitrogen on the regulation and circadian expression of tobacco nitrate reductase. *Plant Physiol. Biochem.* **29**, 1–9.

De Vries, S. E., Dirk, R., Mendel, R. R., Schaart, J. G., and Feernstra, W. J. (1986). Biochemical characterization of nitrate reductase deficient mutants of *Nicotiana plumbaginifolia*. *Plant Sci.* **44**, 105–110.

Dirk, R., Negrutiu, I., Sidorov, V., and Jacobs, M. (1985). Complementation analysis by somatic hybridization and genetic crosses analysis of nitrate reductase-deficient mutants of *Nicotiana plumbaginifolia*. *Mol. Gen. Genet.* **201**, 339–343.

Dorbe, M. F., Caboche, M., and Daniel-Vedele, F. (1992). The tomato *Nia* gene complements a *Nicotiana plumbaginifolia* nitrate reductase-deficient mutant and is properly regulated in heterologous transgenic plants. *Plant Mol. Biol.* **18**, 363–375.

Duke, S. H., and Duke, S. O. (1984). Light control of extractable nitrate reductase activity in higher plants. *Physiol. Plant.* **62**, 485–493.

Ekes, M. (1981). Ultrastructural demonstration of ferricyanide reductase (diaphorase) activity in the envelopes of the plastids of etiolated barley (*Hordeum vulgare* L.) leaves. *Planta* **151**, 439–446.

Evans, D. A. (1989). Somaclonal variation—Genetic basis and breeding applications. *Trends Genet.* **5**, 46–50.

Fankhauser, H., Bucher, F., and King, P. J. (1984). Isolation of biochemical mutants using haploid mesophyll protoplasts of *Hyoscyamus muticus*. IV. Biochemical characterization of nitrate non-utilizing clones. *Planta* **160**, 415–421.

Faure, J. D., Vincentz, M., Kronenberger, J., and Caboche, M. (1991). Co-regulated expression of nitrate and nitrite reductases. *Plant J.* **1**, 107–113.

Feenstra, W. J., and Jacobsen, E. (1980). Isolation of a nitrate reductase deficient mutant of *Pisum sativum* by means of selection for chlorate resistance. *Theor. Appl. Genet.* **58**, 39–42.

Fernandez, E., Schnell, R., Ranum, L. P. W., Hussey, S. C., Silflow, C. D., and Levebvre, P. A. (1989). Isolation and characterization of the nitrate reductase structural gene of *Chlamydomonas reinhardtii*. *Proc. Natl. Acad. Sci. U.S.A.* **86**, 6449–6453.

Fu, Y., and Marzluf, G. A. (1990). *Nit 2*, the major positive-acting nitrogen regulatory gene of *Neurospora crassa*, encodes a sequence-specific DNA-binding protein. *Proc. Natl. Acad. Sci. U.S.A.* **87**, 5331–5335.

Gabard, J., Marion-Poll, A., Chérel, I., Meyer, C., Müller, A. J., and Caboche, M. (1987). Isolation and characterization of *Nicotiana plumbaginifolia* nitrate reductase-deficient mutants: Genetic and biochemical analysis of the NIA complementation group. *Mol. Gen. Genet.* **209**, 596–606.

Gabard, J., Pelsy, F., Marion-Poll, A., Caboche, M., Saalbach, I., Grafe, R., and Müller, A. J. (1988). Genetic analysis of nitrate reductase deficient mutants of *Nicotiana plumbaginifolia*: Evidence for six complementation groups among 70 classified molybdenum cofactor deficient mutants. *Mol. Gen. Genet.* **213**, 206–213.

Galangau, F., Daniel-Vedele, F., Moureaux, T., Dorbe, M. F., Leydecker, M. T., and Caboche, M. (1988). Expression of leaf nitrate reductase genes from tomato and tobacco in relation to light–dark regimes and nitrate supply. *Plant Physiol.* **88**, 383–388.

Garrett, R. H., and Nason, A. (1967). Involvement of a b-type cytochrome in the assimilatory nitrate reductase of *Neurospora crassa*. *Proc. Natl. Acad. Sci. U.S.A.* **58**, 1603–1610.

Grafe, R., and Müller, A. J. (1983). Complementation analysis of nitrate reductase deficient mutants of *Nicotiana tabacum* by somatic hybridization. *Theor. Appl. Genet.* **66**, 127–130.

Grafe, R., Marion-Poll, A., and Caboche, M. (1986). Improved *in vitro* selection of nitrate reductase-deficient mutants of *Nicotiana plumbaginifolia*. *Theor. Appl. Genet.* **73**, 299–304.

Grandbastien, M. A., Spielmann, A., and Caboche, M. (1989). TNT1, a mobile retroviral-like transposable element of tobacco isolated by plant cell genetics. *Nature (London)* **337**, 376–380.

Hamano, T., Oji, Y., Mitsuhashi, Y., Matsuki, Y., and Okamoto, S. (1984). Action of thiol proteinase on nitrate reductase in leaves of *Hordeum distichum L. Plant Cell Physiol.* **25**, 1469–1475.

Hoff, T., Stummann, B. M., and Henningssen, K. W. (1991). Cloning and expression of a gene encoding a root specific nitrate reductase in bean (*Phaseolus vulgaris*). *Physiol. Plant.* **82**, 197–204.

Hyde, G. E., and Campbell, W. H. (1990). High-level expression in *Escherichia coli* of the catalytic active flavin domain of corn leaf NADH:nitrate reductase and its comparison to humain NADH:cytochrome b_5 reductase. *Biochem. Biophys. Res. Commun.* **168**, 1285–1291.

Hyde, G. E., Wilberding, J. A., Meyer, A. L., Campbell, E. R., and Campbell, W. H. (1989). Monoclonal antibody-based immunoaffinity chromatography for purifying corn and squash NADH:nitrate reductases. Evidence for an interchain difulfide bond in nitrate reductase. *Plant Mol. Biol.* **13**, 233–246.

Ingemarsson, B. (1987). Nitrogen utilization in *Lemma*. I. Relations between net nitrate flux, nitrate reduction, and *in vitro* activity and stability of nitrate reductase. *Plant Physiol.* **85**, 856–859.

Jacobsen, E., Braaksma, F. J., and Feenstra, W. J. (1984). Determination of xanthine dehydrogenase activity in nitrate reductase deficient mutants of *Pisum sativum* and *Arabidopsis thaliana. Z. Pflanzenphysiol.* **113**, 183–188.

Johnson, J. L., Bastian, N. R., and Rajagopalan, K. V. (1990). Molybdopterin guanine dinucleotide, a modified form of molybdopterin identified in the molybdenum cofactor of dimethylsulfoxide reductase from *Rhodobacter sphaeroides. Proc. Natl. Acad. Sci. U.S.A.* **87**, 3190–3194.

Kaiser, W. M., and Brendle-Behnisch, E. (1991). Rapid modulation of spinach leaf nitrate reductase activity by photosynthesis. I. Modulation *in vivo* by CO_2 availability. *Plant Physiol.* **96**, 363–367.

Kaiser, W. M., and Spill, D. (1991). Rapid modulation of spinach leaf nitrate reductase activity by photosynthesis. I. Modulation *in vitro* by ATP and AMP. *Plant Physiol.* **96**, 368–375.

Kamachi, K., Amemiya, Y., Ogura, N., and Nakagawa, H. (1987). Immuno-gold localization of nitrate reductase in spinach (*Spinacia oleracea*) leaves. *Plant Cell Physiol.* **28**, 333–338.

Kennedy, I. R., Rigaud, J., and Trinchant, J. C. (1975). Nitrate reductase from bacteroids of *Rhizobium japonicum*: Enzyme characteristics and possible interaction with nitrogen fixation. *Biochim. Biophys. Acta* **397**, 24–35.

King, P., and Khanna, V. (1980). A nitrate reductase-less variant isolated from suspension culture of *Datura innoxia* (Mill). *Plant Physiol.* **66**, 632–636.

Kinghorn, J. R., and Campbell, E. I. (1989). Amino acid sequence relationships between bacterial, fungi, and plant nitrate reductase and nitrite reductase proteins. *In* "Molecular and Genetic Aspects of Nitrate Assimilation" (J. Wray and J. Kinghorn, eds.), pp. 385–403. Oxford Science Publications, Oxford.

Kleinhofs, A., Warner, R. L., Muehlbauer, F. J., and Nilan, R. A. (1978). Induction of specific gene mutations in *Hordeum* and *Pisum*. *Mutat. Res.* **51**, 29–35.

Kleinhofs, A., Warner, R. L., Hamat, H. B., Juricek, M., Huang, C., and Schnorr, K. (1988). Molecular genetics of barley and rice nitrate reductases. *Curr. Top. Plant Biochem. Physiol.* **7**, 35–42.

Kleinhofs, A., Warner, R. L., Lawrence, J. M., Melzer, J. M., Jeter, J. M., and Kudrna, D. A. (1989). Molecular genetics of nitrate reductase in barley. *In* "Molecular and Genetic Aspects of Nitrate Assimilation" (J. Wray and J. Kinghorn, eds.), pp. 197–211. Oxford Science Publications, Oxford.

Kleinhofs, A., Warner, R. L., and Melzer, J. M. (1989). Genetics and molecular biology of higher plant nitrate reductases. *Recent Adv. Phytochem.* **23**, 117–155.

Kramer, S. P., Johnson, J. L., Ribeiro, A. A., Millington, D. S., and Rajagopalan, K. V. (1987). The structure of the molybdenum cofactor. Characterization of di(carboxamidomethyl)molybdopterin from sulfite oxidase and xanthine oxidase. *J. Bacteriol. Chem.* **262**, 16357–16363.

Kubo, Y., Ogura, N., and Nakagawa, H. (1988). Limited proteolysis of the nitrate reductase from spinach leaves. *J. Biol. Chem.* **263**, 19684–19689.

Kudla, B., Kaddick, M. X., Langdon, T., Martinez-Rossi, N. M., Benett, C., Sibley, S., Davies, W., and Arst, H. N. (1990). The regulatory *are A* mediating nitrogen metabolite repression in *Aspergilus nidulans*: Mutations affecting specificity of gene activation alter a loop residue of a putative "zinc finger." *EMBO J.* **9**, 1355–1364.

Làzàr, G. B., Fankhauser, H., and Potrikus, I. (1983). Complementation analysis of a nitrate reductase-deficient *Hyoscyamus muticus* cell line by somatic hybridization. *Mol. Gen. Genet.* **189**, 359–364.

Lillo, C., and Henrikson, A. (1984). Comparative studies of diurnal variations of nitrate reductase activity in wheat, oat and barley. *Physiol. Plant.* **62**, 89–94.

Lu, J. L., Ertl, J. R., and Chen, C. M. (1990). Cytokinin enhancement of the light induction of nitrate reductase transcript level in etiolated barley leaves. *Plant Mol. Biol.* **14**, 585–594.

Lycklama, J. C. (1963). The absorption of ammonium and nitrate by perennial rye-grass. *Acta Bot. Neerl.* **12**, 361–423.

MacDonald, D. W., Cove, D. J., and Coddington, A. (1974). Cytochrome-*c* reductases from wild type and mutant strains of *Aspergillus nidulans*. *Mol. Gen. Genet.* **128**, 187–199.

Marion-Poll, A., Chérel, I., Gonneau, M., and Leydecker, M. T. (1991). Biochemical characterization of *cnx* nitrate reductase-deficient mutants of *Nicotiana plumbaginifolia*. *Plant Sci.* **76**, 201–209.

Martino, S. J., and Smarrelli, J. (1989). Nitrate reductase synthesis in squash cotyledons. *Plant Sci.* **61**, 61–67.

Marzluf, G., and Fu, Y. H. (1989). Genetics, regulation and molecular studies of nitrate assimilation in *Neurospora crassa*. *In* "Molecular and Genetic Aspects of Nitrate Assimilation" (J. Wray and J. Kinghorn, eds.), pp. 314–327. Oxford Science Publications, Oxford.

McEwan, A. G., Jackson, J. B., and Ferguson, S. J. (1984). Rationalization of properties of nitrate reductases in *Rhodopseudomonas capsulata*. *Arch. Microbiol.* **137**, 344–349.

Meltzer, J. M., Kleinhofs, A., and Warner, R. L. (1989). Nitrate reductase regulation: Effects of nitrate and light on nitrate mRNA accumulation. *Mol. Gen. Genet.* **217**, 341–346.

Mendel, R. R., and Müller, A. J. (1985). Repair *in vitro* of nitrate reductase-deficient tobacco mutants (*cnx*A) by molybdate and by molybdenum cofactor. *Planta* **163**, 370–375.

Mendel, R. R., Màrton, L., and Müller, A. J. (1986). Comparative biochemical characterization of mutants at the nitrate reductase/molybdenum cofactor loci *cnx* A, *cnx* B and *cnx* C of *Nicotiana plumbaginifolia*. *Plant Sci.* **43**, 125–129.

Meyer, C., Chérel, I., Moureaux, T., Hoarau, J., Gabard, J., and Rouzé, P. (1987). Bromphenol blue:nitrate reductase activity in *Nicotiana plumbaginifolia*. An immunological and genetic approach. *Biochimie* **69**, 735–742.

Meyer, C., Levin, J. M., Roussel, J. M., and Rouzé, P. (1991). Mutational and structural analysis of the nitrate reductase heme domain of *Nicotiana plumbaginifolia*. *J. Biol. Chem.* **266**, 20561–20566.

Müller, A. J. (1983). Genetic analysis of nitrate reductase-deficient tobacco plants regenerated from mutant cells. Evidence for duplicated structural genes. *Mol. Gen. Genet.* **192**, 275–281.

Müller, A. J., and Grafe, R. (1978). Isolation and characterization of cell lines of *Nicotiana tabacum* lacking nitrate reductase. *Mol. Gen. Genet.* **161**, 67–76.

Müller, A. J., and Mendel, R. R. (1989). Molecular and genetic aspects of nitrate reductase assimilation. *In* "Molecular and Genetic Aspects of Nitrate Assimilation" (J. Wray and J. Kinghorn, eds.), pp. 166–185. Oxford Science Publications, Oxford.

Nakagawa, H., Kubo, Y., Shiraishi, N., Sato, Y., and Ogura, N. (1990). Functional domains in higher plant nitrate reductase. *In* "Abstracts of Third International Symposium on Nitrate Assimilation, Molecular and Genetic Aspects," p. 78. Bombannes, INRA-Versailles, France.

Nam, H. G., Giraudat, J., Den Boer, B. G. W., Loos, W. D. B., Hauge, B. M., and Goodman, H. M. (1989). Restriction fragment length polymorphism linkage map of *Arabidopsis thaliana*. *Plant Cell* **1**, 699–705.

Napoli, C., Lemieux, C., and Jorgensen, R. (1990). Induction of a chimeric chalcone synthase gene into petunia results in reversible co-suppression of homologous genes in trans. *Plant Cell* **2**, 279–289.

Nelson, S. R., Ryan, S. A., and Harper, J. E. (1983). Soybean mutants lacking constitutive nitrate reductase activity. *Plant Physiol.* **72**, 503–509.

Notton, B. A., Fido, R. J., Whitford, P. N., and Barber, M. J. (1989). Effect of proteolysis on partial activities and immunological reactivity of spinach nitrate reductase. *Phytochemistry* **28**, 2261–2266.

Nussaume, L., Vincentz, M., and Caboche, M. (1991). Constitutive nitrate reductase: A dominant conditional marker for plant genetics. *Plant J.* **1**, 267–274.

Oaks, A., and Hirel, B. (1985). Nitrogen metabolism in roots. *Annu. Rev. Plant Physiol.* **36**, 345–365.

Oaks, A., Poulle, M., Goodfellow, V. J., Cass, L. A., and Deising, H. (1988). The role of nitrate and ammonium ions and light on the induction of nitrate reductase in maize leaves. *Plant Physiol.* **88**, 1067–1072.

Oelmüller, R., Schuster, C., and Mohr, H. (1988). Physiological characterization of a plastidic factor signal required for nitrate-induced appearance of nitrate and nitrite reductases. *Planta* **174**, 75–83.

Oostinder-Braaksma, F. J., and Feenstra, W. J. (1973). Isolation and characterization of chlorate-resistant mutants of *Arabidopsis thaliana*. *Mutat. Res.* **19**, 175–185.

Pelsy, F., and Gonneau, M. (1991). Genetic and biochemical analysis of intragenic complementation events among nitrate reductase apoenzyme-deficient mutants of *Nicotiana plumbaginifolia. Genetics* **127**, 199–204.

Pelsy, F., Kronenberger, J., Pollien, J. M., and Caboche, M. (1991). M2 seed screening for nitrate reductase deficiency in *Nicotiana plumbaginifolia. Plant Sci.* **76**, 109–114.

Poulle, M., Oaks, A., Bzonek, P., Goodfellow, V. J., and Solomonson, L. P. (1987). Characterization of nitrate reductases from corn leaves (*Zea mays* cv W64A × W182E) and *Chorella vulgaris. Plant Physiol.* **85**, 375–378.

Pouteau, S., Chérel, I., Vaucheret, H., and Caboche, M. (1989). Nitrate reductase mRNA regulation in *Nicotiana plumbaginifolia* nitrate reductase-deficient mutants. *Plant Cell* **1**, 1111–1120.

Rajasekhar, V., Gowri, G., and Campbell, W. (1988). Phytochrome-mediated light regulation of nitrate reductase expression in squash cotyledons. *Plant Physiol.* **88**, 242–244.

Remmler, J. L., and Campbell, W. H. (1986). Regulation of corn leaf nitrate reductase. II. Synthesis and turnover of the enzyme's activity and protein. *Plant Physiol.* **80**, 442–447.

Roldan, J. M., Romero, F., Lopez-Ruiz, A., Diez, J., and Verblen, J. P. (1987). Immunological approaches to inorganic nitrogen metabolism. *In* "Inorganic Nitrogen Metabolism" (W. R. Ullrich, P. J. Aparaciao, P. J. Syrett, and F. Castillo, eds.), pp. 94–98. Springer-Verlag, Berlin.

Saux, C., Lemoine, Y., Marion-Poll, A., Valadier, H., Deng, M., and Morot-Gaudry, J. F. (1987). Consequence of absence of nitrate reductase activity on photosynthesis in *Nicotiana plumbaginifolia* plants. *Plant Physiol.* **84**, 67–72.

Scazzocchio, C., and Arst, H. N. (1989). Regulation of nitrate assimilation in *Aspergillus nidulans. In* "Molecular and Genetic Aspects of Nitrate Assimilation" (J. Wray and J. Kinghorn, eds.), pp. 299–313. Oxford Science Publications, Oxford.

Schoenmakers, H. C. H., Koornneef, M., Alefs, S. J. H. M., Gerrits, W. F. M., Van der Kop, D., Chérel, I., and Caboche, M. (1991). Isolation and characterization of nitrate reductase deficient mutants in tomato (*Lycopersicon esculentum* Mill.). *Mol. Gen. Genet.* **227**, 458–464.

Scholten, H. J., de Vries, S. E., Nijdam, H., and Feenstra, W. J. (1985). Nitrate reductase deficient cell lines from diploid cell cultures and lethal mutant M2 plants of *Arabidopsis thaliana. Theor. Appl. Genet.* **71**, 556–562.

Schuster, C., Schmidt, S., and Mohr, H. (1989). Effect of nitrate, ammonium, light and a plastidic factor on the appearance of multiple forms of nitrate reductase in mustard (*Sinapsis alba* L.) cotyledons. *Planta* **177**, 74–83.

Shen, T. C., Funkhouse, E. A., and Guerrero, M. G. (1976). NADH- and NAD(P)H-nitrate reductase in rice seedlings. *Plant Physiol.* **68**, 292–294.

Solomonson, L., and Barber, M. (1990). Assimilatory nitrate reductase: Functional properties and regulation. *Annu. Rev. Plant Physiol. Mol. Biol.* **41**, 225–253.

Solomonson, L., and McCreery, M. J. (1986). Radiation inactivation of assimilatory NADH:nitrate reductase from *Chlorella.* Catalytic and physical sizes of functional units. *J. Biol. Chem.* **261**, 806–810.

Somers, D. A., Kuo, T. M., Kleinhofs, A., Warner, R. L., and Oaks, A. (1983). Synthesis and degradation of barley nitrate reductase. *Plant Physiol.* **72**, 949–952.

Steven, B. J., Kirk, D. W., Bright, S. W. J., and Wray, J. L. (1989). Biochemical genetics of further chorate resistant, molybdenum cofactor defective, conditional-lethal mutants of barley. *Mol. Gen. Genet.* **219**, 421–428.

Strauss, A., Bucher, F., and King, P. J. (1981). A search for biochemical mutants using haploid mesophyll protoplasts of *Hyoscyamus muticus*. 1. A NO_3^- non-utilizing clone. *Planta* **153**, 75–80.

Streit, L., Martin, B. A., and Harper, J. E. (1987). A method for the separation and purification of the three forms of nitrate reductase present in wild-type soybean leaves. *Plant Physiol.* **84**, 654–657.

Sueyoshi, K., Ogura, N., and Nakagawa, H. (1989). Identification of possible intermediates in *in vivo* degradation of spinach nitrate reductase. *Agric. Biol. Chem.* **53**, 151–156.

Suzuki, A., and Gadal, P. (1984). Glutamate synthase: Physicochemical and functional properties of different forms in higher plants and in other organisms. *Physiol. Vég.* **22**, 471–486.

Thibaud, J. B., and Grignon, C. (1981). Mechanism of nitrate uptake in corn roots. *Plant Sci. Lett.* **22**, 279–289.

Tischner, R., Ward, M. R., and Huffaker, R. C. (1989). Evidence for a plasma-membrane-bound nitrate reductase involved in nitrate uptake of *Chlorella sorokiniana*. *Biochem. J.* **250**, 921–923.

Truong, H. N., Meyer, C., and Daniel-Vedele, F. (1991). Characteristics of *Nicotiana tabacum* nitrate reductase protein produced in *Saccharomyces cerevisiae*. *Biochem. J.* **278**, 393–397.

Ulrich-Eberius, C. I., Novacky, A., Fischer, E., and Lüttge, U. (1981). Relationship between energy-dependent phosphate uptake and the electrical membrane potential in *Lemna gibba* GI. *Plant Physiol.* **67**, 797–801.

Van der Krol, A. R., Mur, L. A., Beld, M., Mol, J. N. M., and Stuitje, A. R. (1990). Flavonoid genes in petunia: Addition of a limited number of gene copies may lead to a suppression of gene expression. *Plant Cell* **2**, 291–299.

Vaucheret, H., Vincentz, M., Kronenberger, J., Caboche, M., and Rouzé, P. (1989). Molecular cloning and characterization of the two homeologous genes coding for nitrate reductase in tobacco. *Mol. Gen. Genet.* **216**, 10–15.

Vaucheret, H., Chabaud, M., Kronenberger, J., and Caboche, M. (1990). Functional complementation of tobacco and *Nicotiana plumbaginifolia* nitrate reductase deficient mutants by transformation with the wild type alleles of the tobacco structural genes. *Mol. Gen. Genet.* **220**, 468–474.

Vaughn, K. C., and Campbell, W. H. (1988). Immunogold localization of nitrate reductase in maize leaves. *Plant Physiol.* **88**, 1354–1357.

Vincentz, M., and Caboche, M. (1991). Constitutive expression of nitrate reductase allows normal growth and development of *N. plumbaginifolia* plants. *EMBO J.* **10**, 1027–1035.

Walker-Simmons, M., Kudrna, D., and Warner, R. (1989). Reduced accumulation of ABA during water stress in a molybdenum cofactor mutant of barley. *Plant Physiol.* **90**, 728–733.

Wang, X. M., Scholl, R. L., and Feldman, K. A. (1986). Characterization of a chlorate-hypersensitive, high nitrate reductase *Arabidopsis thaliana* mutant. *Theor. Appl. Genet.* **72**, 328–336.

Wang, X. M., Feldman, K. A., and Scholl, R. L. (1988). A chlorate-hypersensitive, high nitrate/chlorate uptake mutant of *Arabidopsis thaliana*. *Physiol. Plant.* **73**, 305–310.

Warner, R. L., and Huffaker, R. C. (1989). Nitrate transport is independent of NADH and NAD(P)H nitrate reductase in barley seedlings. *Plant Physiol.* **91**, 947–953.

Warner, R. L., Lin, C. J., and Kleinhofs, A. (1977). Nitrate reductase-deficient mutants in barley. *Nature (London)* **269**, 406–407.

Warner, R. L., Kleinhofs, A., and Muehlbauer (1982). Characterization of nitrate reductase-deficient mutants in pea. *Crop Sci.* **22,** 389–393.

Warner, R. L., Narayanan, K. R., and Kleinhofs, A. (1987). Inheritance and expression of NAD(P)H nitrate reductase in barley. *Theor. Appl. Genet.* **74,** 714–717.

Wilkinson, J. Q., and Crawford, N. M. (1991). Identification of the *Arabidopsis CHL3* gene as the nitrate reductase structural gene *NIA2. Plant Cell* **3,** 461–471.

Wray, J. (1986). The molecular genetics of higher plant nitrate assimilation. *In* "A Genetic Approach to Plant Biochemistry" (A. D. Blonstein and P. J. King, eds.), pp. 101–157. Springer-Verlag, New York.

Wray, J. L. (1988). Molecular approaches to the analysis of nitrate assimilation. *Plant, Cell Environ.* **11,** 369–382.

Xuan, L. T., Grafe, R., and Müller, A. J. (1983). Complementation of nitrate reductase-deficient mutants in somatic hybrids between *Nicotiana* species. *In* "Protoplasts" (I. Potrykus, C. T. Harms, A. Hinnen, P. S. King, and R. D. Shillito, eds.), pp. 75–76. Birkhaeuser, Basel.

GENETIC IMPLICATIONS OF SOMACLONAL VARIATION IN PLANTS

Virginia M. Peschke*[1] and Ronald L. Phillips†

*Department of Biology, Washington University, St. Louis, Missouri 63130
†Department of Agronomy and Plant Genetics and the
Plant Molecular Genetics Institute,
University of Minnesota, St. Paul, Minnesota 55108

I. Introduction

Somaclonal variation was defined by Larkin and Scowcroft (1981) as the genetic variation displayed in tissue culture-regenerated plants and their progeny. This variation, along with the corresponding changes observed in tissue cultures per se has been documented by numerous authors. While serving as a resource to the investigator seeking traits such as resistance to disease or stress, somaclonal variation is often a

[1] Present address: Monsanto Company, St. Louis, Missouri 63198.

41

Copyright © 1992 by Academic Press, Inc.
All rights of reproduction in any form reserved.

nuisance (at best) to those attempting to create defined changes in plants through transformation, or who wish to preserve germplasm *in vitro*. As stated by Karp and Bright (1985), "Resolution of this problem may turn out to make a greater contribution to plant science than direct exploitation of the phenotypic variation itself." It is in this spirit that the present review is written.

Throughout this paper, "somaclonal variation" is taken to mean any genetic, cytogenetic, or molecular changes produced during tissue culture or plant regeneration. Changes detected in the sexual progeny of regenerated plants are often considered to be "stable," and contrasted with the "epigenetic" traits sometimes detected in tissue cultures or primary regenerants. However, as will be discussed, this distinction is a blurry one, as not all traits inherited even through one or more sexual generations will continue to behave in a genetically stable fashion. Our scope will be limited to variation known or believed to involve the nuclear genome, although rearrangements and genetic variation involving both the mitochrondrial and chloroplast genomes have been documented (e.g., Brettell *et al.*, 1980; Gengenbach *et al.*, 1981; Chourey and Kemble, 1982; Hanson, 1984; Kemble and Shepard, 1984; Rose *et al.*, 1986; Hartmann *et al.*, 1987; Harada *et al.*, 1991; Shirzadegan *et al.*, 1991).

In this review, we will investigate the topic of somaclonal variation from two perspectives. The first is an examination of the various categories of somaclonal variation and speculation on the underlying mechanisms, including a discussion of how seemingly unrelated phenomena may be interconnected. The second is a description of the unusual genetic outcomes that would be predicted, and the implications of these types of variation for use of tissue culture-derived materials. Emphasis will be placed on the genetic changes that may be detected in regenerated plants and their progeny, because they are of most practical significance. However, changes occurring in the cultures will also be discussed and cannot be ignored; they are likely to provide more data on the underlying mechanisms, because they are the first events to occur, and they have not yet been "screened" by the requirements of morphogenesis.

II. Influence of Genotype/Developmental Stage on the Occurrence and Frequency of Somaclonal Variation

A. EXPLANT SOURCE TISSUE

Relatively few studies have focused on the influence of the explant source on somaclonal variation, possibly because the source often defines whether or not the culture system will work; that is, explants from various tissue sources do not culture equally well. However, two generalizations can be made:

1. Meristems cultured without a state of dedifferentiation will produce little or no variation compared to when a dedifferentiated state is induced (Karp and Bright, 1985; Bayliss, 1980; D'Amato, 1985; Potter and Jones, 1991).

2. Differences in the stability of tissue cultures produced from different explant tissue sources can often be traced to variability preexisting in the explant. The most widely recognized case of this occurrence is polysomaty (wherein diploid and polyploid cells coexist in the same tissue); this condition may be found in over 90% of plant species (D'Amato, 1952, 1985). For example, van den Bulk et al. (1990) found little difference in regenerants from tomato cultures derived from different explants (cotyledon, leaf, and hypocotyl) except for a high degree of polyploidy (58%) in plants regenerated from hypocotyl cultures. They also demonstrated that the hypocotyl is polysomatic whereas the other explants tested (cotyledon and leaf) showed few to no nondiploid cells. Similarly, a comparison of protoplast-derived potato plants showed a higher frequency of chromosomal and phenotypic alterations when the protoplasts were obtained from a chromosomally variable cell suspension than when they were freshly isolated from diploid tissue (Sree Ramulu et al., 1986). In a case apparently unrelated to explant chromosomal instability, some differences in variation for agronomic traits were observed in rice plants regenerated from seed versus plumule culture (Heszky et al., 1989), though no explanation for this was suggested.

B. Culture Type

As indicated above, meristem-derived cultures without a significant dedifferentiation step give rise to normal or nearly normal plants. Among those types of cultures that regenerate after a dedifferentiated (callus) phase, one can distinguish between regeneration by organogenesis (formation of roots or shoots from meristematic regions) or somatic embryogenesis (formation of somatic embryos, which then "germinate"). It is clear from histological studies that organogenesis often involves more than one cell (e.g., Springer et al., 1979); a number of genetic analyses have demonstrated this as well. For example, both sectored and nonsectored maize plants that regenerated via organogenesis from the same explant contained the identical chromosome aberration (Benzion and Phillips, 1988). This result indicated that the variant was produced in callus and that the sectored plant had a multicellular origin. In addition, chimeras produced early in regenerated plant development (based on sector size) have been seen in tobacco (de Boucaud and Gaultier, 1983) and tomato (Sree Ramulu *et al.*, 1976). Additional experiments in which tobacco chimeras were synthesized by mixing cells (Marcotrigiano and Gouin, 1984) provide further evidence for a multicellular origin, though the plants did not survive long enough for genetic testing.

In the case of somatic embryogenesis, the embryoid is often derived from a single cell, although evidence for a multicellular origin has also been obtained. For example, histological evidence for a single-cell origin of somatic embryos has been observed for pearl millet (Vasil and Vasil, 1982), sugarcane (Ho and Vasil, 1983), and *Daucus* (Haccius 1978). In contrast, some maize somatic embryos have been shown to originate from groups of cells (Vasil *et al.*, 1985). Primarily mixoploid celery plants were regenerated via embryogenesis from mixoploid cell cultures, indicating that the plants were derived from more than one cell (Browers and Orton, 1982).

Some authors have argued that plants regenerated from somatic embryoids ought to contain fewer mutations than those regenerated via organogenesis due to the (presumably) more stringent genetic requirements imposed by embryo formation (Swedlund and Vasil, 1985). It is clear that somatic embryogenesis, and plant regeneration in general, do screen out variability; plants derived from tissue culture are usually less variable than the cultures from which they have regenerated. Alfalfa plants that regenerated via embryogenesis from mixoploid cultures (originating from polysomatic explant tissue) were most often diploid, except for one apparently stable aneuploid configuration ($2n = 30$ versus the normal $2n = 32$) (Feher *et al.*, 1989). Similarly, *Citrus sinensis* plants regenerated via embryogenesis from highly variable colchicine-treated cultures were either diploid or tetraploid (Gmitter *et al.*, 1991). However, this "screen" is not complete: variability from plants regenerated via somatic embryogenesis has been reported in a number of studies (e.g., Karp and Maddock, 1984; Browers and Orton, 1982; Ahloowalia and Maretzki, 1983). A comparative study (Armstrong and Phillips, 1988) indicated that maize plants regenerated from embryogenic callus were in fact more variable, both phenotypically and cytologically, than those from organogenic callus from the identical explant. After 4 months, 20.8% of the plants from Type I (organogenic) and 33.3% of the plants from Type II (friable embryogenic) callus contained one or more abnormalities; these frequencies increased to 24.2 and 37.2% at 9 months. This finding may be explained by either an increased mistake rate or a constant mutation rate per cell generation multiplied by an increased number of generations. As expected, a lower (but nonzero) percentage of chimeras was observed from the embryogenic callus, consistent with an origin of fewer cells.

C. Culture Age

Many investigators have noted an increase in variability of all types with an increase in culture age (e.g., McCoy *et al.*, 1982; Benzion and Phillips, 1988; Lee and Phillips, 1987a,b; Armstrong and Phillips, 1988; Muller *et al.*, 1990). Rigorous analysis of culture age effect is limited by the ability to assess growth rates (i.e., number of cell generations/time) as well as the time it may take for a mutant to proliferate sufficiently to be detected in a regenerated plant. However, age effect has generally been attributed to one or a combination of four causes:

1. The culture per se becomes more prone to change as it gets older, i.e., the mutation rate per cell generation increases.
2. The mutation rate is approximately constant over time, so the number of mutations increases over time, provided selection does not eliminate them.
3. A number of mutations take place early but are not detected until much later, when a sufficient number of mutant cells have accumulated to be included in samples of the callus or in a regenerated plant. This problem in sampling will produce a "lag period" of apparent culture stability.
4. Mutations that occur at an early culture age are actively selected, forming an increasingly greater proportion of the cells over time.

Orton (1984) discusses somaclonal variation from the perspective of the above-mentioned fourth cause, indicating that a growing callus has a set of metabolic requirements different than that of a plant (e.g., no need to photosynthesize) and that selection will work to remove excess "baggage." For example, there are many reports of albino plants arising from tissue culture, e.g., brome grass (Gamborg *et al.*, 1970), ryegrass (Ahloowalia, 1975), sugarcane (Ahloowalia and Maretzki, 1983), and barley (Luhrs and Lorz, 1988). In these cases, there is no simple way to determine whether the mutants have been selected or whether they are merely tolerated by the culture environment. However, many genetic mutants detected in cultures and regenerated plants are heterozygous and recessive; they would provide no apparent selective advantage or disadvantage unless gene dosage is important.

A study by Benzion and Phillips (1988) indicates that the mutation rate does not necessarily increase over time but that, rather, mutations occur early that are able to accumulate, because selection does not eliminate them. Pedigrees of a large number of callus lines indicated that cytological aberrations detected in plants regenerated after many months in culture could often be traced back to a common subculture point early in the culture process. Also, though the general frequency of aberrations increased over time, specific aberrations (which could be followed due to being cytologically distinct) often had different patterns (e.g., an increase followed by a decrease), which could be caused by selection or drift. Armstrong and Phillips (1988) saw a decrease in the frequency of chimeric plants over time in culture while the frequency of total variant plants increased, which also supports the hypothesis that many mutations occur after short culture periods but are not detected in regenerated plants at that point due to sampling. These findings indicate that use of short culture periods to decrease variability may not always be effective.

In contrast, mutations detected in rice regenerants appear to have occurred at various times throughout the culture process (Fukui, 1983). Any given regenerant contained from zero to four mutations, based on progeny tests. All plants with at least one mutation carried the same one (early heading). Within this group, a subset also contained a second mutation (albinism). Only in plants with both early heading and albinism were the last two mutations (short culm and sterility) detected, demonstrating that the mutations had occurred in sequence during the lifetime of the callus, rather than all together at a relatively early point.

D. EXPLANT GENOTYPE

1. Preexisting Variation

As described above for the case of polysomaty, it is clear that the explant tissue may not be genetically homogeneous, that heterogeneity may be "magnified" by the proliferation of differing cell types, and that preexisting variation is often reflected in the regenerated plants. Protoplasts from a Su/su (light green) tobacco plant were cultured by Lorz and Scowcroft (1983). In 20/79 cases in which a culture produced variant (dark green or pale) plants, all of the plants from the culture were phenotypically identical. This finding points either to extremely early mutational events or to variation preexisting in the protoplasts. Barbier and Dulieu (1983) observed a similar result, again demonstrating either preexisting variation or mutation in the very first cell generations. Navarro *et al.* (1985) found that *Citrus clementina* plants regenerated from a given nucellus would either be all normal or all abnormal.

The question has been raised whether somaclonal variation is primarily due to such preexisting variation, rather than to new mutations (Morrish *et al.*, 1990). To address this, variation between two lines of *Pennisetum glaucum* (pearl millet) was compared. One line was a green inbred characterized as "stable." The other line, considered to be "genetically unstable," was apparently less inbred and in addition contained several plant color marker alleles (one dominant purple mutation, one non-Mendelian chlorophyll mutation). More color changes were detected in somaclones derived from the marker line than from the stable line. However, statistically significant differences for plant height and seed weight were observed between progeny of regenerants from the genetically stable line and the corresponding noncultured controls, indicating that somaclonal variation could be observed even within the stable line. Though this study and others indicate that preexisting variation may be perpetuated in culture and detected in regenerated plants, this does not disprove the occurrence of *de novo* variation. For example, the study by Lorz and Scowcroft (1983) cited above, which demonstrated preexisting variation in the explant, also produced a number of heterogeneous protoplast-derived colonies (i.e., colonies producing plants of several colors). Because the colonies clearly originated from a single cell, the variability must have originated in culture.

2. Varietal Differences

A number of reports indicate that somaclonal variation can be influenced by the explant variety (cultivar) used. A dramatic example of this has been observed in oats (McCoy *et al.*, 1982), for which cytological stability of regenerants from two cultivars, Lodi and Tippecanoe, was compared. In plants regenerated after 4 months, 12% of the Tippecanoe plants had one or more variants compared to 49% from Lodi; these frequencies increased by 8 months to 48 and 89%, respectively. Variety differences have also been reported for maize (Zehr *et al.*, 1987), wheat (Mohmand and Nabors, 1990), and *Medicago media* (Nagarajan and Walton, 1987). The basis for these differences has not been determined, but in the same way that there are dramatic differences in the ability of various lines to form regenerable cultures, differences may exist in the degree to which the tissue culture environment disrupts the cellular environment of a particular line.

III. Major Types of Somaclonal Variation and Their Genetic Consequences

A. CHROMOSOMAL ABERRATIONS

1. Polyploidy/Aneuploidy

Possibly hundreds of studies of chromosomal variation and behavior in plant tissue culture have been undertaken over the past 30 years and have been described in a number of detailed reviews (Sunderland, 1977; Bayliss, 1980; D'Amato, 1985; Lee and Phillips, 1988). Initial observations focused on ploidy changes, because they can be detected even in mitotic cells with small chromosomes and/or large chromosome numbers. The first species successfully cultured (e.g., tobacco, carrot), generally fell into this category. Polyploidy in tissue culture has been assumed to take place via mechanisms similar to those seen *in vivo;* it has been generally explained as the product of either endopolyploidization or nuclear fusion (Sunderland, 1977; Bayliss, 1980). Aneuploidy is somewhat more difficult to understand, because it is less often seen *in vivo* as a natural part of plant development. Cytological evidence indicates that aneuploidy may be caused by nondisjunction, lagging chromosomes, aberrant spindles, and chromosome breakage producing dicentric and acentric chromosomes (Sunderland, 1971; see Section III,A,2,a). Balzan (1978) proposed that aneuploidy may be due in some cases to polyploidy followed by chromosome elimination, based on the observation of lobed (constricted) nuclei reminiscent of "budding."

It should be remembered that aneuploidy is better tolerated in species that are normally polyploid than in those with a diploid chromosome number, due to the greater imbalance of genetic material in the diploid situation. This "rule" is borne out in analyses of chromosome number in regenerated plants; aneuploid regenerants are common in polyploids but occur rarely in diploids (summarized in D'Amato, 1985). Similarly, whatever aneuploidy does occur in diploid regenerants is much less likely to be transmitted to progeny generations than in the case of regenerants from polyploid species.

2. Chromosome Breakage

a. Detection and Location of Breakpoints. It became apparent to some researchers interested in somaclonal variation that a great deal

mutator (Spm) [also called *Enhancer (En)*]. *Ac* and *Spm* are classified as "autonomous" elements, because they produce the gene product(s) necessary for their own transposition. A number of "nonautonomous" elements related to both *Ac* and *Spm* have been identified; these non-autonomous elements are able to transpose only in the presence of the corresponding autonomous element. Autonomous transposable element activity in regenerated plants or their progeny can be detected by making test crosses to genetic stocks containing the appropriate nonautonomous transposable element controlling easily scored genetic markers. Based on such tests, *Ac* (Peschke *et al.,* 1987) and *Spm* (Peschke and Phillips, 1991) were observed in the progeny of 1–3% of the regenerated plants analyzed. Analysis of noncultured control materials indicated that neither of these elements was present in the explant source. Furthermore, cell lines that produced transposable element-containing plants also produced a number of plants without transposable element activity, indicating that "activation" took place sometime during callus rather than during plant regeneration. Evola *et al.* (1984, 1985) have also reported activation of *Ac* and *Spm* in a small number of regenerated plants. Culley (1986, 1991) has reported the activation of *Ac* in maize endosperm cultures containing an *Ac*-responsive allele as a marker, but such cultures are nonregenerable and could not be tested genetically.

Strong evidence for the activation of a transposable element in tissue culture has also been obtained in alfalfa (*Medicago sativa* L.). An unstable flower color allele (white with purple sectors) at the *C2* locus is most likely to be due to the presence of an autonomous transposable element; when the element is present in the gene its function is lost, and when it excises color is restored (Groose and Bingham, 1986). The frequency of reversion (transposable element excision) is more than 20 times greater *in vitro* than *in planta*. Genetic studies indicate that the factor responsible for the instability behaves as a single gene (Ray and Bingham, 1991). Pouteau *et al.* (1991) have recently demonstrated that the Tnt-1 retrotransposon in tobacco may become active (as assayed by levels of transcription) by protoplast isolation. This response can be induced by the cell wall hydrolases used and does not appear to be a response to wounding or osmotic stress.

The activation of transposable elements is different than many other types of phenotypic somaclonal variation in that it represents a gain of function. The high frequency with which activation occurs indicates that such elements exist in the genome in an inactive or slightly imperfect form. This is known to be the case in maize; most if not all of the maize lines tested with DNA probes for known transposable elements contain multiple copies of sequences homologous to these probes, even though no genetic activity of these elements is detected (Fedoroff *et al.*, 1983; Cone *et al.*, 1986; Chandler *et al.*, 1988; Johns *et al.*, 1985). This was shown to be true as well for the lines giving rise to transposable element activity in tissue culture (Peschke and Phillips, 1991; Peschke *et al.*, 1991).

In many cases not involving tissue culture, transposable element activity has been correlated with reduced methylation at specific sites within the elements (Chomet *et al.*, 1987; Schwartz and Dennis, 1986; Chandler and Walbot, 1986; Bennetzen, 1987; Schwartz, 1989). Tissue cultures were initiated from a maize line containing an *Ac* element known to cycle frequently between activity and inactivity; though the element started out inactive, it became active at 80 times the normal frequency (Dennis and Brettell, 1990). At the same time, a reduction in methylation was detected in the 5' end of the element. Some correlation has been observed between transposable element methylation and activity in our materials from tissue culture, though it is not as clear-cut as in the studies just mentioned (Peschke *et al.*, 1991). Though plants with *Ac* activity often contained *Ac*-homologous sequences hypomethylated at certain restriction sites, this did not appear to be a prerequisite for the activity because some plants with activity did not show the methylation change. In contrast, all *Ac*-homologous sequences in plants without *Ac* activity were methylated at the sites tested.

Culley (1986, 1991) observed that *Ac*-homologous sequences in endosperm callus cultures (which appeared to have *Ac* activity based on phenotype) were much less methylated than plant tissues of the starting genotype. This indicated to us the possibility that activation of transposable elements in tissue culture might result from large-scale demethylation in callus followed by improper remethylation during plant regeneration. However, a comparison of 4-month-old maize callus and seedling DNA indicated that methylation within and surrounding *Ac*-homologous sequences was similar in both tissues (V. M. Peschke, unpublished).

3. Evidence for Transposition in Callus

If transposable elements are involved in producing tissue culture-derived mutations, one would expect to observe their transposition in callus. Gorman and Peterson (1978) demonstrated that maize endosperm cultures initiated from a genotype with an active transposable element showed spotting similar to that observed on kernels of the same genotype. Culley (1986, 1991) used a similar genetic system to observe directly the activation of transposable elements in maize endosperm callus. However, in the absence of genetic markers visible in the callus, transposition can be assayed by following the change in band size on DNA blots probed with element-homologous sequences. James and Stadler (1989) initiated callus containing the maize element *Mutator* (*Mu*) and observed such band shifts in 38% of the sublines; a similar result was obtained by Planckaert and Walbot (1989). No novel fragments were observed in the control cell lines containing methylated (inactive) *Mu* elements, nor was there evidence for any reduction in methylation of these elements. In contrast, some elements in the active lines appeared to become inactive over time (James and Stadler, 1989; Planckaert and Walbot, 1989).

4. Mutational Effects of Transposable Elements

As discussed earlier, one motivation for studying transposable elements in regenerated plants and callus was to learn whether they might be responsible for some of the other types of variability that have been reported. To date, there is little direct evidence to indicate that they are involved in somaclonal variation. At first glance, one would expect many transposable element-induced variants to be unstable. However, there are few reports of such instability; the most noteworthy case is the unstable alfalfa color allele described above (Groose and Bingham, 1986). Unstable transmission of a trait for supernumary heads in wheat has been reported (Ahloowalia and Sherington, 1985), though it is not known whether a transposable element is present in this case.

Though many transposable element-induced mutations are in fact genetically unstable (and recognized in this way), stable mutants can be produced by transposable elements as well. For example, a nonautonomous element will remain stably inactive in the absence of the corresponding autonomous element; whether it has any phenotypic effect depends on its site of insertion. In addition, insertion of a transposable element generally creates a "footprint," a small duplication of host DNA, which is not always correctly repaired on excision (Sachs *et al.*, 1983; Pohlman *et al.*, 1984a,b; Schwarz-Sommer *et al.*, 1985; Saedler and Nevers, 1985; Chen *et al.*, 1986). Some footprints produce no apparent change in phenotype (Sachs *et al.*, 1983; Dooner, 1980), whereas others result in nulls (Dooner and Nelson, 1977) or altered gene products (Dooner and Nelson, 1979; Echt and Schwartz, 1981; Chen *et al.*, 1986). Large deletions (Peacock *et al.*, 1984; Taylor and

Walbot, 1985; Dooner *et al.*, 1988) and chromosome breakage (Mc-Clintock, 1951; Doring and Starlinger, 1984; Robertson and Stinard, 1987) have also been reported. Finally, a transposable element that is active in callus may become inactive before plant regeneration (James and Stadler, 1989; Planckaert and Walbot, 1989), producing a genetically stable insertion.

Williams *et al.* (1991) examined *Ac*-homologous sequences in the DNA of tissue culture-derived plants containing 10 different variants (7 single-gene recessive and 3 of unknown origin) and were unable to detect any changes in the number or location of these sequences. However, the lack of transposable element involvement in a specific case does not imply that the elements never contribute to somaclonal variation. There are over a dozen known transposable element systems in maize (Peterson, 1986), any of which could potentially become active in tissue culture. In any case, transposable element activation can be considered to be a symptom of genetic instability in tissue culture; whether it is a major cause of other types of variation remains to be seen.

As in the case of chromosome breakage, an initial occurrence of transposable element activation may produce yet more activation. For example, the presence of one active *Spm* element in the genome can cause the transient activation of additional elements (McClintock, 1958, 1959, 1971). Molecular analysis indicates that the methylation of inactive *Spm* element-homologous sequences is reduced by the presence of an active element, and that in some cases these sequences will become genetically active (Banks *et al.*, 1988; Fedoroff, 1989).

C. Methylation Changes

1. Relationship between Methylation and Gene Expression

As discussed in the previous section, transposable element activity is correlated in many cases with reduced methylation. This general correlation has been recognized for some time for many animal genes (Yisraeli and Szyf, 1984; Holliday, 1987; Cedar and Razin, 1990). Likewise, some plant gene activity appears to be correlated with methylation (Hepburn *et al.*, 1983; Bianchi and Viotti, 1988) whereas for other genes it does not (Nick *et al.*, 1986). In all cases it is important to remember that correlations do not prove a cause-and-effect relationship. In addition, the ability to change gene regulation by altering methylation *in vitro* (e.g., Hershkovitz *et al.*, 1990) does not prove that this is the plant's normal mechanism for doing so. However, the abundance of information suggesting a relationship between methylation and gene regulation has produced increasing interest in whether tissue culture may cause changes in methylation.

2. Observations on Callus and Regenerated Plants

Methylation changes are most frequently studied by the use of restriction enzymes (having varying sensitivity to methylated bases) to cleave callus or plant DNA, which is then fractionated by gel electrophoresis. Changes affecting the genome as a whole may be estimated by the degree of restriction enzyme cleavage seen in the ethidium bromide-stained gel, whereas effects on individual gene methylation may be analyzed by Southern blotting and hybridization using probes corresponding to specific gene sequences. Two of the first reports on methylation changes in tissue culture (Brown and Lorz, 1986; Brown, 1989) indicated that both increases and decreases in overall methylation could be observed in 16% of the regenerated maize plants tested. These changes persisted through several generations of selfing (Brown, 1989). Such changes were also detected using probes to cloned sequences such as the maize *Waxy* gene (Brown, 1989). No correlation between methylation changes and abnormal phenotypes was observed; however, the probes used bore no apparent relationship to the aberrant phenotypes detected. A similar result has been observed in rice (Muller *et al.*, 1990), wherein methylation changes in tissue culture-derived plants were not obviously correlated with abnormal phenotypes (although they were correlated with genomic rearrangements detected by other probes).

Both increases and decreases in methylation have been observed to result from tissue culture, though there is some trend toward reduced methylation as the more common event. Examination of petunia 18S–25S ribosomal genes indicated that methylation in callus was generally lower than in other plant tissues; this level was stable over 10 months in callus. Shoots regenerated from the callus were similar to the original explant in their methylation levels (Anderson *et al.*, 1990). Soybean 5S ribosomal genes showed reduced methylation early in culture, based on the cleavage of formerly uncleavable *Hpa*II restriction sites. After several years, this methylation could be regained (Quemada *et al.*, 1987). Culley (1986, 1991) observed that *Ac*-homologous DNA sequences in endosperm-derived maize callus were greatly reduced in methylation compared to leaf tissue of the same genotype, though this was not observed in a small sample of embryo-derived maize callus (V. M. Peschke, unpublished observations). Phillips *et al.* (1992) compared methylation of 21 single-copy sequences between control plants and 22 R1 (first progeny generation) families. Of the families, 34% showed varied methylation with one or more of the probes. In all cases, the change was a reduction in methylation; no increases were seen. In addition, 17% of the R1 families containing alterations did not segregate for the change, indicating that the original regenerant was homozygous (Section IV,B).

One common approach to studying methylation effects that requires some caution is the use of 5-azacytidine or other cytosine analogs to inhibit cytosine methylation. Though these analogs do appear to cause activation of transposable elements and other genes (Pan and Peterson, 1988; Jones, 1984), it may not be certain whether the result observed is due to induced hypomethylation or to some other cause. For example, it has been shown that 5-azacytidine may cause point mutations, mitotic recombination (Zimmermann and Scheel, 1984), and chromosome breakage (Jones, 1984) in addition to reducing methylation. In cultures of maize and tobacco, application of 5-azacytidine appeared to have more of an effect on growth than on methylation (Brown *et al.*, 1989).

D. OTHER DNA-LEVEL ABERRATIONS

1. *Single-Base Changes*

Little information on the frequency of changes at the nucleotide level has been reported, primarily because sequence analysis first requires the identification of a mutant presumed to correspond to a previously characterized sequence. Though such data may become increasingly common due to the relative ease provided by the polymerase chain reaction (PCR), which allows sequencing of a variant without first cloning it, one must still define a region of interest to target. It is doubtful that completely random sequence comparisons would yield sufficient information to justify their use at this point.

Two cases wherein sequence analysis has been done involve the maize *Adh1* locus, which encodes the primary form of maize alcohol dehydrogenase. Cultures were initiated from materials of known isozyme constitution; 1382 plantlets were regenerated and their root proteins were analyzed on starch gels. Two variants were isolated: one electrophoretic variant (Brettell *et al.*, 1986a) and one null (Dennis *et al.*, 1987). Both were sequenced and each was found to be due to a single base change (conversion of an A to a T). In the case of the electrophoretic variant, this resulted in an amino acid change, whereas in the case of the null, a premature stop codon was produced.

2. RFLP Alterations

DNA analysis using restriction fragment length polymorphisms (RFLPs) may be used to screen for one of three types of variants: methylation changes, chromosome rearrangements, and single base changes. The use of RFLPs to detect methylation changes was described in Section III,C,2. RFLPs may also be produced following insertions or deletions of >50–100 bp, translocations, inversions, and gene copy number alterations. Though it is possible to detect single base changes using RFLPs, only four to six bases are examined by each enzyme/probe combination. As with many of the phenomena described in this paper, it has been relatively easy to detect polymorphisms produced by tissue culture but much more difficult to understand their significance. For example, Muller et al. (1990) detected RFLP changes in regenerated rice using both methylation-sensitive and -insensitive enzymes; plants showing variability with one type of change also contained the other. However, polymorphisms of either kind could be observed regardless of the plant phenotype (normal or abnormal).

There is some evidence that the types of DNA variation observed may depend on the genomic organization of the species being cultured. Brown et al. (1991) examined maize and rice callus cultures with probes representing functional loci. A wide range of variability in both restriction fragment size and gene copy number was observed in maize callus; less was seen in rice callus. An intermediate amount of variability was detected in regenerated maize plants. According to Dhar et al. (1988), most plant species contain middle-repetitive DNA interspersed with single-copy sequences, whereas rice has long blocks of single-copy and repetitive DNA that are not interspersed. Based on this observation, Brown et al. (1991) hypothesized that changes in the repetitive DNA will have more effect on the single-copy genes of maize than of rice. In another study of rice, significant differences in the frequency of polymorphism were detected depending on the probe used (Brown et al., 1990). However, any given probe revealed approximately the same frequency of polymorphisms whether methylation-sensitive or -insensitive enzymes were used, indicating that DNA methylation changes were not the primary cause of the polymorphism observed.

3. Increases/Decreases in Gene Copy Number

A number of studies have shown that gene copy number may be altered in culture, although some of these changes appear to have little or no relationship to the observed phenotype (e.g., Lapitan *et al.*, 1988; Brown, 1989; Brown *et al.*, 1991). On the other hand, gene amplification "with a purpose" has been observed in cultures undergoing selection, wherein multiple copies of a gene product may afford an advantage to the cell (e.g., Donn *et al.*, 1984). It should also be noted that polyploidy may function in some cases as a mechanism of gene amplification. Selection for resistance to methotrexate, which will cause specific gene amplification in mammalian cultures, did not do so in maize. However, five out of six lines selected as "methotrexate tolerant" contained predominantly tetraploid cells; it is presumed that the polyploidy may have contributed to the tolerance observed (Tuberosa and Phillips, 1986).

Several studies of ribosomal gene copy number have shown decreases in number resulting from tissue culture (potato: Landsmann and Uhrig, 1985; triticale: Brettell *et al.*, 1986b). Cullis (1976) previously demonstrated that decreases in flax ribosomal DNA copy number may occur as a result of an environmental stress (fertility imbalance), and also found ribosomal DNA copy number decreases in callus (Cullis and Cleary, 1986). It should be noted that copy number changes (mostly decreases) using several other probes for repeated DNA were observed in these lines as well (Cullis and Cleary, 1986).

E. PHENOTYPIC VARIATION

1. Qualitative Variation

In general, the most easily characterized and understood variants are those that behave genetically as caused by one or two genes. Examples of these are numerous, e.g., rice (Fukui, 1983), maize (Edallo *et al.*, 1981; Lee and Phillips, 1987b), tomato (Evans and Sharp, 1983), and soybean (Freytag *et al.*, 1989). Though most behave as recessive mutations [e.g., 43/44 phenotypes seen by Lee and Phillips (1987b)], some dominant traits have been observed as well (Evans and Sharp, 1983; Larkin *et al.*, 1984).

To date, only two tissue culture-derived single-gene mutants have been examined at the DNA sequence level (Section III,D,1). Less is known about the majority of single-gene variants, though a few have been mapped to previously identified loci (e.g., Evans and Sharp, 1983).

2. Quantitative Variation

As with qualitative variants, the examples of quantitative inheritance of tissue culture-derived variants are extensive. In general, quantitative changes observed have been deleterious (e.g., reduced yield), but a number of them may be useful in plant improvement programs. Examples of beneficial changes have included early maturity in maize (Lee *et al.*, 1988) and sorghum (Bhaskaran *et al.*, 1987), increased dry matter in potato (Evans *et al.*, 1986), increased submergence tolerance in rice (Adkins *et al.*, 1990), and increased yield (without other changes) in oat (Dahleen *et al.*, 1991). However, it is clear from these and other studies that a large number of lines must be screened due to the preponderance of undesirable changes.

F. CELL CYCLE DISTURBANCE AS A "UNIFYING HYPOTHESIS"

In terms of the "bottom line," i.e., the plants that may be eventually used by plant breeders, other scientists, and growers, the phenotype produced by somaclonal variation is of the greatest interest. Despite the many phenomena that have been discussed in previous sections and that could potentially be responsible for phenotypic variation, there is little evidence for any one of them being of primary importance. In addition, it has been difficult to correlate these occurrences to one another in any simple way. However, it is conceivable that all of the effects of tissue culture trace back to one or a few initial events.

As illustrated in Fig. 1, the primary event causing tissue culture-induced variability may be disturbance in the cell cycle, perhaps caused by imbalance in nucleotide pools (Section III,A,2,c) or the effects of the exogenous hormones required (Bayliss, 1977; Liscum and Hangarter, 1991). In the most extreme case, polyploidy would result from improper completion of mitosis. In a less extreme instance, the unusually late replication of heterochromatin could result in chromosome breakage and lead (directly or indirectly) to aneuploidy. Chromosome breakage may lead to a variety of outcomes that have been discussed, such as the activation of transposable elements (Section III,B), aneuploidy (through loss of chromosome fragments), methylation changes (Grafström *et al.*, 1984), and further chromosome breakage (McClintock, 1939, 1942, 1978). In addition, chromosome breakage in *Escherichia coli* may initiate the so-called SOS response, resulting in single base changes (Walker, 1984); Burr and Burr (1988) discuss evidence for a similar error-prone repair system in plants. Studies of the human fragile-X syndrome have demonstrated a correlation between chromosome breakage at the fragile site in culture cells, the

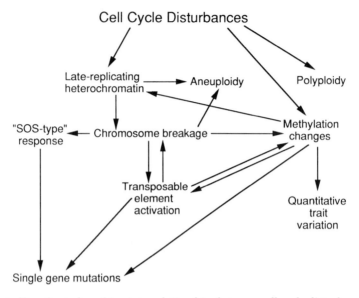

Fig. 1. Hypothesis describing interrelationships between cell cycle disturbances, cytological variation, methylation changes, and genetic variation produced by plant tissue culture (see Section III,F).

presence of chromosome rearrangement *in vivo,* clinical symptoms (presumed to be due to improper X chromosome reactivation in the region), and methylation at the fragile site (Oberle *et al.,* 1991). Though the exact relationship between these events remains to be determined, the phenomenon illustrates the complex interconnections between chromosome structure, methylation, and phenotype.

The occurrence of methylation increases and decreases (whether due to chromosome breakage or directly as a result of cell cycle perturbations) can give rise to yet additional changes. The most significant of these is the potential for quantitative trait variation, because a large number of genes may be affected simultaneously (Phillips *et al.,* 1990). The correlation between methylation changes and activity of transposable elements and other genes has been described; the frequent decreases in methylation seen in culture (Section III,C,2) indicate the potential for gains of gene activity. Deamination of 5-methylcytosine results in its conversion to thymine, producing point mutations (Coulondre *et al.,* 1978). Finally, activation of transposable elements may generate single gene mutations (stable or unstable), methylation changes in (and activation of) transposable elements, and additional chromosome breakage.

IV. Unusual Genetic Behavior Resulting from Somaclonal Variation

As described throughout this paper, the variation produced through tissue culture is far from limited to "well-behaved" single gene mutations. In contrast, many cases of unusual genetic behavior result from tissue culture; these are important to recognize in order to utilize fully the variation produced. If anything, it is suspected that such anomalies may be underreported in the literature (especially unstable or poorly transmitted traits) due to the difficulty in studying them.

A. SECTORING

As indicated in Section II,B, sectored regenerated plants (chimeras) may occur regardless of whether their origin in the culture is by organogenesis or embryogenesis. Sectoring may result in a plant with a desirable phenotype that is not transmitted to progeny, if the somatic and germ cells derive from separate lineages. For example, thornless blackberry sports found in nature often result from periclinal chimeras; seeds on such sports produce thorned plants (Darrow, 1931). Sectoring involving some but not all germ cells will result in distorted genetic ratios in the first sexual generation; barring other problems, subsequent generations should segregate normally. An extreme example of such an off ratio is observed in monoecious plants such as maize, wherein chimeras are often detected by the observation that a mutant not seen at all in the R1 (first sexual) generation is observed to segregate in the R2 (second sexual) generation (e.g., McCoy and Phillips, 1982; Lee and Phillips, 1987b; Armstrong and Phillips, 1988). This result is explained by the presence of different genotypes in the ear and tassel such that even a self-pollination does not produce homozygous mutant progeny.

B. HOMOZYGOSITY OF REGENERATED PLANTS

Though most new mutations are present in a heterozygous state in the regenerated plant (based on progeny segregation data), a number of variants are reported to have originated in a homozygous condition. Examples of such variants include seed color in mustard (George and Rao, 1983), awns in wheat (Larkin et al., 1984), plant height in rice (Oono, 1985), and single-locus methylation changes in maize (Phillips et al., 1992). Though it is possible that simultaneous mutation of both homologues could occur, this is extremely unlikely unless mutation is "directed" in some way. In the case of homozygous methylation changes, this could occur if a specific DNA-binding protein recognizes and binds to both homologues, preventing methylation (R. L. Phillips, personal communication). A number of alternatives to simultaneous mutation involve a single mutational event followed by incorporation of the mutation into the homologous chromosome region. This could occur by (1) haploidy or monosomy followed by doubling, (2) somatic crossing-over, or (3) gene conversion. In theory, one could distinguish between the first possibility and the other two if the cultured material was heterozygous for genetic or molecular markers on each chromosome (e.g., a cross between two inbred lines).

C. Unstable Mutations

Mutations that could be classified as "unstable" fall into two broad categories. The first, "mutable alleles," discussed in Section III,B,4, may be caused by insertion and excision of transposable elements. Though this instability is generally manifested as variegation in somatic tissue, transposition events may also take place in the germ line, giving rise to new alleles (Schwarz-Sommer *et al.*, 1985).

An example of a second type of tissue culture-induced unstable trait that has been reported involves the "disappearance" of a dwarf mutant phenotype of rice, which was present in the original regenerated plants, stable when self-pollinated (or when two dwarfs were crossed), but "lost" upon outcrossing to a number of normal lines (Oono, 1985). Two lines of evidence indicate that methylation may be involved in regulating this phenotype. First, a study in rice indicated that dwarfism and hypomethylation may be induced by treatment with 5-azacytidine (azaC); this dwarfism was stable through at least three generations of selfing (Sano *et al.*, 1990). Second, it is known that methylation of at least one gene (the *Mu* transposable element in maize) can be altered by the crossing scheme; methylation is kept low by outcrossing to lines without active *Mu* elements, but increases when *Mu*-containing lines are intercrossed (Bennetzen, 1987). Crosses of azaC-induced rice dwarfs to tall lines would be useful in clarifying the situation.

D. Altered Transmission

There are occasional reports of tissue culture-induced mutant phenotypes that behave genetically as recessive loci, but appear in a smaller proportion of segregating progeny than expected (e.g., Evans and Sharp, 1983; Pugliesi *et al.*, 1991). Some of these turn out to be two-gene segregations, or the result of a sectored parent plant. As indicated in Section IV,A, off ratios produced by sectoring should correct themselves in subsequent generations. Likewise, in the situation of two-gene segregation, ratios in some progeny families should resemble single-gene segregation, once one or the other of the loci becomes homozygous for a mutant allele. However, a number of traits may continue to segregate abnormally even after several generations. Such behavior can be caused by linkage of the trait to a small chromosomal aberration, which may or may not be the actual cause of the mutant. Other expected outcomes of this situation would be pollen sterility or size dimorphism, and the possibility of better transmission through the female than the male.

V. Conclusions

By now, somaclonal variation has been observed at every genetic level, from the mutation of a single DNA base pair to the heritable alteration of traits encoded by multiple loci. Despite the wealth of descriptive information, the basis for somaclonal variation remains elusive. It is hoped that the constant refinement of DNA analysis techniques, especially those that facilitate analysis of very small amounts of tissue, will allow dissection of the earliest events occurring in culture.

The question of whether tissue culture represents a useful means of generating variability for agricultural or biological purposes is still largely unanswered. The primary consideration is whether similar changes may be induced or selected for more easily using other means. Some types of variation (e.g., chromosome breakage) may be generated using a variety of agents, but others (e.g., methylation changes) may be novel. It should be noted, however, that the extensive scrutiny given recently to DNA-level events in tissue culture and regenerated plants has not been applied to plants produced by other mutagenic treatments. Even if a given type of mutational event is not a unique feature of tissue culture, the regeneration of plants from one or a few cells allows representation of the mutation throughout a large portion of the plant (Evans and Sharp, 1986). In addition, tissue culture is unique as a tool for use of selective agents, with or without additional mutagenic treatment.

The other side of the question is whether it will ever be possible to reduce significantly or to eliminate tissue culture-induced variability to allow its use as a stable genetic system. The use of young cultures helps to reduce variability and increase regenerative potential, but it requires continual reinitiation of cultures and still does not entirely eliminate the problem. Cryogenic techniques to preserve cultures long-term in a genetically "younger" form require special equipment and could possibly generate additional variability as well as losses in viability. In some ways, it is difficult to be optimistic. The processes of dedifferentiation and callus growth require complex changes in gene regulation, which have been generally considered to be transient. It is now apparent that mechanisms implicated in gene regulation, such as methylation changes, may persist through many sexual generations. However, the power associated with using the cell as a unit of selection and manipulation surely makes the effort worthwhile.

REFERENCES

Adkins, S. W., Shiraishi, T., McComb, J. A., Ratanopol, S., Kupkanchanakul, T., Armstrong, L. J., and Schultz, A. L. (1990). Somaclonal variation in rice—Submergence tolerance and other agronomic characters. *Physiol. Plant.* **80,** 647–654.

Ahloowalia, B. S. (1975). Regeneration of ryegrass plants in tissue culture. *Crop Sci.* **15,** 449–452.

Ahloowalia, B. S., and Maretzki, A. (1983). Plant regeneration via somatic embryogenesis in sugarcane. *Plant Cell Rep.* **2,** 21–25.

Ahloowalia, B. S., and Sherington, J. (1985). Transmission of somaclonal variation in wheat. *Euphytica* **34,** 525–537.

Anderson, S., Lewis-Smith, A. C., and Smith, S. M. (1990). Methylation of ribosomal RNA genes in *Petunia hybrida* plants, callus cultures, and regenerated shoots. *Plant Cell Rep.* **8,** 554–557.

Armstrong, C. L., and Phillips, R. L. (1988). Genetic and cytogenetic variation in plants regenerated from organogenic and friable, embryogenic tissue cultures of maize. *Crop Sci.* **28,** 363–369.

Ashmore, S. E., and Gould, A. R. (1981). Karyotype evolution in a tumour derived plant tissue culture analysed by Giemsa C-banding. *Protoplasma* **106,** 297–308.

Balzan, R. (1978). Karyotype instability in tissue cultures derived from the mesocotyl of *Zea mays* seedlings. *Caryologia* **31,** 75–87.

Banks, J. A., Masson, P., and Fedoroff, N. (1988). Molecular mechanisms in the developmental regulation of the maize *Suppressor-mutator* transposable element. *Genes Dev.* **2,** 1364–1380.

Barbier, M., and Dulieu, H. (1983). Early occurrence of genetic variants in protoplast cultures. *Plant Sci. Lett.* **29,** 201–206.

Bayliss, M. W. (1977). The effects of 2,4-D on growth and mitosis in suspension cultures of *Daucus carota*. *Plant Sci. Lett.* **8,** 99–103.

Bayliss, M. W. (1980). Chromosomal variation in plant tissues in culture. *Int. Rev. Cytol.* **11A,** 113–144.

Bennetzen, J. L. (1987). Covalent DNA modification and the regulation of *Mutator* element transposition in maize. *Mol. Gen. Genet.* **208,** 45–51.

Benzion, G., and Phillips, R. L. (1988). Cytogenetic stability of maize tissue cultures: A cell line pedigree analysis. *Genome* **30**, 318–325.

Bhaskaran, S., Smith, R. H., Paliwal, S., and Schertz, K. F. (1987). Somaclonal variation from *Sorghum bicolor* (L.) Moench cell culture. *Plant Cell Tissue Org. Cult.* **9**, 189–196.

Bianchi, A., Salamini, F., and Parlavecchio, R. (1969). On the origin of controlling elements in maize. *Genet. Agrar.* **22**, 335–344.

Bianchi, M. W., and Viotti, A. (1988). DNA methylation and tissue-specific transcription of the storage protein genes of maize. *Plant Mol. Biol.* **11**, 203–214.

Brettell, R. I. S., Thomas, E., and Ingram, D. S. (1980). Reversion of Texas male-sterile cytoplasm maize in culture to give fertile, T-toxin resistant plants. *Theor. Appl. Genet.* **58**, 55–58.

Brettell, R. I. S., Dennis, E. S., Scowcroft, W. R., and Peacock, W. J. (1986a). Molecular analysis of a somaclonal mutant of maize alcohol dehydrogenase. *Mol. Gen. Genet.* **202**, 235–239.

Brettell, R. I. S., Pallotta, M. A., Gustafson, J. P., and Appels, R. (1986b). Variation at the *Nor* locus in triticale derived from tissue culture. *Theor. Appl. Genet.* **71**, 637–643.

Browers, M. A., and Orton, T. J. (1982). Transmission of gross chromosomal variability from suspension cultures into regenerated celery plants. *J. Hered.* **73**, 159–162.

Brown, P. T. H. (1989). DNA methylation in plants and its role in tissue culture. *Genome* **31**, 717–729.

Brown, P. T. H., and Lorz, H. (1986). Molecular changes and possible origins of somaclonal variation. *In* "Somaclonal Variation and Crop Improvement" (J. Semal, ed.), pp. 148–159. Martinus Nijhoff, Dordrecht, The Netherlands.

Brown, P. T. H., Yoneyama, K., and Lorz, H. (1989). An investigation into the role of 5-azacytidine in tissue culture. *Theor. Appl. Genet.* **78**, 321–328.

Brown, P. T. H., Kyozuka, J., Sukekiyo, Y., Kimura, Y., Shimamoto, K., and Lorz, H. (1990). Molecular changes in protoplast-derived rice plants. *Mol. Gen. Genet.* **223**, 324–328.

Brown, P. T. H., Gobel, E., and Lorz, H. (1991). RFLP analysis of *Zea mays* callus cultures and their regenerated plants. *Theor. Appl. Genet.* **81**, 227–232.

Burr, B., and Burr, F. (1981). Transposable elements and genetic instabilities in crop plants. *Stadler Symp.* **13**, 115–128.

Burr, B., and Burr, F. (1988). Activation of silent transposable elements. *In* "Plant Transposable Elements" (O. Nelson, ed.), pp. 317–323. Plenum, New York.

Catcheside, D. G. (1939). A position effect in *Oenothera*. *J. Genet.* **38**, 345–352.

Cedar, H., and Razin, A. (1990). DNA methylation and development. *Biochim. Biophys. Acta* **1049**, 1–8.

Chandler, V. L., and Walbot, V. (1986). DNA modification of a maize transposable element correlates with loss of activity. *Proc. Natl. Acad. Sci. U.S.A.* **83**, 1767–1771.

Chandler, V. L., Talbert, L. E., and Raymond, F. (1988). Sequence, genomic distribution, and DNA modification of a *Mu1* element from non-Mutator maize stocks. *Genetics* **119**, 951–958.

Chen, C. H., Freeling, M. L., and Merckelbach, A. (1986). Enzymatic and morphological consequences of *Ds* excisions from maize *Adh1*. *Maydica* **31**, 93–108.

Chomet, P. S., Wessler, S., and Dellaporta, S. L. (1987). Inactivation of the maize transposable element *Activator* (*Ac*) is associated with its DNA modification. *EMBO J.* **6**, 295–302.

Chourey, P. S., and Kemble, R. J. (1982). Transposition event in tissue cultured cells of maize. *Proc. Int. Congr. Plant Tissue Cell Cult., 5th, 1982*, pp. 425–426.

Compton, M. E., and Veilleux, R. E. (1991). Variation for genetic recombination among tomato plants regenerated from three tissue culture systems. *Genome* **34**, 810–817.

Cone, K. C., Burr, F. A., and Burr, B. (1986). Molecular analysis of the maize anthocyanin regulatory locus *C1*. *Proc. Natl. Acad. Sci. U.S.A.* **83**, 9631–9635.

Coulondre, C., Miller, J. H., Farabaugh, P. J., and Gilbert, W. (1978). Molecular basis of base substitution hotspots in *Escherichia coli*. *Nature (London)* **274**, 775–780.

Creissen, G. P., and Karp, A. (1985). Karyotypic changes in potato plants regenerated from protoplasts. *Plant Cell Tissue Org. Cult.* **4**, 171–182.

Culley, D. E. (1986). Evidence for the activation of a cryptic transposable element *Ac* in maize endosperm cultures. *Proc. Int. Congr. Plant Tissue Cell Cult., 6th, 1986* Abstr., p. 220.

Culley, D. E. (1991). Studies of regulated gene expression in early maize endosperm development. Ph.D. Thesis, University of Minnesota, St. Paul.

Cullis, C. A. (1976). Environmentally induced changes in ribosomal cistron number in flax. *Heredity* **36**, 73–79.

Cullis, C. A., and Cleary, W. (1986). DNA variation in flax tissue culture. *Can. J. Genet. Cytol.* **28**, 247–251.

Dahleen, L. S., Stuthman, D. D., and Rines, H. W. (1991). Agronomic trait variation in oat lines derived from tissue culture. *Crop Sci.* **31**, 90–94.

D'Amato, F. (1952). Polyploidy in the differentiation and function of tissues and cells in plants. *Caryologia* **4**, 311–358.

D'Amato, F. (1985). Cytogenetics of plant cell and tissue cultures and their regenerates. *CRC Crit. Rev. Plant Sci.* **3**, 73–112.

Darrow, G. M. (1931). A productive thornless sport of the Evergreen blackberry. *J. Hered.* **22**, 404–406.

de Boucaud, M.-T., and Gaultier, J.-M. (1983). Evidence of a different nuclear population in chimerae of regenerated plants from tobacco aneuploid tissue cultures. Cytophotometric and statistical analysis. *Z. Pflanzenphysiol.* **110**, 97–105.

Dennis, E. S., and Brettell, R. I. S. (1990). DNA methylation of maize transposable elements is correlated with activity. *Philos. Trans. R. Soc. London, Ser. B* **326**, 217–229.

Dennis, E. S., Brettell, R. I. S., and Peacock, W. J. (1987). A tissue culture induced *Adh1* null mutant of maize results from a single base change. *Mol. Gen. Genet.* **210**, 181–183.

Dhar, M. S., Dabak, M. M., Gupta, V. S., and Ranjekar, P. K. (1988). Organisation and properties of repeated DNA sequences in rice genome. *Plant Sci.* **55**, 43–52.

Dimitrov, B. (1987). Relationship between sister-chromatid exchanges and heterochromatin or DNA replication in chromosomes of *Crepis capillaris*. *Mutat. Res.* **190**, 271–276.

Doerschug, E. B. (1973). Studies of *Dotted*, a regulatory element in maize. I. Induction of *Dotted* by chromatid breaks. II. Phase variation of *Dotted*. *Theor. Appl. Genet.* **43**, 182–189.

Dolezel, J., and Novak, F. J. (1986). Sister chromatid exchanges in garlic (*Allium sativum* L.) callus cells. *Plant Cell Rep.* **5**, 280–283.

Dolezel, J., Lucretti, S., and Novak, F. J. (1987). The influence of 2,4-dichlorophenoxyacetic acid on cell cycle kinetics and sister-chromatid exchange frequency in garlic (*Allium sativa* L.) meristem cells. *Biol. Plant.* **29**, 253–257.

Donn, G., Tischer, E., Smith, J. A., and Goodman, H. M. (1984). Herbicide-resistant alfalfa cells: an example of gene amplification in plants. *J. Mol. Appl. Genet.* **2,** 621–635.

Dooner, H. K. (1980). Regulation of the enzyme UFGT by the controlling element *Ds* in *bz-m4*, an unstable mutant in maize. *Cold Spring Harbor Symp. Quant. Biol.* **45,** 457–462.

Dooner, H. K., and Nelson, O. E., Jr. (1977). Controlling element-induced alterations in UDPglucose:flavonoid glucosyltransferase, the enzyme specified by the *bronze* locus in maize. *Proc. Natl. Acad. Sci. U.S.A.* **74,** 5623–5627.

Dooner, H. K., and Nelson, O. E., Jr. (1979). Heterogeneous flavonoid glucosyltransferases in purple derivatives from a controlling element-suppressed *bronze* mutant in maize. *Proc. Natl. Acad. Sci. U.S.A.* **76,** 2369–2371.

Dooner, H. K., Ralston, E., and English, J. (1988). Deletions and breaks involving the borders of the *Ac* element in the *bz-m2(Ac)* allele of maize. *In* "Plant Transposable Elements" (O. Nelson, ed.), pp. 213–226. Plenum, New York.

Doring, H.-P., and Starlinger, P. (1984). Barbara McClintock's controlling elements: Now at the DNA level. *Cell (Cambridge, Mass.)* **39,** 253–259.

Echt, C. S., and Schwartz, D. (1981). Evidence for the inclusion of controlling elements within the structural gene at the *waxy* locus in maize. *Genetics* **99,** 275–284.

Edallo, S., Zucchinali, C., Perenzin, M., and Salamini, F. (1981). Chromosomal variation and frequency of spontaneous mutation associated with *in vitro* culture and plant regeneration in maize. *Maydica* **26,** 39–56.

Evans, D. A., and Sharp, W. R. (1983). Single gene mutations in tomato plants regenerated from tissue culture. *Science* **221,** 949–951.

Evans, D. A., and Sharp, W. R. (1986). Applications of somaclonal variation. *Bio/Technology* **4,** 528–532.

Evans, N. E., Foulger, D., Farrer, L., and Bright, S. W. J. (1986). Somaclonal variation in explant-derived potato clones over three tuber generations. *Euphytica* **35,** 353–361.

Evola, S. V., Burr, F. A., and Burr, B. (1984). The nature of tissue culture-induced mutations in maize. *11th Annu. Aharon Katzir-Katchalsky Conf.* Abstr.

Evola, S. V., Tuttle, A., Burr, F., and Burr, B. (1985). Tissue culture associated variability in maize: Molecular and genetic studies. *Proc. Int. Congr. Plant Mol. Biol., 1st,* 1985 Abstr., p. 10.

Fedoroff, N. (1989). The heritable activation of *cryptic Suppressor-mutator* elements by an active element. *Genetics* **121,** 591–608.

Fedoroff, N., Wessler, S., and Shure, M. (1983). Isolation of the transposable maize controlling elements *Ac* and *Ds*. *Cell (Cambridge, Mass.)* **35,** 235–242.

Feher, F., Hangyel Tarczy, M., Bocsa, L., and Dudits, D. (1989). Somaclonal chromosome variation in tetraploid alfalfa. *Plant Sci.* **60,** 91–99.

Freytag, A. H., Rao-Arelli, A. P., Anand, S. C., Wrather, J. A., and Owens, L. D. (1989). Somaclonal variation in soybean plants regenerated from tissue culture. *Plant Cell Rep.* **8,** 199–202, 312.

Fukui, K. (1983). Sequential occurrence of mutations in a growing rice callus. *Theor. Appl. Genet.* **65,** 225–230.

Gamborg, O. L., Constabel, F., and Miller, R. A. (1970). Embryogenesis and production of albino plants from cell cultures of *Bromus inermis*. *Planta* **95,** 355–358.

Gengenbach, B. G., Connelly, J. A., Pring, D. R., and Conde, M. F. (1981). Mitochondrial DNA variation in maize plants regenerated during tissue culture selection. *Theor. Appl. Genet.* **59,** 161–167.

George, L., and Rao, P. S. (1983). Yellow-seeded variants in *in vitro* regenerants of mustard (*Brassica juncea* Coss var. Rai-5). *Plant Sci. Lett.* **30**, 327–330.

Gmitter, F. G., Jr., Ling, X., Cai, C., and Grosser, J. W. (1991). Colchicine-induced polyploidy in *Citrus* embryogenic cultures, somatic embryos, and regenerated plantlets. *Plant Sci.* **74**, 135–141.

Gorman, M. B., and Peterson, P. A. (1978). The interaction of controlling element components in a tissue culture system. *Maydica* **23**, 173–186.

Grafström, R. H., Hamilton, D. L., and Yuan, R. (1984). DNA methylation: DNA replication and repair. *In* "DNA Methylation: Biochemistry and Biological Significance" (A. Razin, H. Cedar, and A. D. Riggs, eds.), pp. 111–126. Springer-Verlag, New York.

Groose, R. W., and Bingham, E. T. (1986). An unstable anthocyanin mutation recovered from tissue culture of alfalfa (*Medicago sativa*). 1. High frequency of reversion upon reculture. 2. Stable nonrevertants derived from reculture. *Plant Cell Rep.* **5**, 104–110.

Haccius, B. (1978). Question of unicellular origin of non-zygotic embryos in callus cultures. *Phytomorphology* **28**, 74–81.

Hanson, M. R. (1984). Stability, variation, and recombination in plant mitochondrial genomes via tissue culture and somatic hybridisation. *Oxford Surv. Plant Mol. Cell Biol.* **1**, 33–52.

Harada, T., Sato, T., Asaka, D., and Matsukawa, I. (1991). Large-scale deletions of rice plastid DNA in anther culture. *Theor. Appl. Genet.* **81**, 157–161.

Hartmann, C., de Buyser, J., Henry, Y., Falconet, D., Lejeune, B., Benslimane, A.-A., Quetier, F., and Rode, A. (1987). Time-course of mitochondrial genome variation in wheat embryogenic somatic tissue cultures. *Plant Sci.* **53**, 191–198.

Hepburn, A. G., Clark, L. E., Pearson, L., and White, J. (1983). The role of cytosine methylation in the control of nopaline synthase gene expression in a plant tumor. *J. Mol. Appl. Genet.* **2**, 315–329.

Hershkovitz, M., Gruenbaum, Y., Renbaum, P., Razin, A., and Loyer, A. (1990). Effect of CpG methylation on gene expression in transfected plant protoplasts. *Gene* **94**, 189–193.

Heszky, L. E., Nam, L. S., Simon-Kiss, I., Lokos, K., Gyulai, G., and Kiss, E. (1989). Organ-specific and ploidy-dependent somaclonal variation; a new tool in breeding. *Acta Biol. Hung.* **40**, 381–394.

Ho, W.-J., and Vasil, I. K. (1983). Somatic embryogenesis in sugarcane (*Saccharum officinarum* L.) I. The morphology and physiology of callus formation and the ontogeny of somatic embryos. *Protoplasma* **118**, 169–180.

Holliday, R. (1987). The inheritance of epigenetic defects. *Science* **238**, 163–170.

Jacky, P. B., Beek, B., and Sutherland, G. R. (1983). Fragile sites in chromosomes: Possible model for the study of spontaneous chromosome breakage. *Science* **220**, 69–70.

James, M. G., and Stadler, J. (1989). Molecular characterization of *Mutator* systems in maize embryogenic callus cultures indicates *Mu* activity *in vitro. Theor. Appl. Genet.* **77**, 383–393.

Johns, M. A., Mottinger, J., and Freeling, M. (1985). A low copy number, *copia*-like transposon in maize. *EMBO J.* **4**, 1093–1102.

Johnson, S. S., Phillips, R. L., and Rines, H. W. (1987). Possible role of heterochromatin in chromosome breakage induced by tissue culture in oats (*Avena sativa* L.). *Genome* **29**, 439–446.

Jones, P. A. (1984). Gene activation by 5-azacytidine. *In* "DNA Methylation: Biochemistry and Biological Significance" (A: Razin, H. Cedar, and A. D. Riggs, eds.), pp. 165–187. Springer-Verlag, New York.

Karp, A., and Bright, S. W. J. (1985). On the causes and origins of somaclonal variation. *Oxford Surv. Plant Mol. Cell. Biol.* **2**, 199–234.

Karp, A., and Maddock, S. E. (1984). Chromosome variation in wheat plants regenerated from cultured immature embryos. *Theor. Appl. Genet.* **67**, 249–255.

Kemble, R. J., and Shepard, J. F. (1984). Cytoplasmic DNA variation in a potato protoclonal population. *Theor. Appl. Genet.* **69**, 211–216.

Landsmann, J., and Uhrig, H. (1985). Somaclonal variation in *Solanum tuberosum* detected at the molecular level. *Theor. Appl. Genet.* **71**, 500–505.

Lapitan, N. L. V., Sears, R. G., and Gill, B. S. (1984). Translocations and other karyotypic structural changes in wheat × rye hybrids regenerated from tissue culture. *Theor. Appl. Genet.* **68**, 547–554.

Lapitan, N. L. V., Sears, R. G., and Gill, B. S. (1988). Amplification of repeated DNA sequences in wheat × rye hybrids regenerated from tissue culture. *Theor. Appl. Genet.* **75**, 381–388.

Larkin, P. J., and Scowcroft, W. R. (1981). Somaclonal variation—A novel source of variability from cell cultures for plant improvement. *Theor. Appl. Genet.* **60**, 197–214.

Larkin, P. J., Ryan, S. A., Brettell, R. I. S., and Scowcroft, W. R. (1984). Heritable somaclonal variation in wheat. *Theor. Appl. Genet.* **67**, 443–455.

Lee, M., and Phillips, R. L. (1987a). Genomic rearrangements in maize induced by tissue culture. *Genome* **29**, 122–128.

Lee, M., and Phillips, R. L. (1987b). Genetic variants in progeny of regenerated maize plants. *Genome* **29**, 834–838.

Lee, M., and Phillips, R. L. (1988). The chromosomal basis of somaclonal variation. *Annu. Rev. Plant Physiol. Plant Mol. Biol.* **39**, 413–437.

Lee, M., Geadelmann, J. L., and Phillips, R. L. (1988). Agronomic evaluation of inbred lines derived from tissue cultures of maize. *Theor. Appl. Genet.* **75**, 841–849.

Lima-de-Faria, A. (1969). DNA replication and gene amplification in heterochromatin. *In* "Handbook of Molecular Cytology" (A. Lima-de-Faria, ed.), pp. 277–325. North-Holland Publ., Amsterdam.

Liscum, E., III, and Hangarter, R. P. (1991). Manipulation of ploidy levels in cultured haploid *Petunia* tissue by phytohormone treatments. *J. Plant Physiol.* **138**, 33–38.

Lorz, H., and Scowcroft, W. R. (1983). Variability among plants and their progeny regenerated from protoplasts of *Su/su* heterozygotes of *Nicotiana tabacum*. *Theor. Appl. Genet.* **66**, 67–75.

Luhrs, R., and Lorz, H. (1988). Initiation of morphogenic cell-suspension and protoplast cultures of barley (*Hordeum vulgare* L.). *Planta* **175**, 71–81.

Marcotrigiano, M., and Gouin, F. R. (1984). Experimentally synthesized chimeras. I. *In vitro* recovery of *Nicotiana tabacum* L. chimeras from mixed callus cultures. *Ann. Bot. (London)* **54**, 503–511.

McClintock, B. (1939). The behavior in successive nuclear divisions of a chromosome broken at meiosis. *Proc. Natl. Acad. Sci. U.S.A.* **25**, 405–416.

McClintock, B. (1942). The fusion of broken ends of chromosomes following nuclear fusion. *Proc. Natl. Acad. Sci. U.S.A.* **28**, 458–463.

McClintock, B. (1950). The origin and behavior of mutable loci in maize. *Proc. Natl. Acad. Sci. U.S.A.* **36**, 344–355.

McClintock, B. (1951). Chromosome organization and genic expression. *Cold Spring Harbor Symp. Quant. Biol.* **16**, 13–47.

McClintock, B. (1958). The *Suppressor-mutator* system of control of gene action in maize. *Carnegie Inst. Wash. Yearb.* **57**, 415–429.

McClintock, B. (1959). Genetic and cytological studies of maize. *Carnegie Inst. Wash. Yearb.* **58**, 452–456.

McClintock, B. (1971). The contribution of one component of a control system to versatility of gene expression. *Carnegie Inst. Wash. Yearb.* **70**, 5–17.

McClintock, B. (1978). Mechanisms that rapidly reorganize the genome. *Stadler Symp.* **10**, 25–47.

McClintock, B. (1984). The significance of responses of the genome to challenge. *Science* **226**, 792–801.

McCoy, T. J., and Phillips, R. L. (1982). Chromosome stability in maize (*Zea mays*) tissue cultures and sectoring in some regenerated plants. *Can. J. Genet. Cytol.* **24**, 559–565.

McCoy, T. J., Phillips, R. L., and Rines, H. W. (1982). Cytogenetic analysis of plants regenerated from oat (*Avena sativa*) tissue cultures: High frequency of partial chromosome loss. *Can. J. Genet. Cytol.* **24**, 37–50.

Meuth, M., L'Heureux-Huard, N., and Trudel, M. (1979). Characterization of a mutator gene in Chinese hamster ovary cells. *Proc. Natl. Acad. Sci. U.S.A.* **76**, 6505–6509.

Mohmand, A. S., and Nabors, M. W. (1990). Somaclonal variant plants of wheat derived from mature embryo explants of three genotypes. *Plant Cell Rep.* **8**, 558–560.

Morrish, F. M., Hanna, W. W., and Vasil, I. K. (1990). The expression and perpetuation of inherent somaclonal variation in regenerants from embryogenic cultures of *Pennisetum glaucum* (L.) R. Br. (pearl millet). *Theor. Appl. Genet.* **80**, 409–416.

Muller, E., Brown, P. T. H., Hartke, S., and Lorz, H. (1990). DNA variation in tissue-culture-derived rice plants. *Theor. Appl. Genet.* **80**, 673–679.

Murata, M., and Orton, T. J. (1983). Chromosome structural changes in cultured celery cells. *In Vitro* **19**, 83–89.

Murata, M., and Orton, T. J. (1984). Chromosome fusions in cultured cells of celery. *Can. J. Genet. Cytol.* **26**, 395–400.

Nagarajan, P., and Walton, P. D. (1987). A comparison of somatic chromosomal instability in tissue culture regenerants from *Medicago media* Pers. *Plant Cell Rep.* **6**, 109–113.

Navarro, L., Ortiz, J. M., and Juarez, J. (1985). Aberrant citrus plants obtained by somatic embryogenesis of nucelli cultured *in vitro*. *HortScience* **20**, 214–215.

Neuffer, M. G. (1966). Stability of the suppressor element in two mutator systems of the *A1* locus in maize. *Genetics* **53**, 541–549.

Nick, H., Bowen, B., Ferl, R. J., and Gilbert, W. (1986). Detection of cytosine methylation in the maize alcohol dehydrogenase gene by genomic sequencing. *Nature (London)* **319**, 243–246.

Nuti Ronchi, V., Bonatti, S., and Turchi, G. (1986). Preferential localization of chemically induced breaks in heterochromatic regions of *Vicia faba* and *Allium cepa* chromosomes. I. Exogenous thymidine enhances the cytologic effects of 4-epoxyethyl-1,2-epoxy-cyclohexane. *Environ. Exp. Bot.* **26**, 115–126.

Oberle, I., Rousseau, F., Heitz, D., Kretz, C., Devys, D., Hanauer, A., Boue, J., Bertheas, M. F., and Mandel, J. L. (1991). Instability of a 550-base pair DNA segment and abnormal methylation in fragile-X syndrome. *Science* **252**, 1097–1102.

Oono, K. (1985). Putative homozygous mutations in regenerated plants of rice. *Mol. Gen. Genet.* **198**, 377–384.

Orton, T. J. (1984). Genetic variation in somatic tissues: Method or madness? *Adv. Plant Pathol.* **2**, 153–189.

Pan, Y.-B., and Peterson, P. (1988). Co-induction of *Uq* activity and an *mn*-type mutant by 5-aza-2'-deoxycytidine. *In* "Plant Transposable Elements" (O. Nelson, ed.), pp. 374–375 (abstr.). Plenum, New York.

Peacock, W. J., Dennis, E. S., Gerlach, W. L., Sachs, M. M., and Schwartz, D. (1984). Insertion and excision of *Ds* controlling elements in maize. *Cold Spring Harbor Symp. Quant. Biol.* **49**, 347–354.

Peschke, V. M., and Phillips, R. L. (1991). Activation of the maize transposable element *Suppressor-mutator* (*Spm*) in tissue culture. *Theor. Appl. Genet.* **81**, 90–97.

Peschke, V. M., Phillips, R. L., and Gengenbach, B. G. (1987). Discovery of transposable element activity among progeny of tissue culture-derived maize plants. *Science* **238**, 804–807.

Peschke, V. M., Phillips, R. L., and Gengenbach, B. G. (1991). Genetic and molecular analysis of tissue culture-derived *Ac* elements. *Theor. Appl. Genet.* **82**, 121–129.

Peterson, P. A. (1953). A mutable pale green locus in maize. *Genetics* **38**, 682–683.

Peterson, P. A. (1986). Mobile elements in maize. *Plant Breed. Rev.* **4**, 81–122.

Phillips, R. L., Kaeppler, S. M., and Peschke, V. M. (1990). Do we understand somaclonal variation? *In* "Progress in Plant Cellular and Molecular Biology" (H. J. J. Nijkamp, L. H. W. van der Plas, and J. van Aartrijk, eds.), *Proc. 7th Int. Congr. Plant Tissue Cell Cult.,* pp. 131–141. Kluwer Acad. Publ., Dordrecht, The Netherlands.

Phillips, R. L., Plunkett, D. J., and Kaeppler, S. M. (1992). Novel approaches to the induction of genetic variation and plant breeding implications. *In* "Plant Breeding in the 1990's" (H. T. Stalker and J. P. Murphy, eds.), pp. 389–408. CAB Int. Press, Oxon, England.

Planckaert, F., and Walbot, V. (1989). Molecular and genetic characterization of *Mu* transposable elements in *Zea mays*: Behavior in callus culture and regenerated plants. *Genetics* **122**, 567–578.

Pohlman, R. F., Fedoroff, N. V., and Messing, J. (1984a). The nucleotide sequence of the maize transposable element *Activator. Cell (Cambridge, Mass.)* **37**, 635–643.

Pohlman, R. F., Fedoroff, N. V., and Messing, J. (1984b). Nucleotide sequence of *Ac. Cell (Cambridge, Mass.)* **39**, 417.

Potter, R., and Jones, M. G. K. (1991). An assessment of genetic stability of potato *in vitro* by molecular and phenotypic analysis. *Plant Sci.* **76**, 239–248.

Pouteau, S., Huttner, E., Grandbastien, M. A., and Caboche, M. (1991). Specific expression of the tobacco *Tnt1* retrotransposon in protoplasts. *EMBO J.* **10**, 1911–1918.

Pugliesi, C., Cecconi, F., Mandolfo, A., and Baroncelli, S. (1991). Plant regeneration and genetic variability from tissue cultures of sunflower (*Helianthus annuus* L.). *Plant Breed.* **106**, 114–121.

Quemada, H., Roth, E. J., and Lark, K. G. (1987). Changes in methylation of tissue cultured soybean cells detected by digestion with the restriction enzymes *Hpa*II and *Msp*I. *Plant Cell Rep.* **6**, 63–66.

Ray, I. M., and Bingham, E. T. (1991). Inheritance of a mutable allele that is activated in alfalfa tissue culture. *Genome* **34**, 35–40.

Rhoades, M. M., and Dempsey, E. (1982). The induction of mutable systems in plants with the high-loss mechanism. *Maize Genet Coop. News Lett.* **56**, 21–26.

Roberts, L. M. (1942). The effects of translocation on growth in *Zea mays. Genetics* **27**, 584–603.

Robertson, D. S., and Stinard, P. S. (1987). Genetic evidence of *Mutator*-induced deletions in the short arm of chromosome 9 of maize. *Genetics* **115**, 353–361.

Rose, R. J., Johnson, L. B., and Kemble, R. J. (1986). Restriction endonuclease studies on the chloroplast and mitochondrial DNAs of alfalfa (*Medicago sativa* L.) protoclones. *Plant Mol. Biol.* **6**, 331–338.

Sachs, M. M., Peacock, W. J., Dennis, E. S., and Gerlach, W. L. (1983). Maize *Ac/Ds* controlling elements—A molecular viewpoint. *Maydica* **28**, 289–301.

Sacristan, M. D. (1971). Karyotypic changes in callus cultures from haploid and diploid plants of *Crepis capillaris* (L.) Wallr. *Chromosoma* **33**, 273–283.

Saedler, H., and Nevers, P. (1985). Transposition in plants: A molecular model. *EMBO J.* **4**, 585–590.

Sano, H., Kamada, I., Youssefian, S., Katsumi, M., and Wabiko, H. (1990). A single treatment of rice seedlings with 5-azacytidine induces heritable dwarfism and undermethylation of genomic DNA. *Mol. Gen. Genet.* **220**, 441–447.

Schwartz, D. (1989). Gene-controlled cytosine demethylation in the promoter region of the *Ac* transposable element in maize. *Proc. Natl. Acad. Sci. U.S.A.* **86**, 2789–2793.

Schwartz, D., and Dennis, E. (1986). Transposase activity of the *Ac* controlling element in maize is regulated by its degree of methylation. *Mol. Gen. Genet.* **205**, 476–482.

Schwarz-Sommer, Zs., Gierl, A., Cuypers, H., Peterson, P. A., and Saedler, H. (1985). Plant transposable elements generate the DNA sequence diversity needed in evolution. *EMBO J.* **4**, 591–597.

Shirzadegan, M., Palmer, J. D., Christey, M., and Earle, E. D. (1991). Patterns of mitochondrial DNA instability in *Brassica campestris* cultured cells. *Plant Mol. Biol.* **16**, 21–37.

Sibi, M., Biglary, M., and Demarly, Y. (1984). Increase in the rate of recombinants in tomato (*Lycopersicon esculentum* L.) after *in vitro* regeneration. *Theor. Appl. Genet.* **68**, 317–321.

Singsit, C., Veilleux, R. E., and Sterrett, S. B. (1989). Enhanced seed set and crossover frequency in regenerated potato plants following anther and callus culture. *Genome* **33**, 50–56.

Spofford, J. B. (1976). Position-effect variegation in *Drosophila*. In "The Genetics and Biology of Drosophila" (M. Ashburner and E. Novitski, eds.), Vol. 1C, pp. 955–1018. Academic Press, New York.

Springer, W. D., Green, C. E., and Kohn, K. A. (1979). A histological examination of tissue culture initiation from immature embryos of maize. *Protoplasma* **101**, 269–281.

Sree Ramulu, K., Devreux, M., Ancora, G., and Laneri, U. (1976). Chimerism in *Lycopersicum peruvianum* plants regenerated from *in vitro* cultures of anthers and stem internodes. *Z. Pflanzenzuecht.* **76**, 299–319.

Sree Ramulu, K., Dijkhuis, P., Roest, S., Bokelmann, G. S., and de Groot, B. (1986). Variation in phenotype and chromosome number of plants regenerated from protoplasts of dihaploid and tetraploid potato. *Plant Breed.* **97**, 119–128.

Stadler, L. J. (1941). The comparison of ultraviolet and X-ray effects on mutation. *Cold Spring Harbor Symp. Quant. Biol.* **9**, 168–178.

Sunderland, N. (1977). Nuclear cytology. In "Plant Tissue and Cell Culture" (H. E. Street, ed.), 2nd ed., pp. 177–205. Blackwell, Oxford.

Swedlund, B., and Vasil, I. K. (1985). Cytogenetic characteristics of embryogenic callus and regenerated plants of *Pennisetum americanum* (L.) K. Schum. *Theor. Appl. Genet.* **69**, 575–581.

Taylor, L. P., and Walbot, V. (1985). A deletion adjacent to the maize transposable element *Mu1* accompanies loss of *Adh1* expression. *EMBO J.* **4**, 869–876.

Tuberosa, R., and Phillips, R. L. (1986). Isolation of methotrexate-tolerant cell lines of corn. *Maydica* **31**, 215–225.

van den Bulk, R. W., Loffler, H. J. M., Lindhout, W. H., and Koornneef, M. (1990). Somaclonal variation in tomato: Effect of explant source and a comparison with chemical mutagenesis. *Theor. Appl. Genet.* **80**, 817–825.

Vasil, V., and Vasil, I. K. (1982). The ontogeny of somatic embryos of *Pennisetum americanum* (L.) K. Schum. I. In cultured immature embryos. *Bot. Gaz. (Chicago)* **143**, 454–465.

Vasil, V., Lu, C.-Y., and Vasil, I. K. (1985). Histology of somatic embryogenesis in cultured immature embryos of maize (*Zea mays* L.). *Protoplasma* **127**, 1–8.

Walbot, V. (1988). Reactivation of the *Mutator* transposable element system following gamma irradiation of seed. *Mol. Gen. Genet.* **212**, 259–264.

Walker, G. C. (1984). Mutagenesis and inducible responses to deoxyribonucleic acid damage in *Escherichia coli*. *Microbiol. Rev.* **48**, 60–93.

Weinberg, G., Ullman, B., and Martin, D. W., Jr. (1981). Mutator phenotypes in mammalian cell mutants with distinct biochemical defects and abnormal deoxyribonucleoside triphosphate pools. *Proc. Natl. Acad. Sci. U.S.A.* **78**, 2447–2451.

Williams, M. E., Hepburn, A. G., and Widholm, J. M. (1991). Somaclonal variation in a maize inbred line is not associated with changes in the number or location of *Ac* homologous sequences. *Theor. Appl. Genet.* **81**, 272–276.

Yisraeli, J., and Szyf, M. (1984). Gene methylation patterns and expression. *In* "DNA Methylation: Biochemistry and Biological Significance" (A. Razin, H. Cedar, and A. D. Riggs, eds.), pp. 353–378. Springer-Verlag, New York.

Zehr, B. E., Williams, M. E., Duncan, D. R., and Widholm, J. M. (1987). Somaclonal variation in the progeny of plants regenerated from callus cultures of seven inbred lines of maize. *Can. J. Bot.* **65**, 491–499.

Zimmermann, F. K., and Scheel, I. (1984). Genetic effects of 5-azacytidine in *Saccharomyces cerevisiae*. *Mutat. Res.* **139**, 21–24.

THE *Mu* ELEMENTS OF *Zea mays*

Vicki L. Chandler and Kristine J. Hardeman

Institute of Molecular Biology, University of Oregon, Eugene, Oregon 97403

I. Introduction

Thirteen years ago, D. S. Robertson, at Iowa State University, described a Mutator system in maize that caused a 50- to 100-fold increase in the mutation frequency (Robertson, 1978, 1983). Extensive genetic and molecular experiments have demonstrated that the increased mutation frequency is caused by a family of transposable elements, referred to as *Mu* elements. Our major goal in this review is to

ADVANCES IN GENETICS, Vol. 30

Copyright © 1992 by Academic Press, Inc.
All rights of reproduction in any form reserved.

focus on the biology of this system, summarizing what is currently known about the structure and properties of the various *Mu* elements. Specifically, we will summarize the genetic properties of *Mu* elements, including the transmission, timing, and maintenance and loss of *Mutator* activity. We will describe the structure and diversity of the *Mu* elements. In discussing the regulation of this system we will summarize the role methylation plays, and the recent discovery and characterization of a *Mu* element that regulates the transposition of other *Mu* elements. Various studies examining *Mu* element transposition will be reviewed, as well as the effects *Mu* elements have on adjacent gene expression. We only briefly discuss the uses of *Mu* elements in mutagenesis and transposon tagging because a detailed review on this aspect of the *Mu* system is currently being written (Walbot, 1992).

A. DISCOVERY OF MUTATOR STOCKS

Mutator stocks originated from a maize stock, in Dr. Alexander Brink's laboratory (University of Wisconsin), that contained an *Ac(Mp)* element inserted into the *P* locus; *P* is required for phlobaphene synthesis in the cob and pericarp (Coe *et al.*, 1988). A revertant was found in this line and transferred to the inbred W23 background by seven backcrosses. At the sixth backcross the stock was tested for *Ac(Mp)* activity, and no evidence of *Ac(Mp)* was found. A self-pollinated ear of this line, which segregated for a defective kernel mutant, was given to Dr. Jerry Kermicle. Kermicle noticed that one of the self-pollinated progeny from this ear segregated for a pale yellow endosperm mutant. Kermicle sent this stock to Robertson in 1961 because of Robertson's interest in carotenoid synthesis. Dr. Robertson designated it *y9* and began characterizing it.

Because *y9* homozygous plants are not healthy, Robertson routinely propagated this allele as a heterozygote. In the course of linkage tests, Robertson crossed F_1 plants, heterozygous for *y9* and marker genes, as males to stocks homozygous for the markers. He then self-pollinated these progeny, planted the resulting seed, and scored the seedlings for the presence of *y9*. It was in these seedling tests that Robertson noticed an unusually large number of new seedling mutants with phenotypes distinct from *y9* (Robertson, 1978). Subsequent tests have clearly demonstrated that this maize line was producing new recessive mutants at frequencies 50–100 times that observed in standard maize stocks. Robertson and numerous other maize geneticists have crossed these stocks into a variety of backgrounds and isolated mutations in a large number of loci.

Throughout this review, we will refer to Mutator plants as those that derived from Robertson's Mutator stocks. We will distinguish between active and inactive stocks based on whether the activity of the *Mu* elements is detectable by at least one of a number of different tests (reviewed in Section II). Non-Mutator stocks are defined as maize stocks with no known history of crosses to Robertson's Mutator stocks.

One of the first mutants isolated from Mutator stocks was unstable, generating small somatic sectors of revertant tissue. Frequently, newly isolated mutants from Mutator stocks are somatically unstable, a characteristic of transposable element insertions. Further experiments demonstrated that *Mutator* activity was transmitted as a dominant multigenic trait, because ~90% of the progeny from crosses between Mutator and standard maize stocks retain the high mutation frequency (Robertson, 1978). Additional experiments failed to find any evidence of segregation distortion or cytoplasmic inheritance (Robertson, 1981a), and demonstrated that if the Mutator system was a transposable element family, it was distinct from four systems previously described for maize (Robertson and Mascia, 1981). Thus, the simplest hypothesis was that the transposition of a new family of transposable elements was responsible for the increased mutation frequency.

The Mutator system appears to have been independently discovered in 1979 by Peterson and co-workers at Iowa State University. Peterson refers to the activity in his stocks as *Cy*. *Cy* activity was first detected in Peterson's Transposable Element-Laden (TEL) stock, so designated because the TEL population has been shown to contain regulatory elements for five different transposable element families: *Cy, En, Uq, Ac,* and *Dt* (Schnable and Peterson, 1988). *Cy* was identified as an unlinked factor that controlled the activity of an allele of *bronze* (*bz*), *bz-rcy*. Subsequent experiments by Peterson and Schnable showed a correlation between *Mutator* activity and *Cy* activity (Schnable and Peterson, 1988, 1989) and demonstrated that the *bz-rcy* allele contained a *Mu* element insertion (Schnable *et al.*, 1989). Similarities and differences between the properties of *Cy* and *Mu* elements will be discussed in more detail in Section V.

B. Discovery of *Mu* Elements

To investigate the molecular nature of mutations arising in Mutator stocks, Freeling and co-workers used Mutator stocks to isolate mutations in the maize *alcohol dehydrogenase 1* (*Adh1*) gene (Freeling *et al.*, 1982; Strommer *et al.*, 1982). Comparison of the DNA sequence from one of the mutants and its progenitor allele demonstrated that a

1.4-kbp insertion had occurred in the first intron of *Adh1* (Bennetzen *et al.*, 1984). The cloning and sequencing of this insertion, designated *Mu1*, revealed that it has terminal inverted repeats of ~200 bp and had made a 9-bp target site duplication on insertion in *Adh1* (Barker *et al.*, 1984). Subsequent studies from numerous laboratories have demonstrated that there is a family of related *Mu* elements that are transposing in Mutator stocks (see Section IV).

II. Assays to Monitor *Mutator* Activity

Several different genetic and molecular assays have been used to characterize the behavior of *Mu* elements in Mutator plants. In this section we describe how the assays are done and discuss the relative strengths and limitations of each.

A. GENETIC ASSAYS

Two distinct genetic assays have been used to monitor the activity of Mutator stocks. The first assay, developed by Robertson, measures the ability of Mutator plants to generate visible seedling mutations at frequencies higher than that in non-Mutator control plants (Robertson, 1978, 1980). We will refer to this assay as the "forward mutation" assay. This assay has been uniquely applied to *Mutator*, because other families of maize transposable elements do not appear to generate forward mutations at frequencies significantly above spontaneous (Robertson and Mascia, 1981). The second assay measures the somatic reversion of an easily scorable mutation caused by a *Mu* transposon. This "somatic reversion" assay has been extensively used in the study of other transposable element systems in maize (reviewed in Fedoroff, 1983, 1989). Plants showing somatic reversion are referred to as "mutable."

To perform the forward mutation assay, Robertson typically uses the Mutator stock as male and crosses it to one of two hybrid non-Mutator lines, which regularly produce two ears. He then self-pollinates the Mutator plant and the other ear of the non-Mutator plant as controls. F_1 seeds are selected from ears in which neither parent segregated for a visible mutant; the F_1 seeds are planted and self-pollinated. Kernels from these ears are then planted in the sand bench and scored for new seedling traits, such as albino, yellow, yellow–green, pale green, glossy, or dwarf. There are numerous genes that can be mutated to produce these phenotypes, making this a convenient and sensitive assay. Typically, ~10% of the Mutator plants produced F_2 progeny that segregated new seedling mutations, as compared to ~0.2% of the control plants (Robertson, 1980). Robertson has propagated his Mutator stocks in such a way for numerous generations. The outcrossed progeny of Mutator plants that segregated for a mutation when self-pollinated were usually not propagated because the presence of a mutation in the parent made scoring for additional mutations in the progeny difficult. However, this crossing protocol may have selected for stocks with particular levels of activity because plants with mutation frequencies that were very high or very low would be selected against.

Using this assay, Robertson demonstrated that most progeny of crosses between Mutator and non-Mutator plants retain the high mutation frequency. Though the number varies, typically ~90% of the progeny retain the high mutation frequency in the subsequent generation. This transmission pattern is observed after numerous generations of outcrosses to non-Mutator lines. This inheritance pattern is not what would be expected if a single gene was responsible for the high mutation frequency, and is not typical of other well-studied transposable elements in maize. For example, there are usually only one or two active *Ac* or *Spm* elements in active stocks, and these elements segregate as single genes, with occasional progeny showing an increase or decrease in the number of elements (reviewed in Fedoroff, 1983, 1989). Robertson hypothesized that in Mutator stocks most of the progeny received at least one fully functional (autonomous) element, and that the small fraction that failed to show an increased mutation frequency did not receive an autonomous element. To explain the continued transmission to ~90% of the progeny, Robertson hypothesized that at least three copies of the autonomous elements existed in most Mutator stocks, and that the elements increase in number prior to meiosis.

The somatic reversion assay for *Mu* element activity exploits the numerous *Mu* element insertions in genes encoding anthocyanin biosynthetic enzymes (Walbot *et al.*, 1986; Brown *et al.*, 1989; Robertson and Stinard, 1989). These alleles are particularly useful for following transposable element activity and to ask whether somatic reversion is dependent on other factors in the genome. In the absence of an active system, the kernels are colorless or bronze; in the presence of an active system, the kernels have colorless or bronze backgrounds with purple revertant sectors. If one takes a mutant allele and crosses it to a tester, as diagrammed in Fig. 1, the number of mutable kernels (with purple spots) is an indication of the number of kernels retaining *Mutator* activity. If the element inserted into the gene was fully autonomous (not

B
$$\frac{bz1\text{-}m^*\ \ Sh1}{bz1sh1}$$ x $$\frac{bz1sh1}{bz1sh1}$$
spotted, plump   bronze, shrunken

Ratios	Phenotype	Predicted Genotype	Interpretation
1.			
50%	spotted, plump	*bz1-m* Sh1*	Element at bz1-m* fully functional, or multiple
50%	bronze,shrunken	*bz1sh1*	copies of regulatory elements in genome.
2.			
25%	spotted, plump	*bz1-m*Sh1*, +Reg	Element at bz1-m* defective and dependent
25%	bronze, plump	*bz1-m*Sh1*, -Reg	on an unlinked regulatory element.
50%	bronze, shrunken	*bz1sh1*	

FIG. 1. Somatic reversion assay. (A) Photograph of an ear segregating a *bz1* mutation containing a *Mu1* element. The mutable allele, designated *bz1-m** is linked to *Sh1* (plump kernels), whereas the null recessive allele, *bz1*, is linked to *sh1* (shrunken kernels). In this ear, all the plump kernels were spotted. (B) Genotype of parents and predicted ratios, depending on whether the element in the *bz1-m** allele is fully functional or is dependent on an unlinked regulator.

dependent on other factors in the genome), the expectation would be that every kernel receiving this chromosome (detected with adjacent markers) would be mutable. The same result would be expected if there were large numbers of autonomous elements in the genome such that every kernel receiving the mutant allele also received an autonomous element. If the element in the gene was defective and dependent on a single unlinked autonomous element in the genome, only half of the kernels receiving the element should be mutable. Segregation analyses of *Mu1* insertions in several different anthocyanin genes have shown that, typically, most of the kernels receiving the mutant allele are mutable (Walbot, 1986; Robertson and Stinard, 1989; Brown and Sundaresan, 1992). These ratios are consistent with transmission patterns seen with the forward mutation assay. Also consistent was the appearance of exceptional progeny that appear to have lost *Mutator* activity (Walbot, 1986; Robertson and Stinard, 1989; Brown and Sundaresan, 1992).

Both assays have been used extensively to investigate the transmission and timing of *Mutator* activity. However, because it is rare that both assays are performed on the same set of plants, it is important to discuss whether the assays reflect similar or distinct processes. Robertson performed extensive analyses comparing the somatic reversion properties of the *a1-mum3* allele and the forward mutation frequencies in the same lineages (Robertson *et al.*, 1985, 1988). He found that the two assays did not always correlate. Robertson classified the kernels containing *a1-mum3* into four categories based on the number of spots: class 1, no spots; class 2, low number; class 3, medium number; and class 4, high number. Kernels were planted and outcrossed to an *a1 sh2* tester; progeny from the outcrosses were then scored for the frequency of somatic sectors and selfed to determine their forward mutation frequency. In general, the higher the level of somatic reversion, the greater the likelihood of observing a high forward mutation rate. Similarly, the kernels with no or low levels of somatic reversion were most likely to show no forward mutation activity. However, the correlation was not absolute because exceptions were observed. In some cases, plants showing a high level of somatic reversion had no detectable new mutations. There were also occasional plants that showed no somatic reversion activity, but that did show a high forward mutation frequency. Robertson suggested that these two assays measure distinct aspects of *Mutator* activity that are dependent on different properties of the *Mu* system. This is possible, but there are also other plausible explanations for the differences observed, as discussed below.

First, because the Mutator system is unstable, sometimes cycling between active and inactive states, the fact that the two assays measure activity at distinct developmental stages could contribute to the loose correlation. The somatic reversion assay usually measures activity of a particular *Mu1* element in the aleurone, which may not be an accurate indicator of the state of the *Mu1* elements in the embryo. Second, the forward mutation assay is measuring the generation of mutants by all *Mu* elements, whereas the somatic reversion assay is measuring the activity of one *Mu1* element. It is possible that different classes of elements might respond differently to different amounts of transposase. Third, a requirement of the forward mutation assay for more transposase activity, relative to the somatic reversion assay, could explain plants retaining somatic reversion activity but producing no new mutants. The forward mutation assay requires that a sufficient number of elements transpose to produce visible mutations above control levels. The somatic reversion assay, measures the ability of a particular element (*Mu1* in all the published cases) to excise from a gene restoring its function. Therefore, it would not be surprising if these two assays required different thresholds of transposase activity. Fourth, plants that showed no somatic reversion activity, but did show forward mutation activity, could be explained as follows. The somatic reversion assay is measuring the activity of a particular *Mu1* element, which may have undergone a mutation while other *Mu* elements are still active. For example, the *Mu1* element at *a1* may have undergone a rearrangement or deletion, such that it can no longer respond to *Mutator* activity. Alternatively, the *Mu1* element may have excised, creating a deletion of *a1*. Both types of events, which result in stable null phenotypes, are documented with *Mu* elements (Taylor and Walbot, 1985; Levy and Walbot, 1991). Given these caveats, we believe it remains an open question as to whether the forward mutation assay and the somatic reversion assay reflect different aspects of the Mutator system.

B. MOLECULAR ASSAYS

Several laboratories have established a strong correlation between *Mutator* activity, as defined by genetic assays, and several characteristics of *Mu* elements. Active Mutator stocks maintain a high copy number of *Mu1* elements on outcrossing, generate new *Mu1*-homologous restriction fragments at high frequencies, and contain *Mu1* elements that are hypomethylated. In contrast, on outcrossing inactive Mutator stocks with non-Mutator stocks, the number of *Mu1* elements decreases dramatically, new *Mu1*-homologous restriction fragments are rarely detected, and the *Mu1* elements exhibit increased methylation. All three of these features can be easily assayed using DNA blot analyses.

Alleman and Freeling (1986) first showed that the number of *Mu1* elements is increased by self-pollination and is maintained on out-crossing active Mutator plants to non-Mutator stocks (that have zero to a few *Mu1*-like elements). Their approach was to digest genomic DNA with a restriction enzyme that cuts at least two conserved sites within the *Mu1* element. This type of digest results in the production of one or a few *Mu1*-homologous bands, which can then be scanned with a densitometer to estimate copy number. Alleman and Freeling's studies suggested a tight copy number control, because the ratio of the numbers of elements in parents and progeny was close to unity on out-crossing, and close to two on self-pollination. In contrast, when Mutator stocks that have lost their characteristic high forward mutation frequency are outcrossed, the number of *Mu1* elements decreases by approximately one-half each generation. Subsequent experiments by Bennetzen *et al.* (1987) and Walbot and Warren (1988) produced similar results in that active Mutator stocks maintained the copy number of their parents (on average) when outcrossed and this number doubled in the first generation of self-pollination. However, in these latter studies much more variation between plants was observed, relative to the Alleman and Freeling (1986) results.

A second assay that has been used to determine the copy number of *Mu1* is to digest genomic DNA with a restriction enzyme that does not cut within the *Mu1* element. This results in a series of *Mu1*-homologous restriction fragments that represent *Mu1* elements at unique genomic locations. When Alleman and Freeling (1986) analyzed Mutator plants and their outcrossed progeny using this assay, they also observed a conservation in the total number of bands. However, with this assay they could frequently distinguish "parental" bands (same size fragments as in the parent) and new unique restriction fragments at molecular weights not observed in the parent. In the two parent/progeny lineages examined, approximately one-half of the *Mu1*-homologous fragments were "parental" and one-half were new. Assuming the new fragments represent transposition events, ~10–15 new *Mu* insertions per generation were detected during sequential outcrossing of Mutator stocks to non-Mutator lines. The generation of new *Mu1*-homologous restriction fragments correlated with *Mutator* activity (as measured by forward mutation frequency; Alleman and Freeling, 1986; Bennetzen *et al.*, 1987). The total number of *Mu1* elements decreased when inactive Mutator stocks were crossed to non-

Mutator stocks, and the generation of new restriction fragments was rarely observed.

Southern blot analysis demonstrates a correlation between loss of activity and an inhibition of digestion of *Mu1* elements by several restriction enzymes sensitive to 5-methylcytosine (Chandler and Walbot, 1986; Bennetzen, 1987). The loss of *Mutator* activity can be observed when Mutator plants are outcrossed, self-pollinated, or intercrossed (Robertson, 1983, 1986; Walbot, 1986). Although some differences have been reported, the correlation between Mutator loss and methylation is observed using either genetic assay (Chandler and Walbot, 1986; Bennetzen, 1987). The most notable difference is that Bennetzen (1987) observed extensive DNA methylation only in self-pollinated or Mutator intercrossed progeny, whereas Chandler and Walbot (1986) and Brown and Sundaresan (1992) observed extensive DNA methylation in progeny deriving from outcrosses to non-Mutator stocks as well. The increased methylation is specific for *Mu1* elements because the methylation status of other repetitive and single-copy genes remains the same in active and inactive Mutator stocks. In addition, the methylation changes are usually limited to the *Mu1* elements, but in some instances some sites in the flanking sequences also become methylated (Walbot and Warren, 1988; Chandler *et al.*, 1988a; Bennetzen *et al.*, 1988; Martienssen *et al.*, 1990). Though it is quite clear that Mutator stocks with extensively methylated *Mu1* elements are inactive, not all inactive stocks contain hypermethylated *Mu1* elements (Bennetzen, 1987; Bennetzen *et al.*, 1988). Thus, the presence of methylated *Mu1* elements is a good indicator that the line is relatively inactive, but the absence of methylated *Mu1* elements does not guarantee that *Mu* elements are still active. Extensive studies examining whether the methylation status of other *Mu* elements also correlates with *Mutator* activity have not been reported.

III. Characteristics of Mutator Stocks

A. Maintenance and Loss of *Mutator* Activities

Mutator activity is heritable, but unstable. When Mutator plants are outcrossed to non-Mutator plants, ~10% of the progeny lack *Mutator* activity (Robertson, 1978, 1985). *Mutator* activity can also be lost when Mutator plants are intercrossed with other Mutator plants or are self-pollinated (Robertson, 1983; Walbot, 1986; Brown and Sundaresan, 1992). Robertson (1983) showed that when two different Mutator stocks, previously propagated by outcrossing, are intercrossed, the high forward mutation frequency increases in the progeny ~1.5-fold above that of the parents. When the F_1 progeny are intercrossed again with other intercrossed F_1 Mutator plants, no further increase is observed. If the intercrossing is continued for subsequent generations the resulting progeny are unable to generate new mutations. The plants that are intercrossed are from diverse genetic backgrounds with only the *Mutator* activity in common, suggesting that the *Mutator* loss is not simply the result of inbreeding (Robertson, 1983).

Mutator plants that had lost activity when outcrossed (outcross-*Mu*-loss) behaved differently than those that had lost *Mutator* activity after intercrossing (intercross-*Mu*-loss) (Robertson, 1986). When out-cross-*Mu*-loss plants were crossed with active Mutator plants, ~80% of the resulting progeny retained *Mutator* activity. Thus, in this type of cross, outcross-*Mu*-loss plants behaved as non-Mutator plants. In contrast, when intercross-*Mu*-loss plants were crossed with active Mutator plants, the resulting progeny showed no *Mutator* activity, suggesting that the elements within the active parent were inactivated in the progeny (Robertson, 1986). Bennetzen and colleagues have examined the nature of the *Mu1* elements in the *Mu*-loss stocks, examining both copy number and methylation state (Bennetzen *et al.*, 1987; Bennetzen, 1987). Stocks that lost *Mutator* activity retained high copy numbers of *Mu1* elements (10–50 copies), regardless of whether they derived from outcrosses or intercrosses (Bennetzen *et al.*, 1987). No new *Mu1* restriction fragments were found in progeny from intercross-*Mu*-loss plants. In contrast, one plant from an outcross-*Mu*-loss plant did show a new *Mu1* restriction fragment. This suggests that though at least some of the outcross-*Mu*-loss plants may have insufficient activity to be detected by the forward mutation assay, they may not be completely inactive. When Bennetzen (1987) examined the methylation state of the *Mu1* elements in these *Mu*-loss stocks, he found extensive hypermethylation of several sites within *Mu1* elements in the intercross-*Mu*-loss plants, but no detectable methylation of *Mu1* elements in outcross-*Mu*-loss plants. He showed that the hypermethylation was stable on outcrossing to non-Mutator plants.

Similar studies were done with Mutator plants that had lost activity as measured by the somatic reversion assay (Walbot, 1986; Chandler and Walbot, 1986). Walbot (1986) defined three states, active (all progeny receiving the mutable allele are spotted), weakly active (low fraction of progeny receiving the mutable allele are spotted), and inactive (no progeny receiving the mutable allele are spotted). Weakly active and inactive states were observed in both outcrossed and self-pollinated progeny. The methylation status of the *Mu1* elements in inactive progeny was compared to that of *Mu1* elements in the active progeny. Both outcross-loss and self-loss plants exhibit increased DNA methylation at numerous sites within the elements (Chandler and Walbot, 1986). Using a different mutant allele, Brown and Sundaresan (1992) have obtained similar results. This contrasts with Robertson and Bennetzen's findings—when outcross-*Mu*-loss was identified using the forward mutation assay, no hypermethylation was observed. This discrepancy might simply reflect differences in the sensitivity of the assays. Because a noticeable increase in the forward mutation frequency may require the transposition of numerous *Mu* elements, it is possible that more *Mutator* activity is required to give a positive result with the forward mutation assay than with the somatic excision assay. If this line of reasoning is correct, the outcross-loss plants arising from Robertson's experiments would be predicted to have enough activity to maintain the unmethylated state, but not enough to generate a measurable forward mutation frequency. The observation that at least one outcross-loss progeny produced a new *Mu1* fragment is consistent with the idea that at least some of these stocks were not completely inactive (Bennetzen *et al.*, 1987).

Several studies have compared the stability of *Mutator* activity when the Mutator system is transmitted through male versus female gametes. As discussed above, Robertson usually maintained Mutator stocks by crossing them as male to non-Mutator females (Robertson, 1978). Robertson (1985) compared the frequency of mutations at two genes, *wx* and *y1*, and the general forward mutation frequency in seedlings when Mutator plants were crossed as males or as females. He found no difference in the frequency of *y1* mutations [1.7×10^{-4} (female) and 2.2×10^{-4} (male)]. In contrast, there was a significant difference in *wx* mutations [4.0×10^{-6} (female) and 5.6×10^{-5} (male)]. In the first generation that Mutator plants were used as female, fewer seedling mutations were observed than when used as males (5% versus 9.2%). However, this difference was not observed in subsequent generations of transmission through the female germ line.

Using the somatic reversion assay, Walbot (1986) observed differences in *Mutator* activity depending on which parent was male or female. When spotted kernels from active lines (100% of kernels receiving the mutable allele are spotted) are outcrossed as females to non-Mutator plants, most of the progeny ears retained full activity (16/18). In contrast, when spotted kernels from active lines were outcrossed as male, only 4 of 12 ears retained full activity. Differential results in the opposite direction were obtained with weakly active plants (only a few kernels receiving the mutable allele are spotted).

When spotted kernels from weakly active plants were crossed as female to non-Mutator plants, two-thirds of the ears contained 100% inactive progeny and the remaining one-third contained only ~1% spotted kernels. When the same plants were crossed as male, one-third of the ears contained 100% inactive progeny and the remaining two-thirds of the ears contained ~12% spotted kernels. Differences between female and male transmission were also observed in reactivation experiments. Active Mutator plants were much better at restoring activity to weakly active stocks when the active stocks were used as female. Inactive Mutator plants were much better at inactivating active stocks when used as female. From these studies, the activity state of the female parent was postulated to be more important than the male parent in influencing the activity state that the progeny will inherit (Walbot, 1986).

Bennetzen (1987) examined the methylation state of *Mu* elements in Robertson's intercross-*Mu*-loss lines. When the intercross-*Mu*-loss plants were used as females with active Mutator plants, the modification state showed a partial maternal dominance. In six of eight crosses using the intercross-*Mu*-loss plant as female with active Mutator male plants, all the elements in the progeny were methylated. In contrast, in seven of eight crosses using the active line as female and the intercross-*Mu*-loss line as male, less than half the elements in the progeny were methylated. Approximately half of the elements came from the *Mu*-loss parent in which they were methylated, suggesting that demethylation of some of the elements occurred.

To explain their results, both Walbot (1986) and Bennetzen (1987) have postulated that a dominant negative factor is present in plants with inactive, methylated *Mu* elements. Recent experiments on the inactivation and reactivation of *Mu* elements calls into question the generality of these findings (Brown and Sundaresan, 1992). Brown and Sundaresan performed reciprocal crosses with active *Mu* plants and inactive *Mu* plants derived from either outcrosses or self-pollinations. They monitored *Mutator* activity using the somatic reversion assay with an allele of *bz1* that contained a *Mu1* insertion, *bz1-mum9*. Interestingly, they found reactivation to be almost 100% efficient in both males and females and independent of whether the inactivation occurred after outcrossing or self-pollinating. In the plants they investigated, they found that when active plants lost *Mu* activity, the loss was slightly greater in progeny transmitted through the male versus the female flower (8% versus 3%). In all the cases examined, inactivation was associated with gain of methylation and reactivation was associated with loss of methylation of *Mu1* elements. They speculate that the difference in their results versus Walbot's results is that the active parents she used may have been losing *Mu* activity at a higher rate compared with their stocks. As her plants also showed a greater loss of activity in crosses to tester when transmitted through the male, reactivation in crosses with inactive *Mu* stocks using pollen would be less efficient than using the ear of the active parent.

Martienssen *et al.* (1990) have proposed a model to explain this difference between male and female transmission. Using the *hcf106* allele that enables the visualization of *Mutator* loss during plant development, Martienssen *et al.* (1990) saw progressive loss of *Mutator* activity during development. Thus, the tassel is more likely to be "off" than the ear because it arises later in development (Martienssen *et al.*, 1990).

B. TIMING OF *Mutator* ACTIVITIES

Robertson has carried out several studies to determine when in development *Mu* elements are active (Robertson, 1980, 1981b). He examined when new *Mu*-induced mutations occur using three tests: he determined where on the ear new mutations occurred and whether they were clustered; he performed allelism tests between phenotypically similar mutants arising from the same male parent; and he looked for the generation of mutant sectors in somatic tissue by scoring Mutator plants heterozygous for several genes expressed in the plant and endosperm. The ear maps demonstrated that mutants with similar phenotypes (and subsequently shown to be allelic) could be found clustered in small sectors on the ear, suggesting new mutations can arise premeiotically. The small size of the sectors (two to five kernels), representing ~1/100 of the germinal tissue of an ear, indicates the mutations occurred late in development. Allelism tests of phenotypically similar mutants transmitted through the male demonstrated that ~18% were allelic, suggesting that premeiotic mutants can be induced in the male lineage as well.

No visible mutant sectors in the mature plant, endosperm, or aleurone tissues were observed in Robertson's experiments. However, mutant sectors have been observed in plants heterozygous for the *hcf106* allele (R. Martienssen, personal communication). Four of the sectors were large enough to permit DNA blot analysis. In all four cases, the previously wild-type allele contained an insertion. Though gene conversion or mitotic recombination could not be ruled out in three cases, in one mutant sector the wild-type allele has a different *Mu* element than the *hcf106* allele has inserted within it (R. Martienssen, personal communication). Taken together, these results suggest *Mu* elements can transpose throughout development but tend to be most active shortly before or during meiosis. Consistent with this, when screening for mutations visible in the aleurone tissue, most arise as single kernels on an ear, and in most instances the aleurone and embryo share the same mutant phenotype.

Somatic reversion also occurs most frequently late in ontogeny. Analyses of genes expressed in the aleurone that contain *Mu1* insertions revealed that reversion occurs late in the ontogeny of that tissue (Robertson and Stinard, 1989; Levy *et al.*, 1989). Most *Mu1*-induced mutations show a high frequency of small revertant sectors. In the few cases in which the number of cells per sector have been measured, they range in size from 1 to 256 cells, with the majority of sectors containing 10–20 cells (Levy *et al.*, 1989; Levy and Walbot, 1990). When elements are inserted into genes that are expressed in the seedling or mature sporophyte, most somatic sectors are also small (Buckner *et al.*, 1990; Patterson *et al.*, 1991). Nonetheless, occasional large revertant sectors do occur (Martienssen *et al.*, 1989). Some of these large sectors represent excisions, whereas others may represent epigenetic changes, as described in more detail in Sections V and VII (Martienssen *et al.*, 1990; Chomet *et al.*, 1991).

IV. Structure and Diversity of *Mu* Elements

A. CLASSES OF *Mu* ELEMENTS

The *Mu* family is the most heterogeneous element family described in maize thus far. It is much more heterogeneous than the *Spm* family of maize transposable elements, whose autonomous elements are nearly identical and whose defective elements, *dSpm*, all appear to be deletion derivatives of *Spm* (reviewed in Fedoroff, 1989). *Mu* elements are also more varied than the *Ac/Ds* family of transposable elements, which includes the homogeneous autonomous element, *Ac*, and defective elements (*Ds* elements). Some *Ds* elements are deletion derivatives of *Ac* and some contain non-*Ac* sequences (reviewed in Fedoroff, 1989). All *Mu* elements share very similar ~220-bp terminal inverted repeats. However, different classes of *Mu* elements have completely unrelated internal sequences. Each class of *Mu* element is defined by homologies within the internal sequences. However, not all members of a class are identical.

Figure 2 illustrates the structures of the characterized *Mu* elements. As stated above, all of these elements share similar terminal inverted repeats of 220 bp. However, the actual number of base pairs repeated at

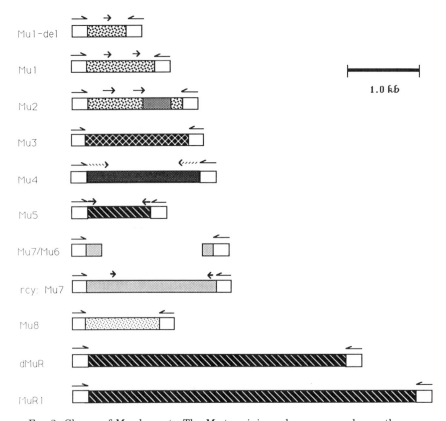

FIG. 2. Classes of *Mu* elements. The *Mu* termini are shown as open boxes; the arrows above them indicate that they are inverted repeats. Additional inverted repeats or direct repeats within the elements are also indicated by arrows. Those *Mu* elements that share sequence identity in their internal regions are diagramed with similarly shaded boxes.

the ends of the elements varies from 185 to 220 bp. In a given element, one end may contain additional sequences homologous to the longest *Mu* terminus even though this sequence is not present in the other end, resulting in a shorter inverted repeat. The homology between the inverted repeats is greater between elements belonging to the same class and less between elements in distinct classes. Finally, the similarity between the termini of *Mu* elements is greatest near the ends, decreasing toward the interior portion (summarized in Walbot, 1991).

The first characterized *Mu* element, *Mu1*, was identified as an insertion into the *Adh1* gene (Bennetzen *et al.*, 1984). *Mu1* has internal direct repeats of 104 bp (Barker *et al.*, 1984). *Mu1* has been found inserted into the following genes: *A1* (O'Reilly *et al.*, 1985), *Bz1* (Taylor *et al.,* 1986; Brown *et al.,* 1989; Hardeman and Chandler, 1989), *Bz2* (McLaughlin and Walbot, 1987), *Kn1* (S. Hake, personal communication), *Sh1* (Sundaresan and Freeling, 1987), *Hcf106* (Martienssen *et al.,* 1989), and *Vp1* (McCarty *et al.,* 1989a). *Mu1*-del, identified as an insertion into the *Bz1* gene (Hardeman and Chandler, 1989), is similar in structure to *Mu1* except it is missing one of the 104-bp direct repeats of *Mu1* and the sequences between the 104-bp repeats (K. J. Hardeman and V. L. Chandler, unpublished).

Mu2, also named *Mu1.7,* was first cloned from a Mutator line by its homology to *Mu1* (Taylor and Walbot, 1987). Sequencing revealed that *Mu2* is similar to *Mu1* except for an additional 385 bp of novel DNA (represented as a differently shaded box in Fig. 2). The novel DNA extends the length of the internal direct repeats by 34 bp. There are only 37 nucleotide differences between the homologous segments of *Mu1* and *Mu2.* Since it was originally isolated, *Mu2* has been found inserted into the *Bz1* (Taylor *et al.,* 1986) and *B* (Patterson *et al.,* 1991) genes.

Mu3 was identified as an insertion into the *Adh1* gene (Chen *et al.,* 1987; Oishi and Freeling, 1987). *Mu3* has also been found inserted into the *Y1* (Buckner *et al.,* 1990) and *Hm1* (S. Briggs, personal communication) genes.

Mu4 and *Mu5* were cloned from a non-Mutator stock (Talbert *et al.,* 1989) by their homology to *Mu1* termini. The inverted repeats of *Mu4* include an extension of 294 bp, making the total length of the inverted repeats 514 bp. The inverted repeats of *Mu5* also include an extension of 138 bp, making the total length of the inverted repeats 358 bp. The extension of the inverted repeats in *Mu4* and *Mu5* shares no sequence identity. To our knowledge, *Mu4* has not been found inserted into genes. However, some Mutator stocks do have increased numbers of *Mu4* elements (Chandler *et al.,* 1988b), suggesting they may be capable of transposition. As discussed below, *Mu5* is closely related to *Mu* elements that can transpose.

Mu6 and *Mu7* were cloned as partial elements from a Mutator stock (V. Chandler, unpublished data); *rcy:Mu7* was isolated as an insertion at *Bz1,* which was regulated by the *Cy* system (Schnable *et al.,* 1989). Sequencing revealed that this 2.2-kbp element contained 203-bp terminal inverted repeats homologous to those of *Mu1* and 65-bp inverted repeats within the internal sequence (Fig. 2). A portion of the internal sequence at one end of *rcy:Mu7* was found to be homologous to *Mu7,* resulting in the naming of the element at *Bz1, rcy:Mu7* (Schnable *et al.,* 1989). A portion of the internal sequence of *rcy:Mu7,* from the opposite end, is homologous to *Mu6* (K. J. Hardeman, unpublished data). Therefore, it is likely that *Mu6* and *Mu7* represent part of a larger *rcy:Mu7*-like element. This element, designated *Mu7*, has also been found inserted into *Knotted* (Veit *et al.*, 1990).

Mu8 was first isolated as an insertion into the *Wx* gene (Fleenor *et al.*, 1990) and has also been found in the *Knotted* gene (Veit *et al.*, 1990).

dMuR was isolated as an insertion into the *Sh1* gene (Chomet *et al.*, 1991). It has sequences at both ends that hybridize to a probe containing the *Mu5* extended inverted repeats. *MuR*, an element of ~5 kbp, was cloned from a Mutator stock using a probe from the internal sequence of *dMuR*. *MuR* was shown genetically to be the regulator of *Mu* elements (Chomet *et al.*, 1991; reviewed in Section V). Results from restriction mapping and hybridization indicate that *MuR* is very similar to *dMuR* except for an additional ~0.9 kb of sequence near one end of the element.

Mu9 (not pictured) was isolated as an insertion into the *Bz2* gene (Hershberger *et al.*, 1991b). It is 4.9 kbp and has been shown to hybridize to internal sequences of *dMuR* (P. Chomet, personal communication). DNA sequence analysis indicated that *Mu9* exhibits regions of homology with *Mu5*. Hershberger *et al.*, (1991b) reported that one end of *Mu9* is homologous to the extended inverted repeats of *Mu5* and ~950 bp of the other end of *Mu9* is ~70% homologous to *Mu5*. We compared the published sequences of *Mu5* and *Mu9* and found that the homology between *Mu5* and *Mu9* extends the complete length of *Mu5* (K. J. Hardeman and V. L. Chandler, unpublished). Therefore, *Mu9*, *MuR1*, *dMuR*, and *Mu5* are all members of the same class of *Mu* elements, with *Mu5* and *dMuR* missing different regions of the larger *MuR/Mu9* elements.

Finally, *MuA* (not pictured in Fig. 2), an element of ~5.5 kbp, was cloned from a Mutator stock by homology to the *Mu1* termini (Qin and Ellingboe, 1990). The restriction pattern of *MuA* is similar, but not identical, to that of *MuR1* and *Mu9*. The *MuA* element was used as a probe to clone a related element designated *MuA2* (Qin *et al.*, 1991). The restriction map of *MuA2* is very similar to that of *MuR1* and *Mu9*. Therefore, it is likely to be a member of the *MuR1/Mu9* class, although to our knowledge its hybridization to these elements has not been directly tested.

B. DISTRIBUTION OF *Mu* ELEMENTS IN *Zea*

The increased mutation rate of Mutator stocks compared to non-Mutator stocks correlates with the transposition of *Mu* elements.

Therefore, it was of interest to determine if *Mu* elements were found in non-Mutator stocks as well as in Mutator stocks. The distribution of *Mu1* has been most thoroughly investigated. Originally, it was reported that most Mutator stocks contained 15–40 fragments that hybridized to *Mu1* internal and terminal sequences, whereas the non-Mutator stocks tested contained no fragments that hybridized to *Mu1* internal or terminal sequences (Bennetzen, 1984; Freeling, 1984). In later studies, it was found that non-Mutator stocks, some of which were the same inbred lines used in the previous studies, contained 1–3 fragments that hybridized to the *Mu1*-internal sequence (Chandler *et al.*, 1986) and ~20–40 fragments that hybridized to the *Mu1*-terminal sequence (Chandler *et al.*, 1986; Talbert *et al.*, 1989). The discrepancy between the number of fragments that hybridized to the *Mu1*-internal sequences could be due to a difference in the non-Mutator stocks used. For example, it was found that in two separately maintained lines of the inbred A188, one contained 2–4 fragments homologous to *Mu1*-internal sequences whereas the other did not contain any fragments homologous to *Mu1*-internal sequences (Masterson *et al.*, 1988). The difference in the number of fragments that hybridized to a *Mu1*-terminal probe could also be due to a difference in hybridization conditions used. As stated earlier, the termini of different *Mu* elements are similar but not identical.

Regardless of reasons for the published differences, it is now clear that many, if not most, non-Mutator stocks do contain sequences homologous to *Mu1*-internal and -terminal sequences and, therefore, the presence of *Mu*-homologous sequences is not a sufficient criterion for the high mutation rate. To determine whether these *Mu1*-homologous fragments were structurally similar to *Mu1*, one of these *Mu1*-homologous sequences was cloned from B37 (a non-Mutator line) and named *Mu1.4-B37*. DNA sequence analysis revealed that it was identical to the *Mu1* element isolated from *Adh1* (Chandler *et al.*, 1988a). Furthermore, this *Mu1* element from B37 was flanked by 9-bp direct repeats, indicative of a target site duplication generated during *Mu* element transposition. Therefore, it would seem that at one time this element transposed to its current location in B37.

The distribution of *Mu4*- and *Mu5*-homologous sequences has also been studied. In non-Mutator lines, only two to three restriction fragments homologous to the *Mu4*-internal probe were found, whereas in these same lines three to six restriction fragments hybridized to the *Mu4*-specific inverted repeat sequence (Talbert *et al.*, 1989). Because the restriction enzyme used in this experiment cut within the cloned *Mu4* element, most of the *Mu4*-specific repeats could be associated with *Mu4*-internal sequences. Two to four sequences in non-Mutator lines were found to hybridize to a *Mu5*-internal probe whereas 12–20 fragments hybridized to a larger probe that included the *Mu5*-specific inverted repeat sequence (Talbert *et al.*, 1989). Therefore, some of the *Mu5*-specific inverted repeats are not associated with *Mu5*-internal sequences. One type of sequence that would fit into this category is the *dMuR* element (Chomet *et al.*, 1991), because it is deleted for most of the *Mu5*-internal sequences (K. J. Hardeman, unpublished data).

Though the distribution of the other *Mu* elements in Mutator stocks has not been thoroughly investigated, a common pattern is that each class usually exists in lower numbers than *Mu1*. In stocks in which an element of a particular class has inserted into a gene, and is therefore assumed to be active, its numbers are usually higher than in non-Mutator stocks (Schnable, *et al.*, 1989; V. L. Chandler, unpublished results). It is unclear at this time whether representatives of all classes of *Mu* elements have been identified, or if more classes remain to be found. It is possible that some of the sequences that hybridize to the *Mu* termini in maize might be isolated *Mu* termini that are not part of a complete element.

A sequence homologous to the internal sequences of *Mu1* and *Mu2*, but not to the *Mu* termini, is present in one or two copies in all maize lines tested (Talbert and Chandler, 1988). This sequence was termed *MRS-A* (for *Mu*-related sequence) and it is transcribed in many different tissues. A similar sequence, which hybridizes to both *Mu2* and to sequences flanking *MRS-A*, is present in a species within the genus most closely related to *Zea, Tripsacum dactyloides*. However, *T. dactyloides* does not contain any sequences that hybridize to the *Mu* termini at high stringencies. These results led to the suggestion that *MRS-A* represents an old nontransposon sequence in maize and that perhaps a similar sequence was encompassed by *Mu* termini to create *Mu2*. This hypothesis assumes that at one time *Mu* termini could transpose in the maize genome. The observation that *Mu* termini contain small imperfect inverted repeats is consistent with this hypothesis (Talbert and Chandler, 1988; Talbert *et al.*, 1989).

The finding that sequences capable of hybridizing with *Mu* termini were not found in *T. dactyloides* suggests that *Mu* elements are more recent events in evolution than the *Ac* family of transposable elements, because sequences that hybridize to *Ds1* were identified and cloned from *T. dactyloides* (Gerlach *et al.*, 1987).

V. Regulation of Mu Element Activities

A. ELEMENTS ENCODING REGULATORY FUNCTIONS

In most Mutator stocks, *Mutator* activity segregates as a multigenic trait. The identification of exceptional lines and their characterization have led to the identification of *Mu* elements that regulate the activities of themselves and other *Mu* elements (Robertson and Stinard, 1989; Chomet *et al.*, 1991; Qin *et al.*, 1991). In a study of two *Mu1* insertions into the *a1* gene (*a1-Mum2* and *a1-Mum3*), stocks were obtained that segregated a factor required for somatic reversion of the *Mu1* element at *a1* (Robertson and Stinard, 1989). In many testcrosses this factor segregated 1:1, suggesting the presence of a single regulatory factor. Interestingly, most *a1-Mum2* and *a1-Mum3* stocks that contained a single segregating regulatory factor did not produce a high forward mutation rate (Robertson and Stinard, 1989). Using a similar stock, Chomet *et al.* (1991) independently identified a factor regulating *a1-Mum2*, which they designated *Mu-Regulator1* (*MuR1*). They demonstrated that the genetic factor was capable of transposition and could confer activity on several different classes of *Mu* elements (Chomet *et al.*, 1991; P. Chomet, D. Lisch and M. Freeling, personal communication).

A novel 4.0-kbp insertion into *shrunken* (*sh1-A83*) was shown to be related to the internal sequences of the larger *MuR1* element (~5.0 kbp), enabling the cloning and characterization of the *MuR1* element (Chomet *et al.*, 1991). The 4.0-kbp insertion into *sh1-A83* appears to contain an ~1-kbp deletion relative to *MuR1* (Chomet *et al.*, 1991) and is designated *dMuR*, for derivative of *MuR*. Three lines of evidence convincingly demonstrate that the cloned element is the genetically identified *MuR1* element (Chomet *et al.*, 1991). First, *MuR1* genetic activity and a *MuR*-cross-hybridizing restriction fragment cosegregate (0 recombinants with 94 chromosomes examined). Second, when *MuR1* undergoes a duplicative transposition, an additional *MuR*-cross-hybridizing restriction fragment is observed. This new cross-hybridizing restriction fragment cosegregates with *MuR1* genetic activity in subsequent crosses. Third, *MuR1* genetic activity and the cosegregating *MuR*-related fragment are simultaneously lost within clonal somatic sectors. DNA was prepared from revertant sectors that appeared in leaves of *a1-Mum2* plants and from adjacent mutant tissue. DNA blot analysis demonstrated that the restriction fragment that segregated with *MuR1* genetic activity was missing from the revertant sectors. In addition, the *MuR* element hybridizes to transcripts that are found specifically in plants with *Mutator* activity.

Qin and Ellingboe (1990) isolated a ~5.1-kbp element by randomly cloning *Mu* elements using the *Mu* terminus as a probe. This element hybridizes with a transcript specifically found in Mutator stocks, leading the authors to speculate that the element (designated *MuA*) was homologous to elements controlling *Mutator* activity. Subsequently, the investigators used a probe from *MuA* to examine Robertson's stocks that were segregating 1:1 for *Mutator* activity. They identified and cloned a *MuA*-homologous restriction fragment cosegregating with *Mutator* activity. The newly cloned element was designated *MuA2*, and the previously cloned *MuA* element was renamed *MuA1* (Qin *et al.*, 1991). The restriction maps of *MuA2* and *MuR1* are very similar, and it is quite likely they represent very similar if not identical elements. Comparison of the sequences flanking the two cloned elements will determine whether they were located at the same or different sites in the genome of the *a1*-*Mum2* stocks.

A third example of the *MuR*/*MuA* class of elements has been characterized and designated *Mu9* (Hershberger *et al.*, 1991b). *Mu9* is ~4.9 kbp and was isolated as an insertion in the *bz2* gene. As shown for *dMuR* and *MuA1*, probes from the internal region of this element recognize transcripts specific for active Mutator stocks and a restriction fragment that cosegregates with *Mutator* activity in Robertson's *a1*-*Mum2* stocks. The restriction map of *Mu9* is very similar to that of *MuR1* and *MuA2*, and *Mu9* hybridizes with probes from *dMuR* (P. Chomet, personal communication), indicating that *Mu9* is closely related to these elements. Because *Mu9* inserted into the *bz2* gene, it is capable of transposition. However, no evidence was presented to indicate that *Mu9* can activate the transposition of itself or unlinked *Mu* elements (Hershberger *et al.*, 1991b). The sequence of *Mu9* reveals multiple open reading frames. However, given the large number of introns in other types of maize elements, it will be necessary to isolate and sequence the cDNAs to determine the element's protein-coding capacity.

Cy was genetically identified as an unlinked factor that controls the activity of an allele of *bronze*, *bz-rcy*, which was subsequently shown to contain a *Mu* element related to *Mu7* (Schnable *et al.*, 1989). Mutability of the *a1*-*Mum2* allele cosegregates with *Cy* (0 recombinants out of 46 chromosomes; Schnable and Peterson, 1989). In contrast to *Mu* regulatory elements in most Mutator lines, in Peterson's stocks *Cy* usually segregates as a single genetic locus. The current data are consistent with the hypothesis that *Cy* represents another example of a

MuR1-like element. However, molecular characterization will be required to substantiate this hypothesis.

Mutator stocks, which show a high forward mutation rate, have increased copy numbers of *MuA2*-like elements (Qin *et al.*, 1991). Robertson may have selected for an increased number of regulatory elements by propagating only those plants that demonstrated a mutation rate significantly above background. Consistent with this idea is the finding that lines with only one to a few copies of the regulator rarely show a high forward mutation rate (Robertson and Stinard, 1989). As Peterson did not impose a similar selection on his *Cy*-containing stocks, this could explain why *Cy* remains in low copy numbers relative to regulatory elements in Mutator stocks.

MuR1, MuA2, and *Mu9* contain *Sst*I sites in their termini, which, when cut, generate a 4.7-kb *Sst*I fragment. Active Mutator stocks contain multiple copies of this fragment and several non-Mutator stocks do not have this fragment (Qin *et al.*, 1991; Hershberger *et al.*, 1991b). This finding is consistent with the hypothesis that non-Mutator stocks have no copies of intact regulatory elements. An alternative explanation is that the regulatory elements are methylated in non-Mutator stocks such that they are not digested and are not active. More extensive experiments with additional restriction enzymes that are not sensitive to methylation should be able to distinguish between these models.

B. CYCLING OF *Mutator* ACTIVITY

As has been discussed, *Mutator* activity is heritable, but unstable. Active lines lose activity, both on outcrossing and on self-pollination. In most instances, this loss of activity correlates with extensive methylation of *Mu1* elements. Examples have been noted wherein a fully active line became inactive within one plant generation (Chandler and Walbot, 1986). However, it is more common to see a gradual loss of activity, with active lines becoming weakly active, followed by complete inactivity (Walbot, 1986). Weakly active lines frequently contain a mixture of methylated and unmethylated *Mu1* elements (Walbot, 1986; Chandler and Walbot, 1986). When inactive lines are propagated, the methylation state is heritable, but occasionally reversible. Plants with methylated *Mu1* elements usually give rise to progeny with inactive methylated elements. However, occasionally progeny with active, hypomethylated *Mu1* elements arise. This has been observed spontaneously (Hardeman and Chandler, 1989), after treatment of inactive seeds with γ-irradiation (Walbot, 1988), or after crosses with active *Mu* stocks (Martienssen *et al.*, 1990; Brown and Sundaresan, 1992). Now that probles are available, it will be important to characterize the methylation state of the regulatory elements in active, weakly active, and inactive Mutator stocks.

VI. Properties of *Mu* Element Transposition

A. GERMINAL AND SOMATIC EXCISION PRODUCTS

Many studies have been undertaken to determine the sequence left behind when transposable elements excise. Most of these studies in maize have focused on germinal revertants of transposable element-induced mutations. However, because germinal revertants of *Mu* insertions are quite rare (Schnable *et al.*, 1989; Brown *et al.*, 1989; Levy *et al.*, 1989), other strategies have also been employed to study the excision products of *Mu* elements. Two groups have used the polymerase chain reaction (PCR) to amplify the somatic excision products of two mutable alleles of *Bz1*, both of which are due to the insertion of *Mu1* (Britt and Walbot, 1991; Doseff *et al.*, 1991). These products were then sequenced and analyzed. An advantage of this strategy, compared to examining germinal revertants, is that it does not bias the sample by including only the excision products that restore gene function. A disadvantage, however, is that the number of times a particular event is isolated may not be representative of the number of times this event occurs in the plant. Therefore, any particular excision event recovered more than once in a given experiment is considered as only one independent event.

Another approach to studying *Mu* excision products was undertaken by Levy and Walbot (1991). In this study, 118 homozygous *bz2::mu1* kernels that were spotted were planted and their cob DNA was analyzed on DNA blots to determine if any of the plants had detectable changes in DNA sequence or modification relative to the progenitor allele. Six of the plants had changes in the *bz2* DNA digestion pattern that could not be attributed to methylation. Four of these alleles were then sequenced after using PCR to clone the DNA. Two of these had an identical deletion and are likely to be due to a single premeiotic event. We refer to these four alleles as somatic genetic changes.

In all the studies mentioned above, a total of 36 excision events were analyzed. These include 4 independent germinal revertants (Schnable *et al.*, 1989; Britt and Walbot, 1991), 29 independent somatic excision products (Doseff *et al.*, 1991; Britt and Walbot, 1991), and 3 independent somatic genetic changes (Levy and Walbot, 1991).

The most frequent excision products isolated from these studies were those in which all or part of one copy of the 9-bp target site duplication was missing, as well as all of the *Mu* sequence. Some of these products were also missing additional sequences adjacent to the target site duplication. However, none of the products in this class had any additional changes. Because these events are simple to describe they are not diagramed in Fig. 3. Twenty-two excision products, including three germinal revertants, fell into this class. Two of the germinal revertants contained imprecise excision events that did not restore the reading frame of the gene. Because they were recovered as revertants, the

FIG. 3. DNA sequences of somatic and germinal excision products. Bold letters indicate new bases. The clone name refers to the name given the excision events in the particular reference in which they were described. (A) DNA sequences of the excision products from *bz::mu1* (Britt and Walbot, 1991), with the sequence of the *bz::mu1* allele shown on top for reference. (B) DNA sequences of the excision products of *bz-Mum9* (Doseff *et al.*, 1991), with the sequence of the *bz-Mum9* allele shown at top for reference.

gene was at least partially functional, and the authors speculate that function of the gene was restored via alternative splicing (Schnable *et al.*, 1989). The other germinal revertant in this class was missing 21 bp, which would restore the reading frame of the gene. Of the somatic events, 18 fell into this category. Four of these were missing the element, along with one copy of the target site duplication, and therefore were identical to the progenitor. The possibility that contamination of the PCR reactions with the progenitor DNA was the source of these four excision products could not be eliminated. The remainder of the somatic excision products in this class were missing from 5 to 44 bp of *Bz1* DNA sequences. In these experiments, changes beyond the primers used for PCR would not be identified. Finally, one of the somatic genetic changes was a deletion of 8 of the 9 bp of one copy of the target site duplication as well as loss of the *Mu* sequences.

A second class of excision products also lacked *Mu* sequences and some flanking *Bz1* sequence, but had insertions of a few (1–5) extra base pairs at the deletion junction. Eight somatic excision products and one germinal revertant belong in this category. Six of these events (1, 4, 5, 6, and 7 in Fig. 3A; 1 in Fig. 3B) contained a single additional base pair. In two of these (4 and 5, Fig. 3A), the extra base pair could be considered a small inverted repeat of the adjacent base pair. Potential mechanisms for how this may have arisen will be discussed below. One excision product (2, Fig. 3B) had an additional 3 bp that could also be considered a small inverted repeat of adjacent sequences. Another excision product (3, Fig. 3A) had an additional 4 bp (CCCT) that could be considered inverted sequences of the 3′ end of the target site duplication (AGGG); however, in the excision product, these additional base pairs were not adjacent to these target site duplication sequences. Finally, one excision event (2, Fig. 3A) had an additional 5–8 bp that are identical to the *Mu1* sequences adjacent to the target site duplication (the exact number is unknown because three of the base pairs were identical in *Mu1* and *Bz1*).

The next class of excision products includes three somatic events. These products were similar to those of the last class except that the number of additional base pairs was much greater. One event (3, Fig. 3B) had an additional 16 bp. This event can be explained as a crossover between seven identical base pairs that are found in both *Mu1* and *Bz1*. Two events had additional sequences that appear to be part of the *Bz1* gene, one had an additional 16 bp (5, Fig. 3B), and the other had an additional 19 bp (4, Fig. 3B). Finally, two of the somatic genetic changes were missing 75–77 bp at the junction between *Bz2* and *Mu1*. The authors ascribe these events to abortive transportation.

The results of the studies of *Mu* element excision revealed that many of the excision products were similar to those resulting from the excision of other plant transposable elements. Similarities include the loss of transposon sequences, deletions of part or all of the target site duplication and surrounding sequences, and the presence of short inverted duplications of sequences adjacent to the deletion. Two models to explain these data have been postulated. Saedler and Nevers (1985) hypothesized that single-stranded nicks at the ends of the target site duplications followed by exonuclease digestion could account for the loss of the transposon and the generation of deletions. They further hypothesized that strand switching by repair enzymes was responsible for the inverted duplication of sequences (Saedler and Nevers, 1985). Coen, *et al.* (1986) hypothesize that after the DNA is cut to release the transposon, the free ends are ligated together to form hairpin structures. These hairpins must then be resolved by a single-stranded nick and in some cases DNA synthesis. Resolution of the hairpins with DNA synthesis would lead to inverted repeats.

B. Properties of Targets

1. Target Site Preference

Many transposable elements, including Tn7, Tn10, and Tc1 (Lichtenstein and Brenner, 1981; Halling and Kleckner, 1982; Mori *et al.*, 1988), show insertion site specificity. In studies of Tn10 it was found that hot spots for Tn10 insertion conformed better to the consensus site for Tn10 insertion than did non-hot spots. There are two sites where *Mu* insertions have been found inserted more than once. One of these sites is in *Bz1*, where *Mu1*, *Mu7*, and *Mu1*-de1 have all inserted. The other site is in *Knotted*, where both *Mu1* and *Mu8* have inserted. As many *Mu* insertion sites have now been sequenced, we were interested in assessing whether *Mu* elements might also show insertion site specificity. A compilation of the 9-bp target site duplications of 23 *Mu*-induced mutations is shown in Fig. 4. Because the target site sequence may diverge after an element inserts, only the target sites of newly isolated *Mu* insertions are presented.

The 9-bp target site duplications are lined up in a 5′ to 3′ direction. However, because the target site duplications are direct repeats, there are two possible 5′ to 3′ orientations for each sequence. Because all *Mu* elements do not contain the same internal sequences, it is impossible to line up the target site duplication sequences according to the direction the *Mu* element inserted. Considering the *Bz1* and *Knotted* sites as potential hot spots (where more than one element inserted), we

Allele	5' Sequence 3'	Element	Reference
bz: :mu1	CAAACAGGG	*Mu1*	Britt and Walbot (1991)
bz-rcy	CAAACAGGG	*Mu7*	Schnable *et al.* (1989)
bz1-A58	CAAACAGGG	*Mu1*-del	K. J. Hardeman and V. L. Chandler (unpublished)
Kn1-mum8	AGTACACTG	*Mu1*	Walko and Hake (personal communication)
Kn1-mum1	AGTACACTG	*Mu8*	Walko and Hake (personal communication)
Kn1-mum6	GTTCCCATC	*Mu8*	Walko and Hake (personal communication)
Kn1-mum3	TTCCCAAAC	*Mu8*	Walko and Hake (personal communication)
Kn1-mum4	GACAACAGG	*Mu1*	Walko and Hake (personal communication)
Kn1-mum2	AGAATCTAT	*Mu8*	Walko and Hake (personal communication)
Kn1-mum7	GTTGTGGTG	*Mu1*	Walko and Hake (personal communication)
Kn1-mum5	GATCTGGCC	*Mu8*	Walko and Hake (personal communication)
bzMum9	GTTTGCGGC	*Mu1*	Doseff *et al.* (1991)
bz: :mu2	GCCCAACTG	*Mu2*	Britt and Walbot (1991)
sh1-A83	AGAGAAACC	*dMuR*	Chomet *et al.* (1991)
sh 9026	CCCAGAGAA	*Mu1*	Sundaresan and Freeling (1987)
wx-mum5	TCGGCAGGC	*Mu8*	Fleenor *et al.* (1990)
Adh1-S3034	TTTTGGGGA	*Mu1*	Barker *et al.,* (1984)
bz2: :mu1	GCCAGACAC	*Mu1*	Nash *et al.* (1990)
a-Mum2	GTTGAATGG	*Mu1*	Shepherd *et al.* (1988)
Adh1-3F1124	ATATAAATC	*Mu3*	Chen *et al.* (1987)
hcf106	GTTGGAGAT	*Mu1*	Barkan and Martienssen (1991)
vp1-mum	ACAGCCAAC	*Mu1*	McCarty *et al.* (1989b)
bz2: :mu9	CCTCCAGGA	*Mu9*	Hershberger *et al.* (1991a)

	1	2	3	4	5	6	7	8	9
A	6	5	7	9	5	15	6	6	3
C	5	6	5	5	10	5	4	2	9
G	9	4	1	6	5	3	11	9	9
T	3	8	10	3	3	0	2	6	2

FIG. 4. Target site duplications of *Mu* element insertions. The DNA sequences of 23 *Mu* insertion sites are shown. For each site, the strand listed is the one that matches the sites where more than one *Mu* element has been found inserted; these are represented by the first five sites. The allele name of insertions is shown as well as the element name in the gene. The matrix below the 23 sites lists the number of times a given base is found at the particular position of the target site.

lined up these sequences such that they would share the most bases in common. In this alignment, four of the nine bases are shared between these two sequences. The remaining 18 target site duplications were aligned in the orientation in which they shared the most bases with the two putative hot spots. When both alignments for a particular sequence were equivalent, the alignment was chosen at random. In comparing the alignment of the 23 target site duplications, no strong consensus sequence is found (Fig. 4).

It is possible that the proteins that cut the target DNA before insertion recognize a site near, but distinct from, the target site duplication. If such a sequence were located near the sites where multiple insertions took place, this would explain the hot spot phenomenon. It is also possible that multiple insertions at a particular sequence could result from the way insertions are selected. Because an insertion must reduce or eliminate gene function in order to be recovered as a new mutation, this may limit the number of target sites where insertions can be recovered in particular genes.

2. Linked versus Unlinked Target Sites

Maize *Ac* elements have been found to move to linked locations more often than to unlinked sites (Greenblatt, 1984). It is unknown if *Mu* elements move to linked locations. The high copy numbers of *Mu* elements in Mutator stocks makes it difficult to determine the origin of new *Mu* elements. One approach used to assess whether *Mu* elements insert into linked sites was to examine the genetic linkage of *Mu1* elements in two sets of progeny from crosses between Mutator and non-Mutator plants (Ingels *et al.*, 1992).

In this study, it was found that in one set of progeny, of the 21 *Mu1* elements, there were three clusters of linked *Mu1* elements, including two clusters with two elements and one cluster with three elements. In the second set of progeny it was found that of 29 *Mu1* elements, there was one cluster of five elements, one cluster of three elements, and two clusters of two elements. The amount of clustering found in the two families is significantly greater than would be expected if *Mu* elements were distributed randomly throughout the genome.

Clusters of *Mu1* elements might be caused by *Mu1* elements moving to linked locations, thereby creating clusters of *Mu1* elements. However, other explanations for this result are also possible. For example, *Mu1* elements may insert preferentially into areas of the genome that are hypomethylated. If hypomethylated regions are clustered, this could lead to a disproportionate amount of *Mu1* elements in some locations.

C. REPLICATIVE VERSUS CONSERVATIVE TRANSPOSITION?

Studies of numerous transposable elements have revealed two major mechanisms of transposition, replicative and conservative. In replicative transposition, replication occurs as an integral part of the transposition mechanism. The donor molecule is replicated and the copy inserted into a new site, with the donor remaining at its original location. The classical example of an element using this mechanism is Tn3 (Sherratt, 1989), which undergoes an intermediate known as the cointegrate, containing two copies of the element. The cointegrate is then resolved by a site-specific recombination event into two molecules, each with one element. In conservative replication, the element is cut from its donor site, followed by reinsertion into a new site. The chromosome containing the donor site may then be repaired or lost. Given the mechanism, a more accurate name for this mechanism is "cut-and-paste." The best characterized element using this mechanism is Tn10 (Kleckner, 1989). If the free ends undergo double-strand gap repair, the element or wild-type sequences may be restored, depending on the genotype of the chromosome serving as template for repair. Alternatively, the free ends generated by excision of the element could be religated. This outcome may not restore the wild-type sequence as some or all of the target site duplication may remain. If the chromosome containing the element is replicated prior to conservative transposition, or if excision is repaired restoring the element to the original site, cut-and-paste transposition will involve some DNA replication and should lead to an increase in element numbers in the genome. Thus, an increase in element copy number could result from either replicative or cut-and-paste mechanisms. Without sophisticated genetic and biochemical experiments to reveal detailed mechanisms of transposition, it is difficult to distinguish which mechanism a particular element is using. In fact, with the bacterial elements, it is becoming clear that both conservative and replicative transposition events share a common intermediate. How this intermediate is resolved determines whether transposition is conservative or replicative (for review, see Chaconas and Surette, 1988).

It is unknown whether *Mu* elements transpose via a cut-and-paste or a replicative mechanism. The following facts are known about *Mu* element transposition. First, *Mu* elements can dramatically increase in copy number from one generation to the next (Hardeman and Chandler, 1989). Stocks containing a low number (three to six) of *Mu1* elements were used to isolate *bz1* mutants. When the mutants were examined, it was found that the *Mu1*-related elements had increased to ~10–20 copies. Alleman and Freeling (1986) have estimated ~10–15 new *Mu1* insertions per generation in their stocks. Second, in plants segregating new *Mu* elements, all the parental elements can also be observed segregating if enough progeny are examined (Alleman and Freeling, 1986; Bennetzen *et al.*, 1987). Third, both somatic excision and *Mu* element insertion into new sites are dependent on the regulatory *MuR* element (Chomet *et al.*, 1991). Fourth, somatic revertant sectors occur at high frequencies, usually late in development. Fifth, the isolation of germinal revertants is rare. Thus, if most somatic revertant sectors are caused by element excision, somatic excision is much more frequent than germ line excision. Sixth, extrachromosomal *Mu* circles have been found in active Mutator stocks (Sundaresan and Freeling, 1987).

The first three facts are consistent with either a replication-linked cut-and-paste mechanism (as utilized by *Ac;* reviewed in Fedoroff, 1989) or replicative transposition. At first glance, the lack of correlation between somatic excision and germ line reversion appears inconsistent with a replication-linked cut-and-paste mechanism. As the germ line in plants is derived from the soma, somatic excisions should produce germinal revertants if they occur in cells that are progenitors to the gametes. If transposition was by a cut-and-paste mechanism, the simplest expectation would be that higher excision rates would lead to higher germinal reversion rates. Consistent with this prediction, it has been observed that higher *Mu* germinal reversion rates can be obtained by selecting kernels with more and larger spots (V. Walbot, personal communication). One could explain the paucity of germinal revertants in Mutator stocks as at least partly due to the timing of *Mu* activity, which is usually late in development. However, new insertions that presumably represent transposition events occur frequently in the germ line (Alleman and Freeling, 1986). If excision was required to produce new insertions, we would expect to recover revertants more frequently in the germ line.

One possible way to explain this potential paradox is to postulate that repair mechanisms differ in somatic and germ line tissues. Recent studies of P element transposition are consistent with the hypothesis that the choice of template for double-strand gap repair determines whether the element is restored to the donor chromosome or is excised (Engles *et al.*, 1990). When the homologue was wild type, germinal revertants were much more abundant than when the homologue contained the insertion. If, in the germ line, the donor chromosome is repaired at high frequency by its sister chromatid, the element would be restored to the donor chromosome, and revertants would be rare. If, in somatic tissues, the donor ends were simply religated, excision would be the outcome and excision would most likely be imprecise. It is also possible that the donor chromosome could be repaired using the homologous chromosome as a template. Repair from the homologous chromosome in a heterozygote should lead to restoration of the wild-type sequence (precise excision), whereas repair in a homozygote would restore the element to the donor. If repair from the homologous chromosome were a frequently used mechanism, the prediction would be that somatic excision would be more frequent in heterozygous relative to homozygous individuals. To our knowledge this has not been observed.

Alternatively, the lack of correspondence between germinal reversion and somatic excision rates may indicate that somatic excision represents an activity other than transposition. There is no strong evidence available at this time that either supports or refutes a true replicative model. Extrachromosomal circles are observed in active *Mu* lines (Sundaresan and Freeling, 1987). However, the generation and function of these circles is unknown. They could represent replicative transposition intermediates, excision products that are transposition intermediates, or nonfunctional molecules that are not transposition intermediates. The combination of more sophisticated genetic experiments with biochemical assays will likely be required to definitely demonstrate the mechanism of *Mu* element transposition.

Few biochemical experiments have been used to investigate *Mu* element transposition. Nuclear proteins that interact specifically with sequences within the *Mu1*-terminal inverted repeats have been described (Zhao and Sundaresan, 1991). These proteins are likely to be host encoded as they are observed in both active Mutator and non-Mutator stocks. The DNA sequences recognized by these factors are conserved in the termini of the other classes of *Mu* elements, and are also found in the unrelated maize element *Ac*. Further experiments should be able to determine whether these factors play an important role in *Mu* element transposition.

VII. Effects of *Mu* Elements on Adjacent Gene Expression

A. Alterations in Transcription

It has been well documented that the insertion of transposable elements within or near genes often alters the transcription of the target gene. Insertions can place the target gene under negative or positive control of the element, or can alter the target gene's developmental or tissue-specific pattern of expression. These phenotypes usually result from a combination of regulatory sequences within and flanking the transposable element. Though only a few mutant alleles have been investigated in detail, it is clear *Mu* insertions can result in similar alterations of the expression of the adjacent gene. We will review the effects of each of three mutations in which a *Mu* element is inserted near the transcription start site of the affected gene.

The insertion of a *Mu1* element in the *hcf106* allele within sequences encoding the 5' untranslated leader produces a mutant phenotype when the elements are active, but a normal wild-type phenotype when the *Mu* elements are inactive (Martienssen *et al.*, 1990). When the *Mu* system is active, leaves of mutant seedlings lack detectable *Hcf106* mRNA (Barkan and Martienssen, 1991). When *Mu* becomes inactive, the mutant phenotype is suppressed and mRNA accumulates that is similar in size and abundance to that of the normal allele. This RNA arises from the activation of a promoter near the end of *Mu1* that directs transcription into the adjacent *Hcf106* gene (Barkan and Martienssen, 1991). The transcripts generated are close to the normal size and contain the entire protein-coding region. Primer extension and S1 nuclease protection experiments demonstrated that the 5' ends of these transcripts are heterogeneous, mapping to ~13 different sites within a 70-nucleotide region. These sites start within the *Mu1* terminus and extend into the flanking *Hcf106* sequences downstream. Interestingly, the activity of this promoter is conditional not only on *Mu* activity but also on signals that regulate the normal allele (Barkan and Martienssen, 1991). Though this is the most detailed characterization of a mutation whose phenotype is modulated by the phase of *Mu* activity, the genetic behavior of alleles of *a1, Kn,* and *vp1* is similar (Chomet *et al.*, 1991; S. Hake, personal communication; R. Martienssen, personal communication). Therefore, this may represent a rather common class of *Mu*-induced mutations.

The insertion of a *Mu3* element into the TATA box located 31 bp 5' of the transcription start site of *Adh1* creates an additional TATA box by duplicating the 9-bp genomic sequence—ATATAAATC—at the site of insertion (Chen *et al.*, 1987). This mutant allele, *Adh1-3F1124*, produces normal levels of ADH1 protein in the pollen but only ~6% of normal levels in the roots. The level of ADH1 in the root still increases under anaerobic conditions, as it does in the wild type. The mature *Adh1* mRNA accumulates to normal levels in pollen but to a reduced level in anaerobically induced roots (Chen *et al.*, 1987; Kloeckener-Gruissem *et al.*, 1992). The structures of the protein produced from the mutant allele in roots appear to be the same as that produced in the progenitor based on ADH1 allozyme activity gels (Chen *et al.*, 1987). The insertion of *Mu3* creates two TATA boxes, separating important 5' controlling sequences required for anaerobic induction from the transcriptional start site (Walker *et al.*, 1987; Ferl, 1990). Recent experiments have isolated *Adh1* mRNA from pollen and aerobically and anaerobically grown roots of plants containing the *ADH1-3F1124* allele; the 5′ ends of the transcripts were determined (Kloeckener-Gruissem *et al.*, 1992). These experiments demonstrate that the proximal TATA box (closest to the normal initiation site) is used in both tissues, whereas only in anaerobically grown roots is the distal TATA box utilized. It is only in anaerobic roots that a 3.5-kb hybrid *Adh1* and *Mu3* transcript accumulates, with its 5′ end mapping to ~30 bp downstream of the distal TATA box. Thus, the two putative TATA boxes appear to function differently in the various tissues and under different environmental conditions.

In the *sh9026* mutation, a *Mu1* element is inserted into the *Sh1* gene, just upstream from the transcription initiation site (Ortiz *et al.*, 1988). The 9-bp target site duplication is 2 bp upstream of the initiation site. This places the normal initiation site 1.4 kb downstream of the *sh1* 5′ promoter sequences. Primer extension analyses and S1 mapping of RNA from the mutant allele have identified novel 5′ ends that begin within the most 5′ *Mu* terminus, a distance downstream from the *sh1* TATA box similar to that observed in the progenitor. Transcripts initiating at these sites and continuing through the *Mu1* element and the *sh1* gene would produce a 4.4-kb transcript. The amount of RNA containing *sh1* sequences in *sh9026* roots is reduced to ~4% of that found in the wild-type allele, whereas transcription assays with isolated nuclei indicate transcription is reduced about sixfold (Ortiz *et al.*, 1988; Strommer and Ortiz, 1989). This finding suggests that both transcription and posttranscriptional processes are altered. RNA blot analyses of developing kernels and anaerobically induced roots do, in fact, reveal a 4.4-kb transcript that hybridizes to both *sh1* and *Mu1* probes. This is likely to arise from transcription starting just upstream of the element and continuing through the *Sh1* gene. A 3.0-kb transcript of a size similar to that observed in the wild-type allele is also observed. It is not clear if the 3.0-kb transcript arises from alternative splicing or alternative transcription initiative events. No 5′ mapping experiments using probes spanning the downstream *Mu* terminus and normal *Sh1* initiation site were presented in the published work (Ortiz *et al.*, 1988; Strommer and Ortiz, 1989). Thus, it is possible that the 3.0-kb transcript results from initiation events from within the downstream *Mu* terminus, as has been observed in *hcf106* (Barkan and Martienssen, 1991). Alternatively, the 3.0-kb transcript could arise from initiation events at the normal site, followed by splicing of the element from the 4.4-kb transcript. Analysis of cDNA clones and additional 5' mapping experiments should distinguish between these models.

B. ALTERATIONS IN mRNA PROCESSING

Insertion of members of the *Ac* or *Spm* transposon families into the transcribed region of a gene can result in altered mRNA processing, including alternative polyadenylation and alternative splicing (for reviews, see Fedoroff, 1989; Weil and Wessler, 1990). Characterization of *Mu1* insertions into the first intron of *Adh1* demonstrated that *Mu1* elements can also alter mRNA processing (Ortiz and Strommer, 1990; Luehrsen and Walbot, 1990).

Three *Adh1* alleles that contain *Mu1* elements within the first intron have reduced levels of ADH1 activity. All three insertions are in the same orientation but are in different locations. *Adh1-S3034* contains a *Mu1* insertion 72 nucleotides downstream from the first exon–intron boundary (Bennetzen *et al.*, 1984); *Adh1-S4477* contains a *Mu1* insertion 413 nucleotides downstream from the same junction, and *Adh1-S4478* has a *Mu* element inserted near the site of *Mu* insertion in *S4477* (Rowland and Strommer, 1985). Results of transcription assays with *Adh1-S3034* nuclei isolated from anaerobically induced roots suggested that more transcription occurs 5' to the insertion than 3' to the insertion (Rowland and Strommer, 1985; Vayda and Freeling, 1986). These results led to the suggestion that there was a block to transcription or polyadenylation occurring within the *Mu* element in *S3034*. Analysis of *Adh1* cDNA clones derived from *S3034* anaerobically induced roots demonstrated that alternative processing of *S3034* pre-mRNA also occurs (Ortiz and Strommer, 1990). Three different poly(A) addition sites within the *Mu1* element were identified, two within the central AT-rich region and one in the 3' terminal inverted repeat. Alternative splicing sites were also identified. *Mu1* contains a donor splice site in the 5' terminal inverted repeat that was joined to the *Adh1* exon 2 acceptor. This splice removes most of the *Mu1* sequences from the mRNA. The same *Mu1* donor site was observed spliced to an acceptor 89 bp downstream within *Mu1*. Luehrsen and Walbot (1990) have also examined the effects of *Mu1* insertions in introns. They inserted *Mu1* elements into a cloned *Adh1-S* intron at similar sites relative to the naturally occurring insertion mutants described above and also at novel sites. Both orientations of the element were constructed for comparison. The orientation that is observed in the naturally occurring insertions is referred to as "a," and the opposite orientation is "b." These constructs were then placed between the

Adh1 promoter and the luciferase reporter gene. Expression of the chimeric genes was tested using transient assays in maize protoplasts from a non-Mutator BMS line. As was observed with the natural *Mu1* insertions (Rowland and Strommer, 1985), these *Mu1* insertions showed a polarity effect, the more 5' insertions causing a greater reduction in luciferase expression (2–3.5% of wild type) than more 3' insertions (10–20% of wild type). Reduced luciferase activity correlated with decreased levels of luciferase transcript. This result was obtained independent of the orientation of the element. Similar results were obtained with *Mu1* insertions in the actin intron 3 (Luehrsen and Walbot, 1990), suggesting that *Mu1* insertions into other introns are also likely to reduce the level of transcript from the affected gene.

To begin to identify the mechanisms contributing to the altered expression, deletion derivatives of the *Mu1* element were inserted between the *Adh1* promoter and luciferase reporter gene, and luciferase expression was assayed by enzyme assays and RNase mapping in transiently transformed protoplasts. These experiments indicated that the central region of *Mu1* was largely responsible for the decrease in luciferase expression, as its deletion restored luciferase expression to levels only about twofold below wild type (Luehrsen and Walbot, 1990). Consistently, the addition of the central region alone reduced luciferase activity to levels lower than that observed with the complete *Mu1* element. This occurs when the element is inserted in either orientation. RNase mapping using probes 5' and 3' to the insertion demonstrated that when the element is in the "a" orientation, only ~28% of the transcripts include sequences downstream of the element insertion site. In contrast, when the element is in the "b" orientation, ~70% of the transcripts include sequences downstream of the element insertion site. The reduction of transcripts that extend through *Mu1* in the "a" orientation may result from a putative polyadenylation site that was mapped near to, but distinct from, the polyadenylation sites characterized by Ortiz and Strommer (1990). Because the ATG is in the first exon of the chimeric construct, intron 1 must be correctly spliced for luciferase expression. The authors hypothesize that the *Mu1* insertion in the "b" orientation may cause reduced luciferase expression by interfering with this splicing event (Luehrsen and Walbot, 1990).

VIII. Applications of *Mu* Elements

A. MUTAGENESIS

The Mutator system is extremely useful for isolating mutations. The high mutation frequency of the Mutator system enables the efficient scoring of mature plant phenotypes as well as seedling and kernel traits. Several general approaches have been used to isolate recessive mutations. The first is to cross a Mutator stock containing the wild-type dominant allele of a gene of interest to a tester containing the recessive allele. This type of approach allows an accurate estimation of the frequency at which the mutations occur. The reported frequencies have ranged from 10^{-3} to 10^{-6} (Robertson, 1985; Walbot *et al.*, 1986; Patterson *et al.*, 1991). Most of the mutations that have been characterized at the molecular level have contained *Mu* element insertions (see Section IV), although several exceptions have been reported (Patterson *et al.*, 1991; McCarty *et al.*, 1989a). Another class of maize element, *Spm*, was found inserted in several mutants isolated from Mutator stocks (Patterson *et al.*, 1991). It is clear that *Mu* elements are unrelated to *Spm* elements (Robertson and Mascia, 1981). However, it is also clear that several different classes of maize elements can become activated in maize stocks (reviewed in Fedoroff, 1989). Thus, it is not safe to assume that the element inserted into the gene of interest is related to *Mu* just because active Mutator stocks were used in the experiment. Until recently, there was no genetic test to confirm that the mutation of interest contained a *Mu* element. The isolation and characterization of regulatory *Mu* elements should improve this situation in the future.

Another approach that has been extensively used to isolate recessive mutations is to self-pollinate active Mutator stocks and score the resulting progeny for phenotypes associated with mutations of interest. This approach has been used to isolate numerous defective kernel, embryo lethal, high chlorophyll fluorescence, and morphological mutations (Clark and Sheridan, 1991; Martienssen *et al.*, 1989; A. Barkan, personal communication; M. Freeling, personal communication).

Mutator stocks are unstable and can completely lose activity within one generation. Thus, when beginning mutagenesis experiments, it is important to begin with Mutator stocks that are known to be active. This can be accomplished by either using the forward mutation assay, the somatic excision assay, or both.

B. TRANSPOSON TAGGING

Once a gene has been "tagged" with a transposable element insertion, it is theoretically possible to then clone the gene using element sequences as a hybridization probe. Multiple genes have been cloned using the Mutator system, demonstrating its utility for transposon tagging. These include *a1* (O'Reilly *et al.*, 1985), *bz2* (McLaughlin and Walbot, 1987), *vp1* (McCarty *et al.*, 1989a), *hcf106* (Martienssen *et al.*,

1989), *y1* (Buckner *et al.*, 1990), *iojap* (R. Martienssen, personal communication), and *hm1* (S. Briggs, personal communication). In this section we summarize current approaches and potential pitfalls to be aware of when using these elements to tag and clone genes.

The most straightforward method for identifying the element responsible for the mutant phenotype is to look for an element-homologous restriction fragment that cosegregates with the mutant phenotype. This is done by examining the DNA from mutant progeny arising from outcrosses or self-pollinations, using DNA blot analyses. Outcrosses to non-Mutator stocks have the advantage of potentially diluting the number of *Mu* elements. It is useful to outcross the mutant with different non-Mutator backgrounds and then self-pollinate (if working with recessive alleles) to generate several families, each of which can then be scored for cosegregating restriction fragments. The discovery of suppressible alleles that produce the wild-type phenotype when the Mutator system becomes inactive (Martienssen *et al.*, 1990) could complicate cosegregation analyses. For this reason we recommend using only DNA from mutant individuals to screen for the presence of a cosegregating fragment. Looking for the absence of a particular fragment in individuals with the wild-type phenotype could result in erroneously concluding that a particular fragment is not cosegregating with the mutation because it is also in normal sibs. Another potential problem is that an endogenous *Mu* element may be tightly linked to the gene of interest and cosegregate with the mutant phenotype. This can be readily eliminated as a concern by examining the parental lines used to generate the original mutation.

Mu1 is frequently the most abundant element found in Mutator stocks, and to date is responsible for generating the largest number of mutations. Thus, when one isolates a mutation from Mutator stocks, a reasonable first step is to probe a blot containing DNA from mutants with the *Mu1* internal probe. Often one will observe 20–40 copies of *Mu1* elements in the stocks. Use of different restriction enzymes can often allow adequate resolution of the fragments. However, subsequent outcrosses may be necessary to reduce the number of *Mu1* elements to a tractable number. Outcrossing does not reduce the number of elements if the Mutator system remains active, and can take several generations even with inactive stocks. Thus, we recommend hybridizing the blots with the internal probes for the other elements. Most of the other *Mu* elements are not found in high copy number in Mutator stocks. Thus, frequently with just one or two DNA blots, it is possible either to eliminate a particular class of *Mu* element as being responsible for the mutation of interest, or to identify a candidate restriction fragment cosegregating with the mutation, which can be verified by additional analyses. As discussed above, we know essentially nothing about potential target site preferences for different classes of *Mu* elements. Multiple alleles of particular genes have been isolated and examined for the nature of the insertion within them. Some genes, such as *bz1*, have contained a high frequency of *Mu1*-related elements (Taylor *et al.*, 1986; Brown *et al.*, 1989; Hardeman and Chandler, 1989), whereas other genes, such as *Kn*, also contained numerous insertions of other classes of *Mu* elements, such as *Mu7* and *Mu8* (Veit *et al.*, 1990). Thus, when working with a new gene, it is not safe to assume that *Mu1* will have generated the mutations being characterized.

IX. Future Directions

Many questions remain unanswered about the regulation of the *Mu* elements. The recent discovery and cloning of the "master" *Mu* element (*MuR1* and the related *MuA2* and *Mu9* elements) provide the basis for examining many of these questions (Chomet *et al.*, 1991; Qin *et al.*, 1991; Hershberger *et al.*, 1991b). For example, it will be important to determine the fate of the master *Mu* element(s) when Mutator stocks cycle between active and inactive states. It is possible that the master *Mu* elements are lost when *Mutator* activity is absent. These stocks might then be similar to non-Mutator stocks, which initial experiments have suggested may lack the master element (Qin *et al.*, 1991; Hershberger *et al.*, 1991b). However, the findings that inactive Mutator stocks can spontaneously become active (Hardeman and Chandler, 1989) and can be reactivated by γ-irradiation (Walbot, 1988) suggest that at least in these stocks the element is there in a quiescent state. This proposed quiescent state of the master *Mu* element might be associated with extensive DNA methylation of the master *Mu*, as has been observed for quiescent *Ac* and *Spm* elements (for review, see Fedoroff, 1989) and for *Mu1* elements (Chandler and Walbot, 1986; Bennetzen, 1987).

Another question regarding the regulation of *Mu* elements is whether there might be a dosage effect of the master *Mu* element. That is, it is possible that increasing the numbers of the master *Mu* elements will result in higher *Mutator* activities. Tools are now available to test the hypothesis that Mutator stocks with a high forward mutation frequency have multiple copies of the master *Mu*, to investigate how the numbers of master *Mu* elements are maintained on outcrossing, and to determine if and when during development the element numbers change.

Many crucial questions remain regarding *Mu* element transposition. Further characterization of the *MuR*-encoded proteins and the development of *in vivo* and *in vitro* transposition assays will allow the dissection of the cis- and trans-acting factors. The various classes of *Mu* elements have similar but nonidentical terminal inverted repeats. It will be interesting to investigate if all classes of *Mu* elements respond equally well to the trans-acting factors. The development of *in vitro* transposition assays may be required to settle the issue of whether *Mu* elements transpose by a replicative or cut-and-paste mechanism. The isolation of *MuR*-encoded cDNAs and the expression and purification of these proteins should provide the basis for eventually developing *in vitro* assays using purified proteins, defined templates, and nuclear extracts from maize to provide any necessary host factors. Candidates for host factors interacting specifically with *Mu* termini have been described (Zhao and Sundaresan, 1991).

The introduction of *MuR* into heterologous plants, which contain no endogenous *Mu* elements, will enable the investigation of whether *Mu* elements favor transposition to linked over unlinked target sites. If *Mu* elements show no preference for linked sites, they may prove particularly useful for tagging genes in other plants, relative to the maize elements that transpose most frequently to nearby target sites. The availability of stocks containing single copies of *MuR* will also help transposon tagging strategies in maize, as these stocks provide a means of establishing that the mutation of interest is associated with a *Mu* element.

In summary, the elements of the *Mu* family have properties that suggest they have several unique features relative to the other well-characterized transposons of maize. These features include their association with high mutation frequencies, the diversity between the different classes of *Mu* elements, and potentially, their mechanism of transposition. Further investigation into the properties and regulation of these elements should enable their more efficient use as mutagens and transposon tags, as well as provide important information on how an organism regulates the activity of transposable elements in its genome.

ACKNOWLEDGMENTS

We are very appreciative of the critical comments and suggestions we received on this review from Alice Barkan, Jette Foss, Garth Patterson, and Manuel Sainz. We wish to thank the following *Mu* researchers for providing us with their unpublished data: A. Barkan, S. Briggs, P. Chomet, M. Freeling, D. Lisch, R. Martienssen, V. Sundaresen, and V. Walbot. We also would like to thank John Korte for his assistance with the figures. The Mutator work ongoing in the Chandler laboratory is supported by a grant from N.I.H. (GM35971) to V. Chandler.

REFERENCES

Alleman, M., and Freeling, M. (1986). The *Mu* transposable elements of maize: Evidence for transposition and copy number regulation during development. *Genetics* **112,** 107–119.

Barkan, A., and Martienssen, R. A. (1991). Inactivation of maize transposon *Mu* suppresses a mutant phenotype by activating an outward-reading promoter near the end of *Mu1*. *Proc. Natl. Acad. Sci. U.S.A.* **88,** 3502–3506.

Barker, R. F., Thompson, D. V., Talbot, D. R., Swanson, J., and Bennetzen, J. L. (1984). Nucleotide sequence of the maize transposable element *Mu1*. *Nucleic Acids Res.* **12,** 5955–5967.

Bennetzen, J. L. (1984). Transposable element *Mu1* is found in multiple copies only in Robertson's *Mutator* maize lines. *J. Mol. Appl. Genet.* **2,** 519–524.

Bennetzen, J. L. (1987). Covalent DNA modification and the regulation of *Mutator* element transposition in maize. *Mol. Gen. Genet.* **208,** 45–51.

Bennetzen, J. L., Swanson, J., Taylor, W. C., and Freeling, M. (1984). DNA insertion in the first intron of maize *Adh1* affects message levels: Cloning of progenitor and mutant *Adh1* alleles. *Proc. Natl. Acad. Sci. U.S.A.* **81,** 4125–4128.

Bennetzen, J. L., Fracasso, R. P., Morris, D. W., Robertson, D. S., and Skogen-Hagenson, M. J. (1987). Concomitant regulation of *Mu1* transposition and *Mutator* activity in maize. *Mol. Gen. Genet.* **208,** 57–62.

Bennetzen, J. L., Brown, W. E., and Springer, P. S. (1988). The state of DNA modification within and flanking maize transposable elements. *In* "Plant Transposable Elements" (O. Nelson, ed.), pp. 237–250. Plenum, New York.

Britt, A. B., and Walbot, V. (1991). Germinal and somatic products of *Mu1* excision from the *bronze-1* gene of *Zea mays*. *Mol. Gen. Genet.* **227,** 267–276.

Brown, J., and Sundaresan, V. (1992). Genetic study of the loss and restoration of *Mutator* transposon activity in maize: Evidence against dominant negative regulator associated with loss of activity. *Genetics* **130,** 889–898.

Brown, W. E., Robertson, D. S., and Bennetzen, J. L. (1989). Molecular analysis of multiple *Mutator*-derived alleles of the *Bronze* locus of maize. *Genetics* **122,** 439–445.

Buckner, B., Kelson, T. L., and Robertson, D. S. (1990). Cloning of the *y1* locus of maize, a gene involved in the biosynthesis of carotenoids. *Plant Cell* **2,** 867–876.

Chaconas, G., and Surette, M. G. (1988). Mechanism of *Mu* DNA transposition. *BioEssays* **9,** 205–208.

Chandler, V. L., and Walbot, V. (1986). DNA modification of a maize transposable element correlates with loss of activity. *Proc. Natl. Acad. Sci. U.S.A.* **83,** 1767–1771.

Chandler, V. L., Rivin, C., and Walbot, V. (1986). Stable non-Mutator stocks of maize have sequences homologous to the *Mu1* transposable element. *Genetics* **114,** 1007–1021.

Chandler, V. L., Talbert, L. E., and Raymond, F. (1988a). Sequence, genomic distribution and DNA modification of a *Mu1* element from non-Mutator maize stocks. *Genetics* **119,** 951–958.

Chandler, V. L., Talbert, L. E., Mann, L., and Faber, C. (1988b). Structure and DNA modification of endogenous *Mu* elements. *In* "Plant Transposable Elements" (O. Nelson, ed.), pp. 339–350. Plenum, New York.

Chen, C.-H., Oishi, K. K., Kloeckener-Gruissem, B., and Freeling, M. (1987). Organ-specific expression of maize *adh1* is altered after a *Mu* transposon insertion. *Genetics* **116**, 469–477.

Chomet, P., Lisch, D., Hardeman, K. J., Chandler, V. L., and Freeling, M. (1991). Identification of a regulatory transposon that controls the *Mutator* transposable element system in maize. *Genetics* **129**, 261–270.

Clark, J. K., and Sheridan, W. F. (1991). Isolation and characterization of 51 embryo-specific mutations of maize. *Plant Cell* **3**, 935–951.

Coe, E. H., Jr., Neuffer, M. G., and Hoisington, D. A. (1988). The genetics of corn. *In* "Corn and Corn Improvement" (G. F. Sprague and J. W. Dudley, eds.), pp. 146–170. Am. Soc. Agron., Madison, WI.

Coen, E. S., Carpenter, R., and Martin, C. (1986). Transposable elements generate novel spatial patterns of gene expression in *Antirrhinum majus. Cell (Cambridge, Mass.)* **47**, 285–296.

Doseff, A., Martienssen, R., and Sundaresan, V. (1991). Somatic excision of the *Mu1* transposable element of maize. *Nucleic Acids Res.* **19**, 579–584.

Engles, W. R., Johnson-Schlitz, D. M., Eggleston, W. B., and Sved, J. (1990). High-frequency P element loss in *Drosophila* is homolog dependent. *Cell (Cambridge, Mass.)* **62**, 515–525.

Fedoroff, N. V. (1983). Controlling elements in maize. *In* "Mobile Genetic Elements" (J. Shapiro, ed.), pp. 1–63. Academic Press, New York.

Fedoroff, N. V. (1989). Maize transposable elements. *In* "Mobile DNA" (D. Berg and M. Howe, eds.), pp. 375–411. ASM Press, Washington, DC.

Ferl, R. (1990). ARF-B2: A protein complex that specifically binds to part of the anaerobic response element of maize *Adh1. Plant Physiol.* **93**, 1094–1101.

Fleenor, D., Spell, M., Robertson, D., and Wessler, S. (1990). Nucleotide sequence of the maize *Mutator* element, *Mu8. Nucleic Acids Res.* **18**, 6725.

Freeling, M. (1984). Plant transposable elements and insertion sequences. *Annu. Rev. Plant Physiol.* **35**, 277–298.

Freeling, M., Cheng, D., and Alleman, M. (1982). Mutant alleles that are altered in quantitative, organ-specific behavior. *Dev. Genet.* **3**, 179–196.

Gerlach, W. L., Dennis, E. S., Peacock, W. J., and Clegg, M. T. (1987). The *Ds1* controlling element family in maize and *Tripsacum. J. Mol. Evol.* **20**, 329–334.

Greenblatt, I. M. (1984). A chromosome replication pattern deduced from pericarp phenotypes resulting from movements of the transposable element, *Modulator,* in maize. *Genetics* **108**, 471–485.

Halling, S. M., and Kleckner, N. (1982). A symmetrical six-base-pair target site sequence determines Tn10 insertion specificity. *Cell (Cambridge, Mass.)* **28**, 155–163.

Hardeman, K. J., and Chandler, V. L. (1989). Characterization of *bz1* mutants isolated from Mutator stocks with high and low numbers of *Mu1* elements. *Dev. Genet.* **10**, 460–472.

Hershberger, J., Warren, C. A., and Walbot, V. (1991). *Mu9*, a new 5 kbp member of the Mutator family of transposable elements. *Maize Genet. Coop. Newsl.* **65**, 98.

Hershberger, R. J., Warren, C. A., and Walbot, V. (1991). *Mutator* activity in maize correlates with the presence and expression of *Mu9*, a new *Mu* transposable element. *Proc. Natl. Acad. Sci. U.S.A.* **88**, 10198–10202.

Ingels, S. C., Bennetzen, J. L., Hulbert, S. H., Qin, M.-M., and Ellingboe, A. H. (1992).

Mutator transposable elements that occur in clusters in the maize genome. *J. Hered.* **83,** 114–118.

Kleckner, N. (1989). Transposon Tn10. *In* "Mobile DNA" (D. E. Berg and M. M. Howe, eds.), pp. 227–268. ASM Press, Washington, DC.

Kloeckener-Gruissem, B., Vogel, J. M., and Freeling, M. (1992). The TATA box promoter region of maize *Adh1* alleles affects its organ-specific expression. *EMBO J.* **11,** 157–166.

Levy, A. A., and Walbot, V. (1990). Regulation of the timing of transposable element excision during maize development. *Science* **248,** 1534–1537.

Levy, A. A., and Walbot, V. (1991). Molecular analysis of the loss of somatic instability in the *bz2::mu1* allele of maize. *Mol. Gen. Genet.* **229,** 147–151.

Levy, A. A., Britt, A. B., Luehrsen, K. R., Chandler, V. L., Warren, C., and Walbot, V. (1989). Developmental and genetic aspects of *Mutator* excision in maize. *Dev. Genet.* **10,** 520–531.

Lichtenstein, C., and Brenner, S. (1981). Site-specific properties of *Tn7* transposition into the *E. coli* chromosome. *Mol. Gen. Genet.* **183,** 380–387.

Luehrsen, K. R., and Walbot, V. (1990). Insertion of *Mu1* elements in the first intron of the *adh1-S* gene of maize results in novel RNA processing events. *Plant Cell* **2,** 1225–1238.

Martienssen, R. A., Barkan, A., Freeling, M., and Taylor, W. C. (1989). Molecular cloning of a maize gene involved in photosynthetic membrane organization that is regulated by Robertson's *Mutator. EMBO J.* **8,** 1633–1639.

Martienssen, R. A., Barkan, A., Taylor, W. C., and Freeling, M. (1990). Somatically heritable switches in the DNA modification of *Mu* transposable elements monitored with a suppressible mutant in maize. *Genes Dev.* **4,** 331–343.

Masterson, R. V., Biagi, K., Wheeler, J. G., Stadler, J., and Morris, D. W. (1988). An embryogenic cell line of maize from A188 (Minnesota) contains *Mu1*-like elements. *Plant Mol. Biol.* **10,** 273–279.

McCarty, D. R., Carson, C. B., Stinard, P. S., and Robertson, D. S. (1989a). Molecular analysis of *viviparous-1*: An abscisic acid-insensitive mutant of maize. *Plant Cell* **1,** 523–532.

McCarty, D. R., Carson, C. B., Lazar, M., and Simonds, S. C. (1989b). Transposable element-induced mutations of the *viviparous-1* gene in maize. *Dev. Genet.* **10,** 473–481.

McLaughlin, M., and Walbot, V. (1987). Cloning of a mutable *bz2* allele of maize by transposon tagging and differential hybridization. *Genetics* **117,** 771–776.

Mori, I., Benian, G. M., Moerman, D. G., and Waterston, R. H. (1988). Transposable element *Tc1* of *Caenorhabditis elegans* recognizes specific target sequences for integration. *Proc. Natl. Acad. Sci. U.S.A.* **85,** 861–864.

Nash, J., Luehrsen, K. R., and Walbot, V. (1990). *Bronze2* gene of maize: Reconstruction of a wild-type allele and analysis of transcription and splicing. *Plant Cell* **2,** 1039–1049.

Oishi, K. K., and Freeling, M. (1987). A new *Mu* element from a Robertson's Mutator line. *In* "Plant Transposable Elements" (O. Nelson, ed.), pp. 289–291. Plenum, New York.

O'Reilly, C., Shepherd, N. S., Pereira, A., Schwarz-Sommer, Z., Bertram, I., Robertson, D. S., Peterson, P. A., and Saedler, H. (1985). Molecular cloning of the *a1* locus of *Zea mays* using the transposable elements *En* and *Mu1. EMBO J.* **4,** 877–882.

Ortiz, D. F., and Strommer, J. N. (1990). The *Mu1* maize transposable element induces tissue-specific abberant splicing and polyadenylation in two *adh1* mutants. *Mol. Cell. Biol.* **10**, 2090–2095.

Ortiz, D. F., Rowland, L. J., Gregerson, R. G., and Strommer, J. N. (1988). Insertion of *Mu* into the *Shrunken1* gene of maize affects transcriptional and post-transcriptional regulation of *Sh1* RNA. *Mol. Gen. Genet.* **214**, 135–141.

Patterson, G. I., Harris, L. J., Walbot, V., and Chandler, V. L. (1991). Genetic analysis of *B-Peru,* a regulatory gene in maize. *Genetics* **127**, 205–220.

Qin, M., and Ellingboe, A. H. (1990). A transcript identified by *MuA* of maize is associated with *Mutator* activity. *Mol. Gen. Genet.* **224**, 357–363.

Qin, M., Robertson, D. S., and Ellingboe, A. H. (1991). Cloning of the *Mutator* transposable element *MuA2*: A putative regulator of somatic mutability of the *a1-Mum2* allele in maize. *Genetics* **129**, 845–854.

Robertson, D. S. (1978). Characterization of a mutator system in maize. *Mutat. Res.* **51**, 21–28.

Robertson, D. S. (1980). The timing of *Mu* activity in maize. *Genetics* **94**, 969–978.

Robertson, D. S. (1981a). Tests of two models for the transmission of the *Mu* mutator in maize. *Mol. Gen. Genet.* **183**, 51–53.

Robertson, D. S. (1981b). Mutator activity in maize: Timing of its activation in ontogeny. *Science* **213**, 1515–1517.

Robertson, D. S. (1983). A possible dose-dependent inactivation of Mutator (*Mu*) in maize. *Mol. Gen. Genet.* **191**, 86–90.

Robertson, D. S. (1985). Differential activity of the maize mutator *Mu* at different loci and in different cell lineages. *Mol. Gen. Genet.* **200**, 9–13.

Robertson, D. S. (1986). Genetic studies on the loss of *Mu* mutator activity in maize. *Genetics* **113**, 765–773.

Robertson, D. S., and Mascia, P. N. (1981). Tests of 4 controlling-element systems of maize for mutator activity and their interaction with *Mu* mutator. *Mutat. Res.* **84**, 283–289.

Robertson, D. S., and Stinard, P. S. (1989). Genetic analyses of putative two-element systems regulating somatic mutability in Mutator-induced aleurone mutants of maize. *Dev. Genet.* **10**, 482–506.

Robertson, D. S., Stinard, P. S., Wheeler, J. G., and Morris, D. W. (1985). Genetic and molecular studies on germinal and somatic instability in Mutator-induced aleurone mutants of maize. *In* "Plant Genetics" (M. Freeling, ed.), pp. 317–332. Liss, New York.

Robertson, D. S., Morris, D. W., Stinard, P. S., and Roth, B. A. (1988). Germ line and somatic Mutator activity: Are they functionally related? *In* "Plant Transposable Elements" (O. Nelson, ed.), pp. 17–42. Plenum, New York.

Rowland, L. J., and Strommer, J. N. (1985). Insertion of an unstable element in an intervening sequence of maize *Adh1* affects transcription but not processing. *Proc. Natl. Acad. Sci. U.S.A.* **82**, 2875–2879.

Saedler, H., and Nevers, P. (1985). Transposition in plants: A molecular model. *EMBO J.* **4**, 585–590.

Schnable, P. S., and Peterson, P. A. (1988). The *Mutator*-related *Cy* transposable element of *Zea mays* L. behaves as a near-mendelian factor. *Genetics* **120**, 587–596.

Schnable, P. S., and Peterson, P. A. (1989). Genetic evidence of a relationship between two maize transposable element systems: *Cy* and *Mutator*. *Mol. Gen. Genet.* **215**, 317–321.

Schnable, P. S., Peterson, P. A., and Saedler, H. (1989). The *bz-rcy* allele of the *Cy* transposable element system of *Zea mays* contains a *Mu*-like element insertion. *Mol. Gen. Genet.* **217**, 459–463.

Shepherd, N. S., Sheridan, W. F., Mattes, M. G., and Deno, G. (1988). The use of Mutator for gene-tagging; cross-referencing between transposable element systems. *In* "Plant Transposable Elements" (O. Nelson, ed.), pp. 137–147. Plenum, New York.

Sherratt, D. (1989). Tn3 and related transposable elements: Site-specific recombination and transposition. *In* "Mobile DNA" (D. E. Berg and M. M. Howe, eds.), pp. 163–184. ASM Press, Washington, DC.

Strommer, J. N., and Ortiz, D. (1989). *Mu1*-induced mutant alleles of maize exhibit background-dependent changes in expression and RNA processing. *Dev. Genet.* **10**, 452–459.

Strommer, J. N., Hake, S., Bennetzen, J., Taylor, W. C., and Freeling, M. (1982). Regulatory mutants of the maize *Adh1* gene induced by DNA insertions. *Nature (London)* **300**, 542–544.

Sundaresan, V., and Freeling, M. (1987). An extrachromosomal form of the *Mu* transposons of maize. *Proc. Natl. Acad. Sci. U.S.A.* **84**, 4924–4928.

Talbert, L. E., and Chandler, V. L. (1988). Characterization of a highly conserved sequence related to Mutator transposable elements in maize. *Mol. Biol. Evol.* **5**, 519–529.

Talbert, L. E., Patterson, G. I., and Chandler, V. L. (1989). *Mu* transposable elements are structurally diverse and distributed throughout the genus *Zea. J. Mol. Evol.* **29**, 28–39.

Taylor, L. P., and Walbot, V. (1985). A deletion adjacent to the maize transposable element *Mu1* accompanies loss of *Adh1* expression. *EMBO J.* **4**, 869–876.

Taylor, L. P., and Walbot, V. (1987). Isolation and characterization of a 1.7-kbp transposable element from a Mutator line of maize. *Genetics* **117**, 297–307.

Taylor, L. P., Chandler, V. L., and Walbot, V. (1986). Insertion of a 1.4 kb and 1.7 kb *Mu* elements into the *bronze1* gene of *Zea mays* L. *Maydica* **31**, 31–45.

Vayda, M. E., and Freeling, M. (1986). Insertion of the *Mu1* transposable element into the first intron of maize *Adh1* interferes with transcript elongation but does not disrupt chromatin structure. *Plant Mol. Biol.* **6**, 441–454.

Veit, B., Vollbrecht, E., Mathern, J., and Hake, S. (1990). A tandem duplication causes the *Kn1-0* allele of *Knotted,* a dominant morphological mutant of maize. *Genetics* **125**, 623–631.

Walbot, V. (1986). Inheritance of Mutator activity in *Zea mays* as assayed by somatic instability of the *bz2-mu1* allele. *Genetics* **114**, 1293–1312.

Walbot, V. (1988). Reactivation of the *Mutator* transposable element system following gamma irradiation of seed. *Mol. Gen. Genet.* **212**, 259–264.

Walbot, V. (1991). The Mutator transposable element family of maize. *Genet. Eng.* **13**, 1–37.

Walbot, V. (1992). Strategies for mutagenesis and gene cloning using transposon tagging and T-DNA insertional mutagenesis. *Annu. Rev. Plant Physiol. Plant Mol. Biol.* **43**, 49–82.

Walbot, V., and Warren, C. (1988). Regulation of *Mu* element copy number in maize lines with an active or inactive Mutator transposable element system. *Mol. Gen. Genet.* **211**, 27–34.

Walbot, V., Briggs, C. P., and Chandler, V. (1986). Properties of mutable alleles recovered from mutator stocks of *Zea mays* L. *In* "Genetics, Development and Evolution" (J. P.

Gustafson, G. L. Stebbins, and F. J. Ayala, eds.), pp. 115–142. Plenum, New York.

Walker, J. C., Howard, E. A., Dennis, E. S., and Peacock, W. J. (1987). DNA sequences required for anaerobic expression of the maize alcohol dehydrogenase 1 gene. *Proc. Natl. Acad. Sci. U.S.A.* **84,** 6624–6628.

Weil, C. F., and Wessler, S. R. (1990). The effects of plant transposable element insertion on transcription initiation and RNA processing. *Annu. Rev. Plant Physiol.* **41,** 527–552.

Zhao, Z.-Y., and Sundaresan, V. (1991). Binding sites for maize nuclear proteins in the terminal inverted repeats of the *Mu1* transposable element. *Mol. Gen. Genet.* **229,** 17–26.

MOLECULAR PERSPECTIVES ON THE GENETICS OF MOSQUITOES

Nora J. Besansky*†, Victoria Finnerty†, and Frank H. Collins*†

*Malaria Branch, Division of Parasitic Diseases, Center for Infectious Diseases, Centers for Disease Control, Atlanta, Georgia 30333
†Department of Biology, Emory University, Atlanta, Georgia 30322

I. Introduction

Mosquitoes have been a focus of scientific study since the turn of the century, when they were first linked with human diseases. This review concentrates on the three most intensely studied genera, *Anopheles, Culex,* and *Aedes.* These genera include the principal vectors of three major groups of human pathogens: malaria parasites of the genus *Plasmodium,* filarial worms of the genera *Wuchereria* and *Brugia,* and numerous arboviruses. Anophelines are the only mosquitoes known to

123

Copyright © 1992 by Academic Press, Inc.
All rights of reproduction in any form reserved.

transmit human malaria parasites, a group of organisms that may be responsible for more morbidity and mortality worldwide than any other human pathogen. Anophelines also transmit filarial worms, as do *Culex* and *Aedes* species. Among the 14 or more different mosquito genera known to harbor arboviruses (Mattingly, 1973), the most important are *Culex* and *Aedes,* which include the principal vectors of yellow fever, dengue, and most encephalitis-causing arboviruses.

Although the connection between mosquitoes and human diseases has been recognized for more than a century (see Harrison, 1978), instances of successful control of either vector populations or the disease agents they transmit are limited in both number and scope in non-Western countries. Most notable has been the failure to eliminate malaria from much of the developing world, especially tropical Africa, in spite of an almost 20-year worldwide malaria eradication campaign directed by the World Health Organization. Attempts to control mosquitoes have been frustrated by both their enormous reproductive capacity and their genetic flexibility. The latter is evidenced not only by rapid development of insecticide resistance but also by the multitude of closely related, cryptic species complexes, some of which appear to be undergoing incipient speciation in the process of adapting to human environments (Coluzzi, 1984; Coluzzi *et al.,* 1985). Further impediments to the control of vector-borne diseases include the rapid spread of drug resistance throughout parasite populations, the increasing movement of people into and out of disease-endemic regions, and the limited funds and public health infrastructure in most developing countries. Thus the widespread use of residual insecticides or antiparasitic drugs has not been and will not be a sustainable solution to the problem of vector-borne disease control. New approaches are needed. The enormous impact of recent developments in molecular genetics on questions of basic biology and human disease has stimulated a reexamination of the prospects for genetic manipulation of vector populations as a means for reducing or eliminating vector-borne diseases, especially malaria (World Health Organization, 1991a,b). Though the specific disease control scenarios currently being considered, such as the replacement of competent vector populations with pathogen-refractory genetic strains, may not be realized, the increase in knowledge of basic mosquito biology on which these control scenarios depend will inevitably stimulate novel approaches to the control of mosquito-borne diseases.

The identification of new vector control strategies depends on deepening our basic understanding of all aspects of mosquito biology from molecular to population levels, including the relationship between the mosquito and the pathogens it transmits. Molecular genetic research over the past two decades has led to the development of recombinant DNA technology, providing the tools for detailed study of individual genes, genome organization, population genetics, and evolution. The application of this technology in mosquito research, now underway in a growing number of laboratories, has been greatly accelerated by adapting molecular techniques and drawing on the already extensive molecular data base from *Drosophila melanogaster,* one of the best genetically understood higher organisms (Ashburner, 1989). Many regulatory and structural DNA sequences are sufficiently conserved between *Drosophila* and mosquitoes such that some *Drosophila* gene constructs and transformation tools may function in mosquitoes with little modification. Cloned *Drosophila* DNA will be valuable in isolating relatively conserved homologous coding and regulatory sequences from mosquito genomic DNA and cDNA libraries and directly from polymerase chain reaction (PCR)-amplified genomic DNA and RNA. However, molecular genetic studies are also confirming the perspective of classical taxonomists, who would warn that mosquitoes and *Drosophila* are on two deeply diverged branches of their order. Thus we believe it is appropriate to caution that many homologous genes in *Drosophila* and mosquitoes may be so diverged that cloned *Drosophila* sequences will not hybridize specifically under any conditions to their mosquito homologues.

The formal genetics and cytogenetics of mosquitoes have been reviewed elsewhere (Wright and Pal, 1967; Kitzmiller, 1976; Steiner *et al.,* 1982). The present review attempts to integrate the variety of recent molecular studies performed on mosquitoes of the genera *Anopheles, Culex,* and *Aedes.* Because such studies have been initiated only recently, this review will rely heavily on unpublished information generously contributed by many colleagues and will inevitably highlight the need for more work on all fronts. The first section deals with how molecular techniques have been used to reveal genetic differentiation, both for the identification of cryptic species and for the understanding of population structure and evolution. Next is a discussion of genome organization as revealed by reassociation kinetics, molecular cloning, and genome mapping. Following this, a section covers the molecular biology of specific physiological systems, including insecticide resistance, immune mechanisms, oogenesis, and salivation. Finally, identification of endogenous mobile elements and potential transformation systems for both cell lines and embryos are considered.

II. Genetic Differentiation

A. CLASSIFICATION

Mosquitoes are classified as lower flies within the insect order Diptera. The branching of lineages leading to lower and higher flies is ancient, probably dating from the Triassic (McAlpine and Wood, 1989). Thus, comparison of the coding regions of the α-amylase gene from three Lepidoptera and three Diptera (the lower fly *Anopheles merus* and two higher flies, *D. melanogaster* and *Drosophila virilis*), shows the *Anopheles* to be as diverged from the *Drosophila* as it is from the Lepidoptera (D. Hickey, unpublished). The family Culicidae, to which mosquitoes belong, is considered monophyletic, probably originating from a Chaoborid-like ancestor (White, 1980). Three subfamilies of the Culicidae are generally recognized: Anophelinae, Toxorhinchitinae, and Culicinae. With the single known exception of *Chagasia bathana* (which, like the chaoborids, has a diploid chromosome number of eight), all mosquitoes have a diploid chromosome number of six (White, 1980).

To determine evolutionary relationships among taxa, Rao and Rai (1987a, 1990) analyzed chromosomal rearrangements, differences in amount and position of C-banded heterochromatin, and nuclear DNA amounts in mosquitoes and their close relatives. Although they state that Toxorhinchitinae is closer to Chaoboridae and is primitive within Culicidae, they present a phylogenetic tree in which the anopheline branch is the first to diverge from the culicid line followed later by the Toxorhinchitinae and finally reaching the Culicinae branches. The phylogenetic relationship among representative species from most of the major genera in the family Culicidae and several related lower flies has also been examined by comparison of DNA sequences from a hypervariable domain of the large ribosomal subunit (28S) coding region (C. Porter, D. Wesson, and F. Collins, unpublished). Both phenetic and cladistic analyses of this ribosomal DNA (rDNA) support a monophyletic origin of the family Culicidae but suggest a very early evolutionary divergence between the Anophelinae and the subfamilies Toxorhynchitinae and Culicinae (Fig. 1). Indeed, differences in this rDNA region are greater among the members of the three different *Anopheles* subgenera represented in Fig. 1 than between the subfamilies Culicinae and Toxorhynchitinae. Other evidence from anophelines supports their extensive divergence from other mosquitoes, including significantly smaller chromosomes and lower nuclear DNA content (Rao and Rai, 1990), and their unique possession of dimorphic sex chro-

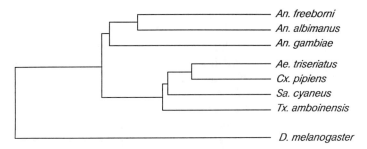

An. freeborni
An. albimanus
An. gambiae
Ae. triseriatus
Cx. pipiens
Sa. cyaneus
Tx. amboinensis
D. melanogaster

FIG. 1. Similarity dendrogram inferred from comparison of D2 expansion domains of 28S rDNA coding regions.

mosomes and long-period interspersion of repetitive sequences in the genome (Black and Rai, 1988). Taken together, the data suggest that Anophelinae is primitive and Toxorhynchitinae intermediate with Culicidae. Interestingly, only Anophelinae and Culicinae contain vectors of human pathogens, because of obligate blood feeding by the adult females. The reproductive strategy of the Toxorhynchitinae does not involve blood feeding, an ability that seems to have been lost.

B. PATTERNS OF GENETIC VARIATION

The ribosomal RNA genes (rDNA) are frequent targets for evolutionary comparisons over a broad range of taxonomic hierarchies because of contrasting levels of variation among their architectural components (reviewed in Beckingham, 1982; Gerbi, 1985). In 20 mosquito species of the subfamilies Anophelinae and Culicinae, 40–1000 rDNA repeat units are tandemly arrayed at a single locus on the sex-determining chromosomes, except for *Aedes triseriatus* and one of two strains of *Aedes albopictus* examined, each of which have an additional rDNA locus on an autosome (Kumar and Rai, 1990a; A. Kumar and K. S. Rai, unpublished). As in other Diptera, each mosquito rDNA repeat unit consists of a transcribed region containing highly conserved 18S, 5.8S, and 28S coding sequences separated by spacers that evolve extremely rapidly (Faria and Leoncini, 1988; McLain and Collins, 1989; McLain *et al.*, 1989; Collins *et al.*, 1989; Beach *et al.*, 1989; Black *et al.*, 1989; Gale and Crampton, 1989). Adjacent to the transcribed region is the nontranscribed intergenic spacer (IGS), which is usually highly variable in length within and between individuals due to different numbers of small internal subrepeats.

An understanding of the underlying architecture of the rDNA IGS allows the use of IGS probes to measure population subdivision. Whereas very little difference in the set of IGS length variants is detected among individuals from laboratory colonies, examination of specimens from field populations has revealed variation that increases with genetic and geographic distance (Collins *et al.,* 1989). In a study of gene flow among *Anopheles gambiae* populations from seven different villages in western Kenya, seven different rDNA spacer genotypes based on length variants were identified (McLain *et al.,* 1989). The frequency distribution of these genotypes revealed restricted gene flow between populations separated by less than 10 km and complete isolation of populations at distances no greater than a few hundred kilometers. Preliminary results from a study of field-collected *Anopheles albimanus* specimens from Panama similarly showed intra- and intermosquito differences in spacer genotypes, with more variation between than within populations. However, no particular variant was characteristic of populations from a given site (Beach *et al.,* 1989; A. M. de Merida, C. Porter, and F. Collins, unpublished). Unlike the situation in *Anopheles,* however, most of the IGS variation in *Aedes* was accounted for by diversity within populations rather than between populations, with no correlation between genetic and geographic distances based on total IGS length (Black *et al.,* 1989; Kambhampati and Rai, 1991c). Among the newly established United States populations of this mosquito, variation at the IGS has occurred rapidly and independently in adjacent populations.

Other rDNA sequences have also been used in evolutionary studies of several species groups within the family Culicidae. The most extensive work to date has involved anopheline mosquitoes in the subgenus *Anopheles,* with a focus on species in the North American *Anopheles maculipennis* group (C. Porter and F. Collins, unpublished). Two regions of the rDNA were analyzed in this work: the rDNA internal transcribed spacer (ITS2) separating the 5.8S and 28S coding regions and a hypervariable domain (D2 expansion domain) near the 5' end of the 28S gene. Relationships among members of closely related species complexes (such as the *Anopheles quadrimaculatus* or *Anopheles punctipennis* complexes) were determined primarily by ITS2 sequence analysis, whereas comparison of the D2 region of the 28S gene proved more useful for more diverged taxa. For example, whereas the ITS2 sequences of the morphologically identical species *Anopheles freeborni* and *Anopheles hermsi* differ by approximately 3.7%, no differences were observed in the D2 region. Among the four species in the *An. quadrimaculatus* complex, differences in the ITS2 regions ranged from

4.5 to 10.7%, whereas D2 region sequence differences ranged from 0.5 to 2.2%. In sequence comparisons among different species complexes in the *An. maculipennis* group, large domains of the ITS2 region could not be aligned. In contrast, most of the D2 region remained taxonomically informative in all comparisons within the subgenus *Anopheles*. Very little within-species variation in the ITS2 sequences has been observed in anopheline mosquitoes, possibly because their rDNA is tandemly arranged at a single locus on either the X or both the X and Y chromosomes and is subject to relatively high levels of within-locus homogenizing processes.

Just as more IGS length variation was found in *Aedes* than *Anopheles* populations, analysis of the ITS2 regions from various *Aedes* species and allied genera revealed more intraspecific sequence variation than was observed in the genus *Anopheles* (D. Wesson, C. Porter, and F. Collins, unpublished). Again, almost all variation in *Aedes* ITS2 sequences was observed in single individuals, and most was due to the presence or absence of short insertions/deletions of between 1 and 6 bp. These differences were not randomly distributed among ITS2 clones but occurred as correlated sets, suggesting the presence of several distinct ITS2 types within one individual. Though the basis of this typology in *Aedes* ITS2 sequences has not yet been established, it is consistent with independent rDNA spacer sequence divergence that could occur at multiple rDNA loci. Indeed, rDNA has been localized by *in situ* hybridization to two different chromosomes in 2 of 13 *Aedes* species examined (Kumar and Rai, 1990a; A. Kumar and K. S. Rai, unpublished). This level of rDNA spacer variation may be characteristic of all mosquitoes on the culicine–toxorhynchitine branches of the Culicidae with homomorphic rather than heteromorphic sex chromosomes. In all mosquitoes examined thus far, the position of all or at least some of the rRNA genes is conserved on the sex-determining chromosomes (Kumar and Rai, 1990a).

Comparison of the 28S D2 regions and ITS2 sequences indicates that the subfamilies Culicinae and Toxorhynchitinae are much more closely related than the *Anopheles* subgenera *Cellia, Anopheles,* and *Nyssorhynchus* (Fig. 1). Though within-species polymorphisms in the ITS2 region complicate its use for phylogenetic analysis of *Aedes* and related genera, overall conservation of ITS2 sequences, even between different genera such as *Aedes* and *Hemagogus,* allows much of this region to be aligned (D. Wesson, C. Porter, and F. Collins, unpublished). By contrast, levels of ITS2 sequence divergence among members of anopheline cryptic species complexes are equivalent to those observed among different *Aedes* subgenera. This may reflect the subjectivity of higher taxonomic groupings of mosquitoes as much as differences in their biology.

Mitochondrial DNA (mtDNA) is another useful tool for genetic analyses. It is commonly used to investigate population structure because it is maternally inherited and different mtDNA genomes recombine only rarely, if at all. The complete mitochondrial genomes of two anophelines, *An. quadrimaculatus* sp. A (Cockburn *et al.*, 1990) and *An. gambiae* (D. Mills Hamm and F. Collins, unpublished), and portions of the *Ae. albopictus* (HsuChen *et al.*, 1984; Dubin *et al.*, 1986) mitochondrial genome have been cloned and sequenced. In sequence, gene order, and orientation of coding regions, these mosquito mtDNAs are similar to those of *Drosophila yakuba* and *D. melanogaster*. As has been observed within all other insect and vertebrate mtDNAs sequenced, the base composition is highly A + T rich and very little noncoding intergenic sequence is present. However, the mtDNA of *Ae. albopictus* and *An. gambiae* (and thus probably all mosquitoes) do differ from *Drosophila* with respect to the order of three tRNA genes (Dubin *et al.*, 1986; D. Mills Hamm and F. Collins, unpublished). Based on sequence data, the *An. quadrimaculatus* and *D. yakuba* mtDNA origins of replication, both extremely A + T rich, have no significant homology (Cockburn *et al.*, 1990). Variation in the size of the mtDNA between the two species has been attributed to length variation in the origin of replication (Cockburn *et al.*, 1990). Restriction analysis of several anophelines also shows some intraspecies length polymorphism in this region (Collins *et al.*, 1990; F. Collins and C. Porter, unpublished; S. Mitchell, S. K. Narang, A. F. Cockburn, J. A. Seawright, and M. Goldenthal, unpublished). The phylogenetic usefulness of sequence-level changes in A + T-rich mtDNA is limited to closely related species because of the rapid saturation of changes at silent codon positions, and the concomitant difficulty in distinguishing identity by descent from identity due to reversion (e.g., DeSalle *et al.*, 1987). On the other hand, structural changes such as the tRNA differences noted above may be useful phylogenetic characters for investigating higher level taxonomic questions outside of Culicidae.

Surveys of mtDNA restriction fragment length polymorphisms (RFLPs) have been conducted among species in the *Aedes scutellaris* and *Ae. albopictus* subgroups (Kambhampati and Rai, 1991a) and among populations of *Ae. albopictus* (Kambhampati and Rai, 1991b) and *Aedes aegypti* (W. J. Tabachnick and D. Zuhn, unpublished). Intraspecies mtDNA RFLPs have also been assessed for sibling species within the *Anopheles albitarsis* (S. K. Narang, T. A. Klein, J. B. Lima, and A. T. Tang, unpublished), *An. gambiae* (F. Collins and C. Porter, unpublished), *An. freeborni* (Collins *et al.*, 1990), and *An. quadrimaculatus* (S. Mitchell, S. K. Narang, A. F. Cockburn, J. A. Seawright, and M. Goldenthal, unpublished) complexes. Although extensive interspecific divergence was revealed in the *Ae. scutellaris* and *Ae. albopictus* subgroups, there was surprisingly little within- or among-population variation in mtDNA haplotypes, even among geographically very distant populations in the case of *Ae. albopictus*. Among 17 worldwide populations of *Ae. albopictus* surveyed, only 3 revealed unique mtDNA RFLPs with a battery of 12 restriction enzymes, and no RFLPs were detected within populations using 4 restriction enzymes (Kambhampati and Rai, 1991b). The result of a survey of mtDNA RFLPs among several populations of *Aedes aegypti* was consistent with low-level intraspecies polymorphism (W. J. Tabachnick and D. Zuhn, unpublished). Where intraspecies polymorphism was detected in the *An. quadrimaculatus* complex, one haplotype usually predominated in all populations surveyed (S. Mitchell, S. K. Narang, A. F. Cockburn, J. A. Seawright, and M. Goldenthal, unpublished). Species-specific haplotypes were detected in the case of two sympatric species in the *An. albitarsis* complex (S. K. Narang, T. A. Klein, J. B. Lima, and A. T. Tang, unpublished) and the *An. quadrimaculatus* complex (S. Mitchell, S. K. Narang, A. F. Cockburn, J. A. Seawright, and M. Goldenthal, unpublished). Only the *An. gambiae* complex showed high levels of intraspecific mtDNA polymorphism (F. Collins and C. Porter, unpublished). For example, analysis of polymorphism among 30 field-collected *Anopheles arabiensis* from Kenya using restriction enzymes that surveyed roughly 7% of the mtDNA genome revealed five distinct haplotypes, four of which were present at frequencies of 0.13 or higher. Similar levels of polymorphism were observed in four other species in this complex, but none of the restriction profiles was unique to a single species. Comparison of within- and between-species mtDNA sequence polymorphism over a 1400-bp region encoding the ND4 and ND5 subunits and several tRNA genes confirmed this relatively high level of intraspecies polymorphism (0.55% difference between three different specimens of *An. gambiae*) relative to interspecies differences (1.3% consistently different bases between *An. gambiae* and *An. merus* and only 0.13% difference between *An. gambiae* and *An. arabiensis*). In fact, the number of shared mtDNA haplotypes and specific nucleotide site polymorphisms suggests recent and perhaps continuing mtDNA introgression between these latter two species.

Though many of these population and species complex studies of mtDNA are preliminary, the widely reported low intraspecific levels of restriction site variation suggest that mtDNA may not be as useful a tool for population genetic analysis of mosquitoes as it has proved to be for other organisms. The only obvious exceptions appear to be the members of the *An. gambiae* complex, which are species with very large and temporally stable geographic distributions in tropical and subtropical Africa. The distributions of both of the *Aedes* species studied have become widespread only relatively recently through association with human commerce (Kambhampati and Rai, 1991a), and the *An. quadrimaculatus* and *An. freeborni* complex mosquitoes are North American species that probably experience substantial fluctuations in overall population size and distribution during winter months.

More recently, a technique known as RAPD-PCR (Williams *et al.*, 1991) has been used to analyze population structure and for species identification. It is based on the use of individual random sequence ten-mers as PCR primers. RAPD-PCR amplification of insect genomes results in a complex pattern of bands, visualized by electrophoresis and ethidium bromide staining of agarose gels. This technique revealed large numbers of polymorphisms between individuals and populations of *Ae. aegypti* (Ballinger-Crabtree *et al.*, 1992), but demands meticulous care in its application and in interpretation of the results (Kambhampati *et al.*, 1992).

C. Identification of Cryptic Species

Until fairly recently, only small numbers of morphologically indistinguishable mosquito species were recognized, most notably the *An. maculipennis* complex discovered in the 1930s followed by the *An. gambiae* complex in the 1960s. The idea that many species cannot be easily identified in the adult stage was difficult for many "cabinet" taxonomists (see Bates, 1949) and for many nontaxonomists reluctant to see their malaria control studies impeded (B. Harrison, personal communication). The existence of large numbers of mosquito species that are nearly identical or that are morphologically indistinguishable as adults is finally gaining wide recognition, and *Anopheles* vector identification in particular is now recognized as a major problem in many areas of the world. Of the 66 *Anopheles* species currently believed to be important vectors of human malaria parasites, at least 55% belong to closely related groups of nearly identical species, or to "cryptic species" complexes (B. Harrison, unpublished). As methods for identifying mosquito species move closer to direct genome analysis, from comparisons of morphological characters through mitotic and polytene chromosome banding patterns to isozymes and, ultimately, DNA sequence, the rate of discovery of new cryptic species has increased. Though interesting from an evolutionary perspective, the existence of cryptic sibling species complexes also poses practical problems in vector control because different complex members may differ in their involvement in disease transmission and their response to various control measures (Coluzzi, 1992). It is therefore important to assess differences between species as they relate to disease transmission, such as biting and resting preferences of adult females, seasonal distribution and abundance, competitive ability, and microhabitat preferences. These data are relevant to the choice and monitoring of vector control strategies and the ability to predict the impact of human activities such as deforestation, irrigation, and desalination on disease transmission (Coluzzi, 1984; 1992).

1. Random Repetitive Probes

Several DNA-based methods have been developed to distinguish among the members of cryptic species complexes. The most widely used approach has been to identify species-specific differences in the undefined highly repeated component of the genome by differential screening of small-fragment genomic libraries with labeled homologous and heterologous genomic DNA. The highly repeated sequences isolated by this strategy typically consist of tandem repeats of 50- to 100-bp monomers present in numbers between 10,000 and 100,000 or more per genome. The high copy numbers make these probes extremely sensitive, enabling species identification in some cases from less than 1 ng of target DNA (or small portions of a specimen). Several laboratories have explored the use of dried or isopropanol-fixed mosquitoes in a squash–blot format, nonisotopic detection systems, and synthetic oligonucleotides based on the probe repeat sequence, all with favorable results. Such modifications should facilitate the use of these assays in studies designed to address population-level questions. Ideally, species diagnosis should be based on probes that are absolutely species specific over the entire species range. However, assays have been devised using a combination of probes that hybridize to different subsets of species in a complex. The species specificity of such probes depends on copy number or sequence variation. These possibilities can be distinguished by varying the stringency of the hybridization conditions, though this step has generally not been taken in published reports. The validity of the probes for species diagnosis should be assessed by careful field testing, especially in areas where they are to be used, because highly repetitive sequences can diverge rapidly in both copy number and sequence between and even within populations (see Section III,B).

DNA dot–blot or mosquito squash–blot assays have been developed for at least four different anopheline complexes, the *An. quadrimaculatus, Anopheles dirus, An. gambiae,* and *Anopheles farauti* complexes. A squash–blot assay to separate species A, B, C, and D of the *An. quadrimaculatus* complex is based on probes that hybridize intensely to one species and sometimes hybridize slightly to one or more other members of the complex (Cockburn, 1990). In the *An. dirus* complex, cloned probes have been developed that uniquely hybridize with sibling species D (Yasothornsrikul *et al.,* 1986; Panyim *et al.,* 1988) or species B (Kertbundit *et al.,* 1986). The probes are sensitive enough to allow positive identifications from as little as 5 ng of DNA, estimated to be about 1/150th the amount of DNA available from a single mosquito. Panyim *et al.* (1988) tested species D probe pMU-D76 against all four species using progeny of females collected from a single geographic area (eight nearby sites). It is unclear whether this probe is valid throughout the range of the species or whether the other probes have been field tested. Another assay for distinguishing the members of this complex has recently been reported (Yasothornsrikul *et al.,* 1988). Preliminary data from the progeny of single field-caught specimens of each species revealed unique restriction fragment patterns of highly repeated DNA superimposed on a background smear of genomic DNA visualized from ethidium bromide-stained gels. Although presented as an alternative method for species diagnosis with the advantages of speed and simplicity, it is unlikely to replace sensitive dot–blot or squash–blot approaches.

DNA dot–blot and mosquito squash–blot protocols have also been developed for the identification of five members of the *An. gambiae* complex, based on the use of cloned probes containing highly repetitive DNA sequences (Gale and Crampton, 1987a,b, 1988). Each probe is very sensitive due to high copy number, but when used alone cannot distinguish all species and must be used in combination with other probes for positive species identification. One probe, pAnal, is male specific and separates female *An. arabiensis* from the other freshwater species only by probing inseminated spermathecae. In spite of these limitations, the technique has been successfully applied to wild-caught female mosquitoes, which were stored in isopropanol and later squashed onto nitrocellulose (G. Yemane, S. Hill, R. Urwin, A. T. Haimanot, and J. Crampton, unpublished). Three probes were used to differentiate mixed populations of *An. gambiae, An. arabiensis,* and

Anopheles quadriannulatus from four locations in Ethiopia. Although 149 of the 153 specimens collected were identified with these probes, none of these identifications were confirmed by an alternative method such as analysis of polytene chromosomes. As is the case in other laboratories, the technique is being improved for field use by incorporating several modifications that increase its speed, simplicity, safety, and economy. These include the substitution of cloned probes by synthetic oligonucleotides 21–26 bases long (Hill *et al.*, 1991a) and the use of nonradioactive labeling and detection protocols (Hill *et al.*, 1991b, 1992).

Similar work with the *An. farauti* complex led to the development of two sets of probes to distinguish three known members, *Anopheles farauti* Nos. 1, 2, and 3 (Booth *et al.*, 1991; Cooper *et al.*, 1991). Identifications using the probes for mosquitoes captured over a broad geographic area were verified by established identification methods. Although the use of these probes resulted in correct identifications most of the time, there were problems with specificity (Booth *et al.*, 1991) and sensitivity (Cooper *et al.*, 1991).

2. mtDNA and rDNA Probes

Species-specific RFLPs in mtDNA genomes or the rDNA multigene family have been described for at least three species complexes, the *An. quadrimaculatus, An. gambiae,* and *An. freeborni* complexes. In the *An. quadrimaculatus* complex, species-specific differences were found in the restriction enzyme map for both mtDNA (S. Mitchell, S. K. Narang, A. F. Cockburn, J. A. Seawright, and M. Goldenthal, unpublished) and rDNA (S. Mitchell, S. K. Narang, A. F. Cockburn, J. A. Seawright, and M. Goldenthal, unpublished). One enzyme, *Cla*I, was found to produce diagnostic mtDNA restriction patterns for each species (Mitchell, 1990); all others tested required separate digestions with two different enzymes to discriminate between all four sibling species (S. Mitchell, S. K. Narang, A. F. Cockburn, J. A. Seawright, and M. Goldenthal, unpublished). Species-specific differences were found in the rDNA IGS or ITS regions of members of the *An. freeborni* (Collins *et al.*, 1990) and *An. gambiae* (Collins *et al.*, 1987, 1989; Finnerty and Collins, 1988) complexes. Diagnostic methods for the *An. gambiae* species complex based on rDNA RFLPs have been validated from field collections by comparison with both cytogenetic (Collins *et al.*, 1988a) and allozyme (Collins *et al.*, 1988b) methods. Interestingly, crossing studies combined with rDNA RFLP analyses of field-collected *An. freeborni* and its sibling species *An. hermsi* not only confirmed the DNA probe method developed to distinguish this cryptic species pair but also invalidated a character, an X chromosome inversion, previously considered to be diagnostic for *An. hermsi* (Fritz *et al.*, 1991).

Detection of species-specific differences in the rDNA spacers has been extended by using the polymerase chain reaction in various *Anopheles* complexes, including *An. gambiae* (Paskewitz, and Collins, 1990; J. Scott, W. Brogdon, and F. Collins, unpublished). *An. freeborni* (Porter and Collins, 1991), and *An. maculatus* (L. G. Chiang, C. Porter, and F. Collins, unpublished). Though this format is relatively expensive in its requirement of a thermal cycler and thermostable polymerase, it is very rapid, relatively simple, and has virtually 100% sensitivity and specificity. Furthermore, the amount of DNA required is so small that living specimens can be identified by using DNA from a single segment of a mosquito leg. The actual identification is accomplished by visualizing the PCR-amplified DNA in an ethidium bromide-stained gel, without the steps of transferring DNA to a membrane and probing. In the *An. gambiae* complex, the method exploits differences in the 5' end of the IGS. In the published work, three primers were used to distinguish *An. gambiae* from *An. arabiensis,* one universal plus-strand primer and two species-specific minus-strand primers. When combined in the same PCR reaction, fragments of unique lengths are produced based on the source of template DNA. This has now been extended to permit identification of five members of the complex using various combinations of one universal plus-strand primer with five species-specific minus-strand primers, each of which amplifies a single fragment of unique length diagnostic of a single species (J. Scott, W. Brogdon, and F. Collins, unpublished). Successful species identification can be achieved from a PCR reaction that incorporates all six primers, but in practice usually no more than three members of the *An. gambiae* complex are sympatric in any part of Africa.

Extensive comparison of the approximately 450-bp ITS2 sequences from 48 different field and colony specimens of five members of the *An. gambiae* complex revealed very few (one to five) species-specific differences from the consensus. By contrast, differences among ITS2 sequences in many other *Anopheles* cryptic species complexes, including the *An. freeborni, An. maculatus, An. quadrimaculatus,* and *An. punctipennis* complexes, are considerably higher and more readily exploited in a PCR-based species-diagnostic assay. For example, an assay used to distinguish *An. freeborni* and *An. hermsi* uses four primers—two universal primers from conserved coding regions that flank the ITS2, and two species-specific primers from within the ITS2, which produce a fragment of characteristic size for each species (Porter and Collins, 1991). Amplification of DNA from other sympatric anopheline species produced much larger fragments that represented extension between the two universal primers across the entire ITS2.

III. Genome Characterization

A. Genome Organization

Genome organization can be described by the haploid DNA content and the amount, types, and dispersion patterns of repetitive DNA sequences. These parameters need to be considered before isolating repeated sequence components from the genome or screening genomic libraries for single-copy sequences. They also form the basis for comparative studies of genome evolution. In general, the biological significance of each of these features and their relationship to one another is far from understood. However, what is clear is the remarkable rate of divergence of these genome characteristics among closely related species and even among populations of the same species.

1. Genome Size

The genome sizes of various *Anopheles, Culex,* and *Aedes* species, including several strains of *Ae. albopictus,* have been measured by various methods (Table 1). Most genome size determinations were made by relative Feulgen cytophotometry, in which the level of Feulgen staining in the test sample is compared with that of a standard of known DNA content (measured by direct chemical analysis). Genome size has also been estimated by reassociation kinetics (see below). Among the many organisms for which haploid DNA content has been measured, there is no simple correlation between genome size and organismal complexity. There does, however, seem to be an overall trend in the family Culicidae, with the most primitive genus, *Anopheles,* possessing an average genome size at least two times smaller than the more recently evolved *Culex* or *Aedes* (Rao and Rai, 1990). Overall, there is about an eightfold variation in haploid DNA content among the mosquito species that have been examined, from 0.23 pg in *Anopheles labranchiae* (Jost and Mameli, 1972) to 1.9 pg in *Aedes zoosophus* (Rao and Rai, 1987a,b). Perhaps more striking is the enormous variation in genome size among different populations of the same species, even those collected from the same geographic location. Thus the threefold variation in DNA content among 23 species of *Aedes* (Rao and Rai, 1987b) is matched by a threefold variation in DNA content among *Ae. albopictus* sampled throughout its range and a 1.7-fold variation between two populations from Singapore (Kumar and Rai,

TABLE 1
Genome Size of Various Anophelines, Culicines, and Aedines[a]

Species/strain	Genome size (pg) ± standard error	Ref.
Anopheles labranchiae	0.23 ± 0.04	Jost and Mameli (1972)
Anopheles atroparvus	0.24 ± 0.04	Jost and Mameli (1972)
Anopheles stephensi	0.24 ± 0.03	Jost and Mameli (1972)
Anopheles freeborni	0.29 ± 0.03	Jost and Mameli (1972)
Anopheles quadrimaculatus (sp. A)	0.24 ± ?; 0.24[b] ± ?	Rao (1985); Black and Rai (1988)
Anopheles gambiae	0.27[b] ± ?	Besansky and Powell (1992)
Culex pipiens	1.02 ± 0.19; 0.54 ± ?; 0.53[b] ± ?	Jost and Mameli (1972); Rao (1985); Black and Rai (1988)
Culex quinquefasciatus	0.54 ± 0.01	Rao and Rai (1990)
Culex restuans	1.02 ± 0.04	Rao and Rai (1990)
Aedes aegypti	0.81 ± 0.03; 0.83[b] ± ?	Rao and Rai (1987b); Warren and Crampton (1991)
Aedes pseudoscutellaris	0.59 ± 0.01	Rao and Rai (1987b)
Aedes triseriatus	1.52 ± 0.06	Rao and Rai (1987b)
Aedes zoosophus	1.90 ± 0.06	Rao and Rai (1987b)
Aedes albopictus		
Kent Ridge (Singapore)	0.75 ± 0.02	Kumar and Rai (1990b)
Amoy (Singapore)	1.29 ± 0.06	Kumar and Rai (1990b)
Koh Samui (Thailand)	0.62 ± 0.02	Kumar and Rai (1990b)
Houston-61/TX (United States)	1.66 ± 0.08	Kumar and Rai (1990b)

[a] All measurements by Feulgen cytophotometry unless otherwise indicated.
[b] Measurement by DNA reassociation kinetics.

1990b). The wide range of intraspecific differences in genome size emphasizes the speed with which the underlying molecular processes of duplication and deletion operate.

The potential significance of changes in genome size at the organismal level has been investigated in *Aedes*. As might be expected, there is a strong correlation between genome size and total length of metaphase chromosomes (Rao and Rai, 1987b). However, no apparent association exists between systematic relationships and genome size. Within *Ae. albopictus,* differences in genome content do not cluster geographically, are not associated with island-dwelling versus continent-dwelling populations, nor are they associated with recent colonization (Kumar and Rai, 1990b). Thus it is unlikely that the DNA sequences involved in rapid changes in genome size are directly responsible for speciation events. Ferrari and Rai (1989) studied the association between DNA content and two phenotypic traits, wing length and development time, related to fitness. In 9 of 10 *Ae. albopictus* laboratory strains compared, they found a significant relationship between DNA content and development time. This implies that differences in DNA content may indirectly affect phenotypic traits subject to selection, at least in the laboratory, but its effect on fitness of natural populations of mosquitoes is still unclear.

2. Repetitive DNA

DNA reassociation studies from the late 1960s led to the discovery of repeated sequences in eukaryotic genomes (reviewed in Britten and Kohne, 1968) and have been used since to determine the amount and organization of highly repeated, moderately repeated, and unique genomic sequences. Reassociation kinetics are estimated by monitoring the rate at which denatured DNA from a given species reassociates in solution. The rate of reassociation depends on the initial concentration of genomic sequences and the time of incubation under defined conditions. Repeated sequences reassociate much faster than single-copy sequences. The data can be expressed as a C_0t curve, the log of the product of initial DNA concentration (C_0) and incubation time (t) plotted against the proportion of DNA remaining single stranded. Nonlinear regression analysis of the C_0t curve allows the relative amounts of highly repeated, moderately repeated, and unique components of the genome to be determined. Comparison of the reassociation rates of long and short DNA fragments reveals the overall pattern of interspersion of repeated and unique sequences—long-period interspersion, the alternation of repeats about 5 kb long with the unique sequence at least 13 kb long; or short-period interspersion, the alternation of repeats about 300 bp long with the unique sequence about 1000 bp long (reviewed in Bouchard, 1982). Total genome size can be estimated from the reassociation rate of the unique component of the curve by comparison with that of an organism of known genome size and little repetitive DNA, commonly *Escherichia coli*. Thus kinetic analysis is an alternative to Feulgen staining for measurement of genome size, and has been applied to mosquito genomic DNA (Black and Rai, 1988; Warren and Crampton, 1991; Besansky and Powell, 1992).

The amounts of repeated sequences in mosquitoes vary from 20% in *An. quadrimaculatus* to 84% in *Ae. triseriatus* (Black and Rai, 1988). From DNA reassociation kinetics, Black and Rai (1988) found that the proportion of the genome occupied by repeated DNA increased as genome size increased in three mosquito species, *An. quadrimaculatus*, two strains of *Ae. albopictus*, and *Ae. triseriatus*. An exception to this trend was *Culex pipiens*, with a genome one-quarter the size of *Ae. triseriatus* but with an equivalent amount of repetitive DNA. The size distribution of repeats from each species was investigated by nuclease S1 digestion of duplexes formed between repeats on 2-kb-long DNA strands, followed by gel chromatography and electrophoresis to measure S1-resistant duplexes. If the duplexes are clustered to a significant extent, as is the case in *D. melanogaster* (Manning *et al.*, 1975), this technique will overestimate the proportion of long repeats. However, kinetic analyses of the four mosquito species do not indicate significant clustering of short repeats (Black and Rai, 1988). From the S1 and kinetic analyses, Black and Rai (1988) conclude that while the absolute amounts of both long (>1000) and short (<1000 bp) repeats increase with genome size, the relative proportion of short repeats increases at the expense of long repeats in all species examined. Therefore, the amplification of short repeats accounts, at least in part, for increased genome size. In light of these results, it is interesting that the *An. quadrimaculatus* genome, smallest of the four species, was found to consist of only two distinguishable components, repetitive and unique, which are organized into a pattern of long-period interspersion, and that genome size variation among anopheline species is minimal (see Table 1). In contrast, the much larger *Culex* and *Aedes* genomes have three distinguishable components, highly repetitive, moderately repetitive, and unique, organized into a pattern of short-period interspersion, and genome size variation is extensive.

Supporting evidence for the role of short repeats in genome size variation emerged from comparisons of two geographic strains of *Ae. albopictus* (Black and Rai, 1988). The larger genome size of the Mauritius strain could be accounted for by 1.4 times more unique sequences,

1.5 times more fold-back repetitive sequences, and 2.7 times more highly repetitive sequences than are present in the Calcutta strain; the amounts of moderately repetitive sequences were equivalent. In addition, the Mauritius strain contained 15% more short repeats than the Calcutta strain. These results strengthen the conclusion that gain (or loss) of short repeats, many of which must be highly repetitive, has played an important role in the evolution of genome size and organization in mosquitoes. In apparent contradiction to this conclusion are the results of a kinetic analysis of *Ae. aegypti* DNA from larvae of the Bangkok strain and from the MOS20 cell line (Crampton *et al.*, 1990a; Warren and Crampton, 1991). In this case, the DNA content of the cell line was twice that of the Bangkok strain, with 2.3-fold more single-copy DNA, 1.6-fold more middle-repetitive DNA, and essentially the same amount of highly repeated DNA. This pattern of variation, however, may reflect processes such as chromosome duplication that are uniquely tolerated by cells growing in culture.

An alternative approach to the analysis of repetitive DNA amount and organization was taken by Cockburn and Mitchell (1989). They prepared libraries in the phage vector EMBL3 with *Sau*3A partial digests of total genomic DNA from *An. quadrimaculatus* spp. A, B, C; *An. freeborni; An. albimanus; Ae. aegypti; Culex quinquefasciatus;* and *Stomoxys calcitrans* (the stable fly). Repetitive DNA amounts and interspersion patterns were inferred by the signal intensity and proportion of 137 plaques randomly selected from each library that hybridized to radiolabeled total genomic DNA from the same species. In general agreement with the results of Black and Rai (1988), they found that anopheline genomes had less repetitive DNA and long-period interspersion patterns, whereas aedine and culicine genomes had more repetitive DNA with short or intermediate interspersion patterns. Though the library hybridization technique is much easier to perform than fluid hybridization studies, results of the technique should be interpreted with caution. Potential errors derive not only from technical considerations such as the degree to which the library represents inverted and tandem repeats, but also from the relatively small part of the cloned genome actually sampled. Because all the cloned fragments were in the 10- to 20-kb size range, interpretation of repeat interspersion pattern is especially difficult. The assumption that all repeated sequences are uniformly distributed throughout the genome is not necessarily justified (e.g., Ananiev *et al.*, 1978). Nevertheless, because the technique is so simple, it can provide a useful preliminary view of genome organization.

Cloned repeats have been used to study the conservation of individual highly repeated sequence families among and within mosquito species. McLain *et al.* (1986) prepared a plasmid library with *Eco*RI-digested genomic DNA from *Aedes malayensis,* from which they selected nine highly repeated sequences for further analysis. Copy number variation for each repeat family was examined by densitometry following hybridization of cloned repeats to DNA dot blots. Each DNA dot contained known amounts of genomic DNA from *Ae. malayensis* and five other sibling species in the *Ae. scutellaris* subgroup. Comparison of signal intensity following low- versus high-stringency hybridization provided additional information on repeat sequence divergence between species. Copy number varied extensively, but six of the nine clones hybridized at high stringency to genomic DNA of all species and the remaining three hybridized strongly at reduced stringency.

Using the same approach, McLain *et al.* (1987) confirmed and extended these results. They isolated eight repeat-containing clones from the Oahu strain of *Ae. albopictus* and compared their sequence and abundance by dot–blot hybridization to species of the *Ae. scutellaris* subgroup, three species of the *Ae. albopictus* subgroup, and 15 strains of *Ae. albopictus* from widely separated geographic areas. Care was taken to ensure that the different clones did not cross-hybridize and that, at least on the basis of restriction enzyme analysis, the clones did not contain tandem repeats. They found that differences in copy number were as large within *Ae. albopictus* as between species. Again, three clones that did not hybridize to some *Ae. albopictus* strains and other species at high stringency hybridized at lower stringency. Their observation of significant intraspecies variation in copy number of some highly repeated DNA sequences clearly suggests caution in the use of highly repeated DNA for species-diagnostic purposes, at least in *Aedes* species.

Kumar and Rai (1991a; 1991b) have expanded on these studies by examining the *in situ* and Southern hybridization patterns of a 5.2-kb cloned repetitive fragment from *Ae. albopictus. In situ* hybridization to mitotic and meiotic chromosomes of several *Aedes* species produced dispersed patterns on all chromosomes of each species, though hybridization intensity was less in *Aedes aegypti.* Southern analysis revealed the presence of homologous sequences in the DNA from representatives of diverse mosquito species as well as a Chaoborid. Nevertheless, the hybridization patterns were unique to each species, although closely related species shared some bands. In contrast, no intraspecific variation was detected. The conservation of this repetitive DNA throughout the evolutionary radiation of mosquitoes in spite of structural reorganizations suggests a function based on nucleotide sequence.

The hybridization pattern of cloned repeats has also been looked at between sibling species in the *An. gambiae* complex (Besansky, 1990b; N. Besansky, unpublished). An *An. gambiae* genomic library was prepared in EMBL3 phage by *Sau*3A partial digestion. Repeat-containing clones were picked at random following hybridization of the library with radiolabeled genomic DNA. Each of 19 repeats was hybridized to Southern blots of genomic DNA from several East and West African laboratory strains of *An. gambiae* and *An. arabiensis* and in some cases, also, one strain of *Anopheles melas* and a congener outside of the complex, *Anopheles stephensi*. All repeats, regardless of copy number, hybridized at high stringency to all members of the complex, whereas none that was tested hybridized to *An. stephensi* (low-stringency hybridization was not attempted). Four clones appeared to contain satellite-type, simple sequence tandem repeats on the basis of (1) regular, additive ladder patterns generated on Southern blots and (2) the insensitivity of the cloned sequence to digestion with any of a panel of seven restriction enzymes recognizing 6-bp targets. The ladder patterns for each clone were identical for *An. gambiae* and *An. arabiensis*. Another 13 clones probably represented dispersed, highly repetitive sequences because they generated smears of varying intensities on Southern blots from members of the complex. By contrast, four unrelated middle repetitive sequences, three of which were isolated from the same EMBL3 insert, produced discrete patterns of bands that differed among members of the complex.

Employing a different strategy, Cockburn and Mitchell (1989) probed phage libraries (described above) from different mosquito species and genera with individual repeat-containing phage clones to determine the proportion and intensity of hybridizing plaques. In general, most anophelines shared repeated DNA families with the exception of two species-specific families identified in the *An. quadrimaculatus* complex. Although cross-generic hybridizations were generally unsuccessful, one *Aedes* clone hybridized to *Culex* and not *Anopheles* and one *Anopheles* clone hybridized to *Aedes* (hybridization to *Culex* was not determined). Apart from the *An. quadrimaculatus* clones, hybridization at reduced stringency was not attempted.

Several cautionary notes apply to all experiments involving the probing of genomic DNA with clones containing uncharacterized repetitive DNA. It is quite possible that any given repetitive clone may contain multiple related or unrelated repeated sequences within it. For example, one clone from an *An. gambiae* EMBL3 phage library proved to contain three unrelated non-LTR retrotransposons (Besansky, 1990a; J. Bedell and N. Besansky, unpublished). This possibility depends upon the interspersion pattern of the genome and insert size of the clone, among other things, but should be tested before firm conclusions are drawn. Additionally, the clone of interest could contain a sequence from a member of one of the many conserved multigene families, such as rDNA, 5S RNA, histones, actins, tubulins, or mtDNA. Finally, hybridization stringency should be very carefully controlled. At moderate stringency, high levels of apparently random hybridization can be observed between sequences with skewed G + C contents or long stretches of simple sequence repeats. Such simple or biased sequence composition may be common not only to many highly repeated sequences that are not homologous, but may also be found in regulatory domains.

In summary, the data suggest that individual families of highly repeated sequences, at least in *Aedes,* may evolve particularly rapidly with respect to copy number, without regard to genome size or phylogenetic relationships. This is underscored by the fact that copy number variation is as extensive within *Ae. albopictus* as between different aedine species. On the other hand, many of these same *Aedes* sequence families apparently evolve quite slowly at the nucleotide sequence level, as suggested by their ability to hybridize at high stringency even to *Anopheles* sequences. This might suggest a sequence-dependent function for such families. Yet, the reciprocal relationship does not seem to hold. That is, most anopheline highly repeated sequence families that have been studied have no homology to sequences outside the genus and limited homology between *Anopheles* species from different series within the same subgenus. With the current limited data, it is difficult to generalize from the above observations. More must be learned about the roles of different types of repeated sequences in the genome, the different factors driving and constraining their evolution, and the phylogenetic relationships among the species examined.

B. Genome Mapping

In some mosquito species, invariably those of medical importance, considerable effort has been directed toward assigning linkage relationships among morphological and isozyme markers and associating these with chromosomes by crossing experiments. In *An. quadrimaculatus,* at least 26 markers have been mapped to specific chromosomes or chromosome arms and another 12 have been assigned only as autosomal (Narang *et al.,* 1990). In *An. albimanus,* at least 51 markers have been assigned to chromosomes and another 12 are only known to be autosomal (Narang and Seawright, 1990). Very extensive linkage maps have also been prepared from *Culex* and *Aedes* mosquitoes, particularly *Ae. aegypti* (e.g., Steiner *et al.*, 1982; Munsterman 1990a,b). Genetic maps can guide the understanding of genome organization and evolutionary relationships among mosquitoes. To the extent that they can be refined, they will also save much time and effort on the part of researchers interested in cloning the genes controlling a variety of phenotypic traits. Molecular cloning techniques have had an important impact on genetic mapping, and have brought the goal of detailed linkage maps of mosquito chromosomes within reach.

All mosquitoes possess polytene chromosomes in certain tissues and developmental stages, including the salivary glands of larvae and the ovarian nurse cells of adult females. Polytene chromosomes represent repeated chromosomal replication in the absence of nuclear division, leading to an increase in DNA content as high as 16,000-fold (Ashburner, 1980). Homologous chromosomes are paired and each chromatid remains aligned with its sisters, causing a thickening of the chromosome and revealing an intricate band/interband pattern that can be mapped and correlated with the genetic map. Unfortunately, the polytene chromosomes of *Aedes* mosquitoes are of very poor quality cytogenetically because they are so difficult to spread, probably due to the fact that their highly repeated dispersed DNA sequences promote extensive ectopic pairing. The chromosomes of *Culex* mosquitoes, whose genomes also contain high levels of repeated DNA, are only slightly better (White, 1980). On the other hand, excellent polytene chromosome preparations can be made from most anophelines, a considerable advantage in linking physical and genetic maps.

With the recent availability of cloned probes and technical improvements of *in situ* hybridization protocols (Reiss, 1990; Graziosi *et al.*, 1990; V. Kumar, V. Petrarca and F. Collins, unpublished), both known and anonymous sequences can be assigned to particular polytene chromosome regions in anophelines. The polytene chromosomes of *An. gambiae* are central to an elegant effort to map the genome of this mosquito (Zheng *et al.*, 1991). This effort relies upon a technique developed for *Drosophila* (Saunders *et al.*, 1989) that involves amplification by PCR of DNA obtained by microdissection of each of the numbered divisions of the *An. gambiae* ovarian nurse cell polytene chromosome complement, thereby creating individual division-specific "libraries." One application of division-specific libraries is the physical mapping of cloned segments of DNA without having to prepare polytene chromosomes, perform *in situ* hybridization, or interpret the banding pattern. Each division-specific library is spotted onto a nylon membrane in order of its chromosomal location. Any cloned sequence, random or known, can then be labeled and used to probe the chromo-somal dot–blot. In addition, each of these libraries can be used as probes to produce division-specific subsets of phage or cosmid genomic libraries as a preliminary step in chromosomal walks across defined genomic divisions (e.g., Sidén-Kiamos *et al.*, 1990).

RFLPs can also be used to map the approximate genetic location of genes. Mapping entails following the segregation of RFLPs in relation to other genetic markers in crossing experiments with inbred strains, or by pedigree analysis as is done in mapping human genes. The RFLPs detected usually result from the gain or loss of a restriction site through base substitution or small insertion/deletions. Variability in the DNA of certain strains of mosquitoes appears to be high enough to support this approach. In ongoing surveys of the G3 colony of *An. gam-biae,* at least 63 *Eco*RI RFLPs have been identified at anonymous loci (P. Romans, unpublished), and RFLPs have also been found for the vitellogenin, dopa decarboxylase, diphenol oxidase, α-amylase, and several other loci (P. Romans, unpublished; Romans, 1991). The cyto-genetic distribution of these RFLPs has been shown to cover all five chromosome arms by *in situ* hybridization.

An advantage of RFLP analysis is that a few crossing experiments can provide an enormous amount of information because (1) the geno-mic DNA extracted from the progeny can be apportioned and digested with a variety of enzymes to maximize the chance of detecting poly-morphisms and (2) a given blot can be stripped and reprobed repeat-edly with different clones. In fact, the only limitations are the amount of interstrain variability and the availability of probes to specific ge-nomic regions. No prior information about the approximate map loca-tion of the gene of interest is required. Both *Ae. aegypti* and *An. gam-biae* have shown considerable variability between strains; 26 of 33 clones examined in the former species and 64 of 72 unique sequence phage clones examined in the latter generated strain-specific RFLPs (D. W. Severson, A. Mori, D. M. Helke, and B. M. Christensen, unpub-lished; P. Romans, unpublished). One serious drawback of an RFLP map of *Ae. aegypti* is the difficulty of obtaining a high-density map. Given a genome size of 780 mb (Rao and Rai, 1987b) and a genetic map length of 220 cM (Munstermann, 1990a), each centimorgan corre-sponds to about 3 mb, but an average resolution of much less than 5 cM may not be very feasible (M. Ashburner, personal communication). Given a map of sufficient density surrounding the gene of interest, the nearest cloned probes can be used in chromosome walks or jumps through libraries to isolate genes defined by features such as chromo-somal deletions, inversions, or translocations. The polytene chromosome division-specific "libraries" for *An. gambiae* will be particularly important in such efforts because they can be used to partition genomic libraries prior to initiating chromosome walks.

Another source of highly polymorphic genetic markers is provided by microsatellite sequences, composed of variable numbers of simple sequence repeats. These are highly abundant and dispersed throughout the genome; $(GT)_n$, the most frequent repeat unit, occurs about once every 30 kb in mice, humans, and rats (see Jacob *et al.*, 1991). Length variation at individual loci can be detected by PCR amplification with primers that anneal to flanking unique sequences. This powerful approach has been successfully used in the mapping of a complex, polygenic trait in the rat (Jacob *et al.*, 1991), with a substantially larger genome size and longer generation time than any mosquito species. If microsatellite sequences behave similarly in mosquitoes, it should be quite feasible to use them to construct a relatively dense linkage map for mosquito genomes. Several clones containing $(GT)_n$ microsatellite sequences have already been isolated from *An. gambiae* (L. Zheng, personal communication), and high priority is being placed on generating high-density (at least 1 cM) microsatellite linkage maps for both *Ae. aegypti* and *An. gambiae*.

IV. Specific Physiological Systems

A. Resistance to Insecticides and Other Toxic Substances

Resistance to various insecticides by *Anopheles, Aedes,* and *Culex* has long been known (Brown, 1967). These traits have served as important functional markers used to assign linkage groups to particular chromosomes or chromosome arms. Genetic and biochemical experiments revealed their mode and pattern of inheritance and suggested a general biochemical or physiological mechanism of resistance without defining its molecular basis. Insects employ several general mechanisms of resistance, including behavioral avoidance, decreased cuticular penetration, decreased sensitivity of the insecticide target site, and increased detoxification (Georghiou, 1986). Recent molecular genetic experiments on resistance mechanisms have documented three types of changes at the DNA level: point mutations in structural genes, changes in regulatory genes, and DNA amplification (Feyereisen *et al.*, 1990).

Among mosquitoes, *Culex* species have been the principal subjects of published molecular studies of insecticide resistance. In *Culex*, a common mechanism of resistance to organophosphate (OP) insecticides is increased detoxification, due to overproduction of one or both of the highly active, nonspecific esterases A and B (Mouches *et al.*, 1987). One esterase B allozyme, B1, was purified from an OP-resistant strain (Tem-R) of *Culex quinquefasciatus* (Fournier *et al.*, 1987). In this strain, esterase B1 was 500 times more abundant than in a susceptible strain (Lab-S) of the same species (Mouches *et al.*, 1987). Antiserum raised to esterases B1 was used to screen a cDNA expression library prepared from Tem-R mRNA, and a clone was isolated and used to show a 250-fold difference in hybridization intensity between the Tem-R and Lab-S DNA, thus demonstrating that the overproduction of esterase B1 was due to gene amplification (Mouches *et al.*, 1986). Since then, independent amplifications of esterase B alleles *B1, B2, B3,* and *B4* in both field and laboratory strains of OP-resistant *Culex* mosquitoes have been shown to be responsible for increased esterase production (Raymond *et al.*, 1989; Pasteur *et al.*, 1990).

The nature of the amplification event has been studied for esterase B1 from the Tem-R strain (Mouches *et al.*, 1990). Overlapping clones, including the esterase B1 gene and its flanking regions, were isolated from a recombinant phage library and were used to construct a restriction map of the amplification unit from Tem-R genomic sequences. The conserved part of the genomic amplification unit, "the core," was identified by the pattern of hybridization; intensely hybridizing single bands were obtained from sequences inside the core, versus multiple bands from sequences outside the 25-kb core. Thus a large segment of flanking DNA had been coamplified along with a single 2.8-kb esterase gene. Restriction mapping of esterase B2 sequences in genomic DNA from six different resistant strains of *C. pipiens* showed that, like the esterase B1 amplification unit, the esterase B2 amplification unit consists of about 30 kb of DNA containing a single esterase gene (Raymond *et al.*, 1991).

Interestingly, the esterase B1 gene as well as the amplification core are bordered by repetitive elements. The 5'-end regions of both core and esterase genes contain a 2- to 3-kb sequence, *CE1*, which is moderately repeated in the genomes of resistant and susceptible strains of *C. pipiens* as well as its sibling species *C. quinquefasciatus,* but is absent in *Culex tarsalis.* Extending downstream of the esterase gene to the 3' end region of the amplification core is another repeated element, *CE2.* It is a composite of a 2.2-kb highly repetitive DNA sequence termed Juan, and a 4.4-kb sequence interrupting Juan. Whereas Juan is found highly repeated in the genomes of resistant and susceptible strains of *C. pipiens, C. quinquefasciatus,* and *C. tarsalis,* the interrupting sequence does not appear to be repeated in susceptible strains.

Whether these elements or other repeated elements play a role in the amplification process awaits the analysis of other amplification units for association with repeated DNA.

The amplification process involving the esterase B alleles is thought to have involved stepwise increases in copy number over time. This is supported by the observation that the level of OP resistance due to esterase B1 in Tem-R mosquitoes has increased significantly over time (cited in Pasteur *et al.*, 1990). Further support is lent by the genomic restriction analysis of the esterase B1 amplification unit. Using cloned fragments from just outside the 5′- and 3′-end boundaries of the conserved amplification core as hybridization probes, Mouches *et al.* (1990) found corresponding genomic sequences amplified 60 and 80 times, rather than the 250-fold amplification characteristic of sequences from within the core. Evidence from *in situ* hybridization to Tem-R meiotic metaphase chromosomes indicates that the esterase B1 amplification units are arrayed tandemly on a single chromosome, and extrachromosomal elements such as double minutes were not detected (Nance *et al.*, 1990). Taken together, the data fit aspects of an "onion skin" model of sequence amplification in which multiple rounds of replication initiate from the same origin within a single replication bubble in a given replication cycle (Stark and Wahl, 1984). This structure can be resolved following recombination between different duplexes into a tandem array. The amount of sequence amplification varies along a bidirectional gradient, with sequences in the middle of the array amplified more than those at the ends. Although a passive role for repeated elements, in which they serve as sites of homologous recombination, can be envisioned in the onion skin model, active transposition plays no part. Until the nature of the amplification units is understood more completely, it is difficult to judge the adequacy of this model, if, in fact, any one model will be adequate.

Although esterase B is not expressed in all tissues or stages of development, every cell type examined, including gametes, contains the amplified gene (Pasteur *et al.*, 1990; Raymond and Pasteur, 1989). Once amplified, the insecticide resistance gene can spread rapidly through populations, even on a global scale (Raymond *et al.*, 1991).

Molecular analysis of other insecticide resistance genes has begun in several different mosquito species. The acetylcholinesterase (AChE) gene is the target site for carbamate and OP insecticides, and insensitive forms of the AChE gene have been found in insecticide-resistant mosquitoes. The AChE gene has been cloned from a phage library of *An. stephensi* DNA, using an AChE cDNA clone from *D. melanogaster* (Hall and Malcolm, 1991). Sequence analysis revealed 69% homol-

ogy at the amino acid level to *D. melanogaster* AChE. *In situ* hybridization to *An. stephensi* polytene chromosomes indicated two locations
on different chromosomes, one near the telomere of chromosome 2R
and the other close to the centromere of chromosome 3L (Malcolm and
Hall, 1990). However, the location is still uncertain because genomic
Southern analysis indicates the presence of only one locus. In *An. albimanus,* the AChE gene maps exclusively to the telomeric end of chromosome 2R (Kaiser *et al.,* 1979), which may be the true location of the
AChE gene in *An. stephensi* as well. Cloning and characterization of
insensitive and sensitive forms of the AChE gene from this mosquito
are in progress.

Cytochrome *P*-450s are a multigene family, some of which are involved in insecticide metabolism. A cytochrome *P*-450-containing clone
was isolated from an *Ae. aegypti* cosmid library using a degenerate
oligonucleotide based on the conserved binding site of rabbit and rat
cytochrome *P*-450s (Bonet *et al.,* 1990). Sequencing confirmed the *P*-
450 identity of a subclone, which was used to probe Northern blots
from pyrethroid-sensitive and -resistant strains of *An. stephensi.* Elevated levels of transcript in the resistant strain led the investigators
to propose that pyrethroid resistance was due to enhanced expression
rather than a change in substrate specificity of *P*-450. The molecular
basis for enhanced expression is under investigation.

Malathion carboxylesterase (MCE), distinct from the nonspecific esterases B discussed above, hydrolyzes the OP insecticide malathion.
MCE activity was found to be increased 55-fold in a malathion-
resistant strain of *C. tarsalis* compared to a susceptible strain (Whyard
et al., 1990). A genomic library from the resistant strain has been prepared and the molecular basis for malathion resistance is currently
under study.

Several dot–blot or microtiter plate immunoassays that have recently been developed or improved can rapidly determine the biochemical insecticide resistance mechanisms in individual mosquito
specimens (Beyssat-Arnaouty *et al.,* 1989; Dary *et al.,* 1990, 1991). Although useful in their own right, they yield no information about the
molecular basis of resistance. There is, however, an assay based on
DNA hybridization, developed for the detection of esterase B sequences, which is capable of specifically detecting the amplification of
specific insecticide-resistance genes (Agarwal *et al.,* 1986). It is based
on the use of a DNA probe derived from the resistance gene being
tested; the probe hybridizes strongly to homologous sequences only in
genomes where amplification has occurred.

Finally, it seems appropriate to include in this section mention of other cases of gene amplification in mosquito cell lines in response to exposure to toxic substances, such as the folate analog methotrexate (mtx). The target protein of this drug is the enzyme dihydrofolate reductase (DHFR), which is involved in purine and pyrimidine synthesis. To study the effect of mtx on the DHFR gene and gene product, *Ae. albopictus* cells were selected for mtx resistance by maintenance in successively increasing mtx concentrations (Shotkoski and Fallon, 1990). Cells resistant to the highest level of mtx (Mtx-5011-256 cells) were 3000 times more resistant and contained about 130 times more DHFR protein than the parental, wild-type cells. The resistant cells also had elevated levels of DHFR mRNA. Southern blots of *Eco*RI-digested DNA from wild-type and resistant cells were probed with cDNAs synthesized from a mRNA fraction enriched for DHFR mRNA. The probe detected an 8.5-kb band in resistant but not wild-type cell DNA, strongly implicating DHFR gene amplification. Cloning of this fragment and subsequent sequencing of the DHFR-related sequences revealed a 614-bp DHFR gene with a 56-bp intron conserved in position between mammalian and mosquito genes (Shotkoski and Fallon, 1991). The copy number of DHFR genes per haploid genome in Mtx-5011-256 cells was estimated at 300 by densitometry of dot–blots probed with the DHFR coding sequence. Because the increase in resistance was accompanied by a conversion from diploid to tetraploid and the additional duplication of a single chromosome (Shotkoski and Fallon, 1990), the number of gene copies per cell is at least 1200. The size and organization of the amplification units are being investigated by pulsed-field gel electrophoresis, *in situ* hybridization, and Southern blotting. Preliminary data suggest that the amplification unit comprises at least 110 kb of sequence, and that it is tandemly arranged on 5 of 13 chromosomes (F. A. Shotkoski and A. M. Fallon, unpublished).

Transformation of the *Ae. albopictus* cell line C6/36 resulted in amplified copies of the exogenous sequences (T. J. Monroe, J. O. Carlson, and B. J. Beaty, unpublished), reminiscent of the amplification of endogenous genes. These cells were transformed with a plasmid containing the *Drosophila* heat-shock promoter driving a bacterial hygromycin-resistance gene and selected with hygromycin B. The cellular DNA of some stable transformants, following restriction endonuclease digestion and electrophoresis through agarose gels, contained discrete bands corresponding to plasmid fragments. Densitometry of these bands led to an estimate of 10,000 plasmid copies per haploid genome. *In situ* hybridization of plasmid DNA to metaphase chromosomes revealed large arrays integrated into different sites on the chromosome arms, centromeres, and telomeres in some cells. In addition, plasmid-carrying minutes and double minutes, some circular, were noted in varying sizes and numbers. Occasionally, whole chromosomes hybridized with plasmid sequences. Furthermore, the rescue of plasmid DNA by transformation of *E. coli* implied that unintegrated plasmids can be maintained. Some of these contained undefined mosquito genomic sequences, resulting from recombination between plasmid and chromosomal DNA. It is not clear if the high copy number of plasmid sequences was selected for by exposure to hygromycin. Monroe *et al.* found a number of stable transformants with only one or a few copies of plasmid per cell, showing that amplification of the plasmid is not always required in the face of the hygromycin selection regimen. The conditions favoring amplification may depend on where in the chromosome the plasmid integrated. An alternative explanation to amplification is suggested by studies of microinjection of mouse embryos with recombinant plasmids. Integration in mammalian cells commonly occurs as a head-to-tail concatemer of multiple copies, although these usually number below 100 as opposed to 10,000 (Gordon and Ruddle, 1985). This process is not well understood. A detailed analysis of more transformants will be helpful in understanding the events resulting in the presence of multiple plasmid sequences in C6/36 cells and how they bear on current molecular models of gene amplification.

B. Oogenesis and Endocrine Regulation

Because of sex, tissue, and stage specificity of oogenesis, its different aspects are being researched as models for regulation of gene expression and as targets for vector control.

The blood meal taken by a female mosquito initiates a complex series of events that culminate in egg production. There is evidence to suggest that distension of the midgut following a blood meal may trigger the activation of the "early" form of the digestive protease trypsin produced by the midgut (Graf and Briegel, 1989). The hydrolysis products from this early trypsin may in turn induce production of the "late" trypsin (C. Barillas-Mury, unpublished). Because late trypsin mRNA is not present prior to the blood meal, and because the increase in message precedes an increase in protein levels, it appears that late trypsin is primarily regulated at the transcriptional level (Barillas-Mury *et al.*, 1991). However, translational regulation may play a role, because the amount of trypsin produced depends on the amount of protein in the blood meal (Briegel and Lea, 1979). Cloning and sequencing of a cDNA and a genomic clone of late trypsin from *Ae. aegypti* revealed that the gene lacks introns, and the results of Southern blot analysis suggest that there is polymorphism in the regions flanking the coding sequence (Barillas-Mury *et al.*, 1991; C. Barillas-Mury, unpublished). Analysis of the regulatory elements, in progress, has uncovered a region similar to the amino acid regulatory element of yeast, which may respond to the availability of amino acids (C. Barillas-Mury, unpublished).

One of the first events following the blood meal is the increased synthesis and accumulation of ribosomes in the fat body in response to an unknown hormonal signal. In *Ae. aegypti*, within 18 hours after the blood meal, the ribosomes are four times more abundant than before the blood meal (Hotchkin and Fallon, 1987).

Coinciding with the increase in ribosome production is the synthesis of large amounts of the major yolk protein, vitellogenin (reviewed in Raikhel *et al.*, 1990), reaching a peak 24–36 hours after a blood meal in *Ae. aegypti* (Gemmill *et al.*, 1986).

To study the regulation of vitellogenesis at the molecular level, the sequences encoding vitellogenin have been cloned from *Ae. aegypti* (Gemmill *et al.*, 1986) and *An. gambiae* mosquitoes (Romans, 1990). Similar strategies were used in both cases, namely, differential screening of a phage genomic library with first-strand cDNA probes prepared from females 24 hours after a blood meal versus males (*Ae. aegypti*) or non-blood-fed females (*An. gambiae*). From *Ae. aegypti*, two clones specific to vitellogenic females were mapped and cross-hybridized (Gemmill *et al.*, 1986). Although homologous throughout the putative 5.5-kbcoding region, the restriction maps were conserved only in a central 2-kb region. Restriction mapping also indicated that these two genes are located in different regions of the genome. Genomic Southern blots suggest the presence of five different vitellogenin genes in *Ae. aegypti*, though it is not clear if all are expressed. In *An. gambiae*, a given individual contains either four or five vitellogenin genes, all of which map to the same site on chromosome 2R by *in situ* hybridization (Romans, 1990; P. Romans, unpublished). At least four of the genes are in a tandem array with the 6.3-kb coding regions separated by about 3-kb spacers. In contrast to the *Aedes* genes and probably as a consequence of their tandem arrangement, the *An. gambiae* genes are virtually identical at the sequence level throughout the entire coding region. At the amino terminus of genes from both *Aedes* (vgA1) and *Anopheles* is a 16-amino acid signal peptide interrupted by a short intron in the eleventh codon (P. Romans and H. Hagedorn, unpublished; P. Romans, unpublished). The promoter sequences from both genera contain a consensus TATA box, duplicate mRNA cap sites, and several sequence motifs similar to those found for *Drosophila* yolk protein genes, including hormone response elements and one (*Aedes*) or two (*Anopheles*) fat body enhancers (P. Romans, unpublished). In addition, *An. gambiae* contains an ovarian enhancer-like sequence. Transient expression from promoter/*LacZ* fusions will be used to study stage- and tissue-specific transcription in transgenic *Drosophila* and *An. gambiae.*

In addition to vitellogenin, the fat body of *Ae. aegypti* synthesizes another protein of 50 kDa that is secreted into the hemolymph and is internalized by the oocytes (Hays and Raikel, 1990). Like vitellogenin, its synthesis peaks about 24 hours after a blood meal and is undetectable by 48 hours. Both synthesis and secretion of the protein are stimulated by 20-hydroxyecdysone. The entire cDNA was cloned using a combination of immunoscreening of a cDNA expression library in phage, and PCR (W.-L. Cho, K. W. Deitsch, and A. S. Raikhel, unpublished). The cDNA clones hybridize to a 1.5-kb mRNA present only in vitellogenic fat bodies, showing the tissue and stage specificity of the message. Sequence analysis revealed an open reading frame with homology to carboxypeptidases. On this basis the protein, named vitellogenic carboxypeptidase (VCP), is thought to play a role in protein hydrolysis associated with embryogenesis.

As vitellogenin expression terminates, the secretory granules and ribosomes of the fat body are degraded in lysosomes (see Raikhel *et al.,* 1990). A lysosomal enzyme has been purified and characterized from *Ae. aegypti* (Cho and Raikhel, 1990). This enzyme exists as a dimer of two 40-kDa subunits. Based on its substrate utilization and N-terminal sequencing, it has been identified as a mosquito cathepsin D, and is being used to study lysosomal regulation.

C. IMMUNE MECHANISMS

Like other insects, mosquitoes possess various defenses against "foreign" invaders; these defenses, by analogy to vertebrate immune systems, have been broadly classified as humoral and cellular immunity, even though the distinction is not always clear (Christensen and Nappi, 1988). Humoral immunity involves the induction of antimicrobial proteins or the melanotic encapsulation of parasites without direct participation by hemocytes. Cellular immunity involves phagocytosis or encapsulation directly mediated by hemocytes. (For recent reviews of insect immunity, see Christensen and Nappi, 1988; Lackie, 1988; Christensen and Tracy, 1989.)

Variation in the ability of different mosquito species or geographic strains to support parasite infections has been recognized for years, and in many cases the mosquito's system for defense against foreign invaders appears to be responsible for this variation. However, in spite of genetic selection studies suggesting a single major locus controlling susceptibility to infection in several parasite/vector associations, most recently with arboviruses in mosquitoes (Miller and Mitchell, 1991), the molecular bases of these differences in immune response have not been elucidated. The possibility of interrupting parasite transmission by genetic engineering is fuelling molecular studies of mosquito immunity.

1. Humoral Responses

One major component to humoral immunity in insects is the arsenal of hemolymph proteins induced by infecting microbes, wounding, or injection of bacterial cell wall components or peptidoglycan degradation products (Boman and Hultmark, 1987). These proteins include the antibacterial diptericin, attacins, and cecropins, the digestive enzyme lysozyme, and hemolin, thought to initiate an immune response by binding to the bacterial cell wall (Sun *et al.*, 1990). Such antibacterial proteins were deemed ineffective against eukaryotic cells (Boman and Hultmark, 1987). However, recent findings may be challenging this view. For example, *Ae. aegypti* responds not only to *E. coli* and laminarin, a bacterial lipopolysaccharide, but also to the filarial nematode *Brugia pahangi* by producing proteins whose molecular weights in SDS–PAGE gels are close to lysozyme, diptericin, attacins, or cecropins (Chaithong and Townson, 1990). Moreover, hemolymph transferred *in vivo* following these immunizations partially protects against subsequent infection with *B. pahangi,* although it is not clear if any of the induced proteins are directly responsible (Chaithong and Townson, 1990) nor if any are simply products of a wounding response. The genes or portions of genes encoding the *Ae. aegypti* counterparts of attacin, cecropin, and diptericin have been cloned from a genomic library using oligonucleotide probes synthesized on the basis of sequence data from other insects (Knapp and Crampton, 1990). A preliminary sequence has been obtained and subclones have been used for Northern and Southern analyses to determine gene number, organization, and pattern of expression in *Ae. aegypti* strains refractory and susceptible to infection by filarial parasites.

Parasiticidal effects of cecropin and magainin, a peptide derived from vertebrates, were demonstrated against developing oocysts of various *Plasmodium* species in mosquitoes (Gwadz *et al.*, 1989). Cecropin B or magainin 2 was injected intrathoracically before or at various intervals following an infective blood meal, and the effects of both peptides were comparable. *Plasmodium* oocysts appeared insensitive to injections prior to 5 days postinfection. Injection 5 or more days postinfection aborted the development of most oocysts; small numbers of sporozoites were nevertheless detected in the salivary glands. A second injection at 9 days postinfection arrested most oocysts that had escaped the first treatment and prevented sporozoite invasion of the salivary glands. However, the levels of magainin and cecropin required to achieve parasiticidal effects were quite close to the 50% lethal dosages for the mosquito host.

In the majority of insects studied, the population of circulating hemocytes is sufficiently large that melanotic encapsulation typically involves the surface attachment and flattening of hemocytes in addition to melanin deposition (Christensen and Nappi, 1988). In mosquitoes, however, hemocyte populations are low and the melanotic capsules are cell free. Although hemocytes may be indirectly involved in some instances (see below), an ultrastructural study of the encapsulation response of *An. gambiae* to malarial ookinetes in the midgut showed no evidence of hemocyte involvement (Paskewitz *et al.*, 1989). Instead, the physiological basis for refractoriness in this case involved elevated activity of the melanogenic enzyme, phenol oxidase, in the midgut epithelia of refractory but not susceptible mosquitoes (Paskewitz *et al.*, 1989).

The genetic basis of refractoriness of *An. gambiae* to malaria parasites has been investigated. *Plasmodium*-refractory and -susceptible strains of *An. gambiae* were originally selected from a parental laboratory colony, G3, polymorphic for the refractory phenotype (Collins *et al.*, 1986). A fully refractory phenotype, manifested by melanotic encapsulation of ookinetes as they complete penetration of the mosquito midgut epithelium 16–24 hours after an infective blood meal (Paskewitz *et al.*, 1988), was selected for after a short number of inbreeding steps following infection with the simian malaria parasite *Plasmodium cynomolgi*. This same line also proved highly refractory to other simian, rodent, and avian malaria parasites, and to some human malaria parasites. Refractoriness in this line appears to be cooperatively controlled by at least two major, unlinked genetic loci (Vernick *et al.*, 1989; Vernick and Collins, 1989). Nothing is known about one of the loci, but the other is inseparable from two esterase genes (Collins *et al.*, 1991). Lines selected for parasite refractoriness and susceptibility show alternative esterase genotypes at both of these esterase loci, and this association is also observed in the unselected parental G3 colony (Vernick and Collins, 1989; Vernick et al., 1989). Using specific enzyme inhibitors, one esterase was defined as an arylesterase, the other as a cholinesterase (Collins et al., 1991; A. Crews-Oyen and F. Collins, unpublished). One or both of these esterases may directly control refractoriness, because esterases can function as serine hydrolases, which have been implicated as prophenol oxidase activators in the enzyme cascade leading to melanotic encapsulation (Lackie, 1988).

Alternatively, the esterase genes may simply be structurally linked to an undefined locus controlling refractoriness by virtue of their inclusion within a paracentric inversion. This possibility was suggested by the molecular analysis of an uncharacterized An. gambiae esterase gene, cloned from a G3 library using an oligonucleotide probe designed to hybridize to conserved sequences around the active site of esterases (F. Collins, unpublished). Though not proved to be one of the esterases associated with refractoriness, the gene has been mapped in situ to a location just outside the breakpoint of a paracentric inversion that is polymorphic in An. gambiae G3 (V. Kumar and F. Collins, unpublished). Further molecular genetic and biochemical studies will be needed to identify unambiguously the esterase-associated gene for refractoriness and to determine its role in this phenomenon.

Romans et al. (1991) used RFLP analysis to test the hypothesis that the esterase-associated refractoriness gene was Dox, the gene for diphenol oxidase A2. An An. gambiae G3 genomic clone of Dox revealed that the malaria-refractory and -susceptible lines were fixed for different Dox alleles, distinguished by a SalI RFLP. Backcrosses between these lines showed that the esterase-associated refractoriness gene and Dox segregate independently.

Attempts to select for refractoriness to infection by the human malaria parasite P. falciparum have been hindered by the difficulty of achieving consistently high mosquito infection rates in the laboratory (Warren and Collins, 1981; Graves and Curtis, 1982). However, one recent attempt using a Punjab strain of An. stephensi and the P. falciparum isolate Nf 54 achieved high percentages of infection (84–100%), even though numbers of oocysts per female (7–132) were still quite variable (Feldmann and Ponnudurai, 1989). Refractoriness in the selected line, Pb 3-9A, although incomplete, was manifested as the absence of oocysts. This indicates that parasite development is being interrupted at an earlier stage than that observed by Collins et al. (1986). It has been suggested that the parasite refractoriness in this strain of *An. stephensi* could be due to higher levels of the proteolytic enzyme, aminopeptidase, in the midgut lumen or wall (Feldmann *et al.*, 1990).

2. Cellular Responses

When invading foreign particles such as bacteria or protozoa are sufficiently small, they may be phagocytosed or trapped in nodules of aggregated hemocytes (Lackie, 1988). If the foreign body is too big to be phagocytosed, hemocytes may play other roles in the immune response. For example, melanotic encapsulation of innoculated *Dirofilaria immitis* microfilariae (mff) is a hemocyte-mediated event in *Ae. aegypti* (Christensen and Tracy, 1989). Ultrastructural studies have shown the lysis of hemocytes in the vicinity of these parasites prior to melanin formation, which occurs near the lysed cells and appears to settle onto the parasite surface (Christensen and Forton, 1986). Phenol oxidase activity is increased in hemocytes during the encapsulation response (Li *et al.*, 1989, 1992). In addition, an 84-kDa polypeptide preferentially expressed during melanotic encapsulation and wound healing is produced by the hemocytes and secreted into the plasma (Beerntsen and Christensen, 1990). In fact, SDS–PAGE, ^{125}I labeling, and lectin binding techniques have shown a significant overall increase in protein concentration in immune-activated hemocytes, some of which are exported to the cell surface (Spray and Christensen, 1991). One 200-kDa surface polypeptide was uniquely expressed in hemocytes from saline- or mff-inoculated mosquitoes. These events are associated with a twofold to threefold increase in the total hemocyte population (Christensen *et al.*, 1989) and a fivefold increase in the proportion of hemocytes whose surface properties allow them to bind lectin, properties that presumably facilitate their recognition and adhesion responses (Nappi and Christensen, 1986). The experiments so far have yielded important clues about the biochemical mechanisms of some aspects of the encapsulation response. Yet, the identity and role of most of the substrates and enzymes involved in the complex sequence of events leading from recognition of the parasite through melanotic encapsulation remain undefined. Work for the future includes the isolation and characterization of the genes encoding and regulating the expression of these proteins.

In *Ae. aegypti,* the genetic basis of another system of defense against the filarial parasites *Brugia malayi* and *Wuchereria bancrofti* was discovered by crossing experiments performed 30 years ago (Macdonald, 1962; Macdonald and Ramachandran, 1965). A single sex-linked recessive locus, f^n, appeared to control susceptibility. Distinct from the melanotic encapsulation response, development of the mff is somehow aborted in the thoracic tissue of resistant strains after penetration of the midgut. Polypeptides synthesized in the thoracic tissue were compared between strains of *Ae. aegypti* refractory and susceptible to infection by *B. malayi* (Wattam and Christensen, 1992). SDS–PAGE gels of radiolabeled polypeptides synthesized *in vivo* showed seven polypeptides unique to blood-fed refractory mosquitoes. The presence of parasites in the blood meal was not necessary to induce their synthesis. All were present 3–24 hours after blood feeding but were undetectable after 48 hours. Wattam and Christensen suggest that these refractory strain-associated proteins may mediate the genetic variation in susceptibility, although there is no evidence as yet that any of them are encoded by the f^n gene. To confirm this suggestion, the f^n gene will need to be isolated and characterized. Work toward this goal is in progress, using RFLP mapping (D. W. Severson, A. Mori, D. M. Helke, and B. M. Christensen, unpublished).

D. Salivation

Another aspect of mosquito physiology that is relevant to the host/parasite relationship is salivation, because most parasites must pass through the salivary glands prior to transmission (James *et al.*, 1991). However, saliva plays many other roles in both sugar and blood feeding (reviewed by James and Rossignol, 1991). Inside the mosquito, enzymes in the saliva aid digestion; externally, the platelet antiaggregating and vasodilatory activity of the saliva facilitate the acquisition of blood (Ribeiro, 1987). The elucidation of these basic processes at the molecular level will enhance our understanding of the parasite's relationship with the salivary gland and may provide salivary gland-specific regulatory DNA sequences that will be useful in engineering pathogen-refractory mosquito strains (Marinotti and James, 1990a).

Several different genes that encode secreted salivary proteins and that are expressed specifically in the salivary glands have been isolated and characterized from *Ae. aegypti*. The first was given the name *Maltase-like I* (*Mal I*) for the similarity of its putative product to a yeast maltase (James *et al.*, 1989). It was isolated from a cDNA library from non-blood-fed females by differentially screening with total abdominal RNA and total salivary gland RNA. A fragment from one of the isolated clones was used to probe developmental Northern blots to determine stage specificity of expression. A 2.1-kb poly(A)$^+$RNA was detected only from salivary gland RNA of adult males and females. Sequence analysis of the cDNA clone revealed a 1.8-kb open reading frame followed by a consensus polyadenylation signal. Analysis of the 5'-noncoding sequence from a corresponding genomic clone identified a TATA box as well as two regions of dyad symmetry with the potential for hairpin loop formation. These putative regulatory structures are of particular interest for their potential roles in directing strong stage- and tissue-specific expression in the mosquito. The 67-kDa deduced protein has a possible secretion signal at its N terminus and six possible sites for asparagine-linked glycosylation. Marinotti and James (1990b) have found a 68-kDa soluble glycoprotein, secreted when mosquitoes take a sugar meal, which is an α-glucosidase probably involved in sugar digestion. The correspondence between this enzyme and the properties predicted for the *Mal I* gene product suggest that they are identical.

A salivary gland-specific α-amylase has also been isolated and sequenced (G. L. Grossmann and A. A. James, unpublished). Its expression pattern analyzed by Northern analysis and *in situ* hybridization overlap that seen with the α-glucosidase *Mal I* (James *et al.*, 1989). Interestingly, the 18-bp repeats found at the 5' end of the *Mal I* gene are also adjacent to the α-amylase coding region, suggesting that they may be involved in regulation of genes encoding carbohydrate-metabolizing enzymes.

Another gene, *D7,* expressed specifically in *Ae. aegypti* salivary glands, has been isolated from a cDNA library using total salivary gland RNA as a probe (James *et al.*, 1991). Oligonucleotides synthesized based on the sequence from *D7* cDNAs were used to isolate homologous clones from salivary gland cDNA and genomic libraries. A Northern blot prepared from RNA from different stages, tissues, and sexes showed that the *D7* cDNA hybridized strongly to a 1.2-kb poly(A)$^+$ RNA from adult female salivary glands and faintly to a message of the same size from adult males. Message was detected up to 48 hours after blood feeding. Sequence comparison of cDNA and genomic clones revealed a TATA box, an open reading frame interrupted by four introns with consensus splice junctions, and a polyadenylation signal. Sequence variation was detected among different cDNAs and between cDNA and genomic clones. It is not yet clear if these differences stem from alternate alleles or multiple gene copies. Polyclonal antibody produced from a recombinant *D7* gene product recognized a 37-kDa protein in salivary glands and saliva from adult females but not from males, indicating female-specific production of a secreted protein and suggesting its involvement in blood feeding.

The salivary glands are sexually dimorphic in the mosquito, being both larger and more differentiated morphologically in the female. Although the salivary glands are three-lobed in both sexes, the distal lateral and medial lobes of the female appear to be sex specific and composed of much larger secretory cells than those found in the female proximal lateral lobes or in any of the male lobes (James et al., 1989). The structural dimorphism coincides both with dietary differences between males and females, as adults of both sexes feed on nectar whereas only the female takes a blood meal, as well as with spatially segregated gene expression of salivary enzymes (Marinotti et al., 1990; Grossman and James, 1990). To explore salivary gland expression more closely, antisense RNAs have been used to probe tissue sections from male and female salivary glands and identify those cells expressing individual genes. *Mal I* is one gene whose product is thought to be involved in sugar digestion. Northern analysis detected abundant *Mal I* gene expression in adult salivary glands of both sexes of *Ae. aegypti*. By *in situ* hybridization with antisense RNA probes, expression occurs throughout the male salivary gland, but is restricted to the proximal lateral lobes in the female (James et al., 1989). Similar *in situ* hybridization patterns were seen in female salivary glands using probes for the α-amylase, another gene presumably involved in sugar feeding (G. L. Grossman and A. A. James, unpublished). On the other hand, the *D7* gene product is not even detected in males and probably functions in blood-feeding females. By Northern analysis, *D7* is preferentially expressed in adult female salivary glands and is barely detected in the male. By *in situ* hybridization, *D7* expression is barely detected in the three lobes of the male salivary gland and is detected only in the distal lateral and medial lobes of the female salivary glands (James et al., 1991). Furthermore, an apyrase secreted into the saliva from salivary glands of female *Ae. aegypti* and diverse anophelines prevents platelet aggregation and thus aids in blood location and ingestion (Ribeiro, 1987). This enzyme is restricted to the distal lateral and medial lobes of the female salivary glands (Ribeiro et al., 1984); another bacteriolytic enzyme from both male and female salivary glands is restricted in females to the proximal lateral lobes (Rossignol and Lueders, 1986). These observations suggest that the proximal lateral lobes serve functions common to both female and male sugar feeding whereas the distal lateral and medial lobes of the female salivary glands provide factors unique to the blood-feeding habit of the female (James *et al.*, 1991) and may have evolved together with this capability.

V. Genetic Transformation

Genetic transformation is broadly defined as the uptake and expression of exogenous DNA by cells in culture or in whole organisms. Gene transfer may be transient, reflecting the episomal state of the introduced DNA, or stable following the integration of DNA into the chromosome. Stable integration into embryonic germ line cells can result in a transformed lineage. It is a powerful tool that has allowed the identification and isolation of genes and regulatory sequences as well as a detailed analysis of gene expression, development, and behavior that would be difficult if not impossible to study with classical genetic techniques. Applied to insect vectors of disease by targeting key stages of parasite development in the host (Warburg and Miller, 1991), it could represent the first step in the engineering and ultimate population replacement of a vector by a nonvector strain. However, among insects, *D. melanogaster* is the only species for which efficient methods of stable germ line gene transfer are currently available (reviewed in Ashburner, 1989). This is due to the availability of efficient *Drosophila*-specific transformation vectors, such as those based on the P and hobo transposable elements, as well as a variety of cloned selectable markers and promoter sequences. Initial optimism that P element vectors would function without modification in mosquitoes has faded, and new efforts are focused on a number of alternative approaches. These include efforts to identify and isolate endogenous mobile elements from mosquitoes as well as detailed research on the basis for P and hobo element mobility, with the expectation that by supplying required *Drosophila*-specific cis- or trans-acting factors, these elements could be mobilized in organisms other than *Drosophila*. Another important approach being investigated in mosquitoes is the potential for using the FRT–FLP site-specific recombination system from the 2μm plasmid of the yeast *Saccharomyces cerevisiae* (Meyer-Leon *et al.*, 1984) as a directed recombination system. The system is particularly attractive because this site-specific recombination, which requires only the appropriate FRT recombination site and a source of the FLP recombinase, has been shown to work efficiently in a number of different organisms, including *D. melanogaster* (Golic and Lindquist, 1989) and mammalian cells (O'Gorman *et al.*, 1991). In addition to the basic transformation vectors and recombination tools, a great deal of effort is also being directed toward the isolation of mosquito-specific promoter and enhancer sequences, selectable marker genes, and mutant mosquito strains that will be required to develop efficient genetic transformation systems for mosquitoes. This process builds on previous genetic research with both mosquitoes and *D. melanogaster* and broadens the understanding of not just mosquito molecular biology but of insect molecular genetics.

A. DNA Delivery

Genetic transformation depends on the successful solution to four separate but interdependent problems: (1) the delivery of DNA into the cell or embryo, (2) the efficient and stable integration of DNA into the chromosome, (3) a suitable marker gene that encodes a dominant phenotypic trait, and (4) a promoter to control expression of the gene. The first problem, how to deliver the exogenous DNA, has been solved for both cultured mosquito cells and embryos. In fact, many methods exist for transfecting mosquito cells, including the use of calcium phosphate, polybrene, lipofection, and electroporation (for review, see Lycett, 1990). With any given method, transfection efficiencies vary from one cell line to another and need to be optimized empirically. Optimal conditions for transfecting the C7-10 clone of *Ae. albopictus* cells using polybrene (Fallon, 1989; Gerenday *et al.*, 1989) and calcium phosphate (Kjer and Fallon, 1991) have been defined. Efficiencies range from 1 in 10,000 cells for polybrene-mediated uptake to as high as 1 in 1000 for the calcium phosphate procedure (Kjer and Fallon, 1991). In both cases only transient expression was tested. Stable transformants of the C6/36 clone of *Ae. albopictus* cells were obtained following the lipofection procedure (Monroe *et al.*, 1990; T. J. Monroe, J. O. Carlson, K. E. Olson, D. L. Clements, and B. J. Beaty, unpublished), and of *Ae. aegypti* MOS20A cells using polybrene with glycerol shock, electroporation, and lipofection (Lycett, 1990; G. J. Lycett, P. Eggleston, and J. M. Crampton, unpublished). Studies in MOS20A cells demonstrated stable transformation at efficies as high as 1 per 500 cells following electroporation (Lycett, 1990).

Three independent laboratories have similarly overcome the more difficult problem of introducing DNA into embryos of *An. gambiae* (Miller *et al.*, 1987), *Ae. triseriatus* (McGrane *et al.*, 1988), and *Ae. aegypti* (Morris *et al.*, 1989) using microinjection. Microinjection of *D. melanogaster* embryos is facilitated by the ability to remove the rigid chorion. Because removal of the mosquito chorion results in the death of the embryo, embryos were injected shortly after oviposition, before the chorion fully hardened, with glass needles whose tips were 3–5 μm in diameter. The embryos had to be slightly desiccated to prevent leakage of the innoculum, but because excess dehydration was lethal, they were injected under water-saturated halocarbon or paraffin oil. Even with these precautions, the survival of injected embryos through adulthood averaged only 14% in *An. gambiae,* 15% in *Ae. aegypti,* and 6% in *Ae. triseriatus.*

Alternative methods for mass injection of insect embryos are being tested. One such method, initially developed for use in transforming plant cells, involves the use of a particle gun to shoot DNA-coated microparticles into the egg. This technique proved fatal to embryos of *An. albimanus* and does not appear promising, due to the small size yet thick eggshell of the mosquito (Carlson and Cockburn, 1989). However, the obvious labor intensity and poor efficiency of individual microinjections indicate the value of continued effort to develop more efficient methods for introducing DNA into mosquito eggs.

B. DNA INTEGRATION

Although random integration of circular or linearized recombinant plasmids into chromosomes is successfully used to transform many diverse cell lines, it is a process that is poorly understood and impossible to control (Handler and O'Brochta, 1991). For example, one common occurrence is the integration of head-to-tail concatemers of multiple copies (Gordon and Ruddle, 1985). Mobile element vectors or site-specific recombination systems improve not only the efficiency and consistency of integration, but also allow some control over the manner in which, or the sites at which, DNA integrates.

For germ line transformation of *D. melanogaster,* vectors based on the P element transposon (reviewed in Engels, 1989) are routinely used. P elements encode a transposase that acts on 31-bp inverted repeats at their termini to effect excision and transposition. Correct processing of transposase message, hence transposition, is limited to the germ line in wild-type P elements. P elements have been engineered such that the transposase coding sequence has been replaced by a multiple cloning site and a promoter controlling a selectable marker gene. If coinjected with an immobilized "helper" plasmid providing preprocessed transposase in trans, the transformation vector can transpose and integrate in somatic and germ cells. In *D. melanogaster,* at least one chromosomal integration event is recovered from 100–200 injected eggs (Handler and O'Brochta, 1991).

The fact that the P element vector has been mobilized in all drosophilids tested (O'Brochta and Handler, 1988) has encouraged efforts to transform mosquitoes through P-mediated integration. In each case, foreign DNA was successfully introduced into the germ line and some integration events were stable over multiple generations (Miller *et al.*, 1987; McGrane *et al.*, 1988; Morris *et al.*, 1989). In all cases, however, the efficiency of integration was much lower than expected for a highly efficient vector such as P. For example, of almost 2300 injections, one integration event was recovered from *An. gambiae* (Miller *et al.*, 1987), a rate at least 10-fold lower than expected from *Drosophila* even though survival rates following injection were comparable. Similarly low rates of integration were obtained from *Ae. triseriatus* (McGrane *et al.*, 1988) and *Ae. aegypti* (Morris *et al.*, 1989). Southern analysis indicated that these integration events were not P element mediated. Other transformation experiments using P element constructs and *Ae. albopictus* C6/36 cells or *Ae. aegypti* MOS20A cells also resulted in plasmid integrations independent of P mechanisms, some of which involved large tandem arrays (T. J. Monroe, J. O. Carlson, and B. J. Beaty, unpublished; Lycett, 1990). It is unclear at present what factor(s) block P element mobility in nondrosophilids. As more is learned about P transposase function in heterologous systems, it may be possible to overcome these obstacles. However, a more profitable approach might be to exploit a transposable element endogenous to the mosquito genome rather than to manipulate one adapted to a different organism. Moreover, other classes of transposable elements might be worth exploring. For example, the retrotransposon *copia* from *Drosophila* was deemed unsuitable for use as a transformation vector because of low rates of mobility, but there is precedent for the use of this kind of element for genome manipulation in yeast (Jacobs *et al.*, 1988). A nonviral retrotransposable element from *Drosophila*, the I element, causes hybrid dysgenesis reminiscent of the P–M system, and this type of element has been found in a broad spectrum of eukaryotic organisms.

Any type of transposable element that integrates at high efficiency might make a good candidate for a transformation vector. In *Drosophila*, these elements reside in the middle repetitive component of the genome. Thus, in seeking out endogenous elements from the mosquito, one approach has been the screening of genomic libraries with radiolabeled total genomic DNA to identify repetitive sequences. This approach was used to isolate T1, a non-LTR retrotransposon from *An. gambiae* (Besansky, 1990a). T1 belongs to the LINE-like class of mobile elements that are ubiquitous in eukaryotic genomes, as examples have been found in insects, plants, mammals, frogs, and protozoans (see Xiong and Eickbush, 1990). They transpose through an RNA intermediate based on their putative reverse transcriptase-encoding sequences, frequent 5′-end truncation, and poly(A)-like tails. However, they are distinguished from retroviral-like elements by the absence of long terminal repeats. T1 is dispersed in about 100 copies in the *An. gambiae* genome and is also found in the genomes of other members of the *An. gambiae* complex, yet is absent from very close relatives belonging to the same series (*Pyretophorus*)—*Anopheles subpictus* and *Anopheles vagus* (N. Besansky, unpublished). T1 is dispersed on all chromosome arms with no apparent target site preference. A high rate of mobility has been inferred both from Southern blot analyses of wild populations of *An. gambiae* from Kenya and from *in situ* hybridizations to polytene chromosomes of *An. gambiae* laboratory strains (Besansky, 1990a; V. Kumar and F. Collins, unpublished).

In the course of characterizing rDNA repeat units from members of the *An. gambiae* complex, two other retrotransposons from *An. gambiae* and *An. arabiensis,* called RT1 and RT2, respectively, were discovered (Paskewitz and Collins, 1989; Besansky *et al.*, 1992). The genomic distribution of these two elements, which do not cross-hybridize on genomic blots, is restricted to the same 17-bp target sequence in a fraction of 28S rDNA units. This is a feature that could be exploited, given an understanding about how the target site is recognized by the integrase. Unfortunately, this information is currently lacking, even in the better known R1 and R2 elements of *Bombyx mori*. The mobility properties of T1, RT1, and RT2 are currently being analyzed to determine the feasibility of using these elements as transformation vectors.

Two different approaches for isolating mobile elements have been taken with *Ae. aegypti* (Crampton *et al.*, 1990a). In the first approach, which was based on the assumption that retrovirus-like transposable elements would be represented as circular DNA molecules in the extrachromosomal fraction of the genome, a plasmid library was generated from extrachromosomal DNA isolated from the MOS20 cell line. When used to probe total genomic Southern blots, one clone, pX16, displayed an intraspecific polymorphism in hybridization pattern, suggesting variation in both chromosomal location and copy number typical of mobile elements. The second approach used oligonucleotide primers corresponding to regions conserved among reverse transcriptases, YXDDML and TAFLHG, in PCR amplifications of *Ae. aegypti* larval and cell line DNA as well as *An. gambiae* DNA (A. Warren and J. M. Crampton, unpublished). Partial sequence data from pX16 and from three cloned PCR products revealed the presence of YXDD boxes in each case, although the tyrosine (Y) was replaced by a valine (V) in pX16 and the second aspartic acid (D) with an isoleucine (I) in one PCR product. Sequence data from the termini are needed to determine whether these elements fall within the retrovirus-like classes of retrotransposons. However, the presence of a methionine (M) rather than an alanine (A) at the X position suggests that they are all retroviral-like (Xiong and Eickbush, 1988).

Though strategies for directly isolating and analyzing mosquito elements have clearly been fruitful, it is uncertain whether an element identified by such approaches will prove useful as a transformation vector. Alternative approaches, such as the search for and investigation of hybrid dysgenesis-like phenomena or unstable genetic mutations in mosquitoes, should also be undertaken. Preliminary searches for hybrid dysgenesis-like phenomena have been carried out with *Anopheles* species of two different species complexes, the *An. quadrimaculatus* complex (J. A. Seawright, personal communication) and the *An. gambiae* complex (F. H. Collins and V. Finnerty, personal communication). These efforts have not yet identified any biological phenomena that appear to indicate actively transposing mobile elements.

Preliminary experiments with the yeast FLP–FRT site-specific recombination system in the mosquito *Ae. aegypti* have been very promising (Morris *et al.*, 1991). In a series of elegant experiments involving the coinjection of mosquito embryos with target plasmids containing FRT recombination sites and plasmids containing the FLP recombinase gene under the control of the *Drosophila hsp70* promoter, efficient FLP-mediated recombination was observed between tandem FRT sites within a plasmid and between FRT sites on separate plasmids. Synthetic FRT sites tested were functionally equivalent to target sites isolated from the yeast 2μm plasmid. These experiments encourage continued investigation of this and perhaps other site-specific recombinase-based systems as methods for genetically transforming mosquitoes.

C. Promoters

The detection of transformed cells or embryos is critically dependent on a promoter capable of expressing the introduced gene in cooperation with the transcriptional apparatus of the host. Because of the evolutionary conservation of the heat-shock response and heat-shock regulatory sequences, the promoter (*hsp70*) of the *D. melanogaster* 70-kDa heat-shock protein has been used successfully to control expression of genes in cell types as diverse as yeast and mouse (Gerenday *et al.*,

1989, and references therein). The *hsp70* promoter functions effectively in transiently transfected *Ae. albopictus* C7-10 cells using conditions defined for *Drosophila* (Durbin and Fallon, 1985), in contrast to the promoter from the *Drosophila* transposon *copia,* which does not (Swerdel and Fallon, 1987). By characterizing the endogenous heat-shock response based on maximal expression of the *Aedes* homologue *hsp66,* Gerenday *et al.* (1989) discovered that an incubation at 41°C rather than 37°C improved transient expression from the *hsp70* promoter by 10-fold in C7-10 cells. The heat-shock response was also maximized in *Ae. aegypti* Aag-2 cells (Lan and Fallon, 1990) and MOS20 cells (G. J. Lycett, P. Eggleston, and J. M. Crampton, unpublished) at 41°C. This improvement translates to anophelines, because it has recently been shown that almost no mortality was observed in *An. albimanus* larvae at 40°C (Benedict *et al.,* 1991) and that transcript accumulation from the mosquito *hsp70* promoter is undetectable at 37°C, but is high at 40°C (M. Q. Benedict, A. F. Cockburn, and J. A. Seawright, unpublished). Indeed, survival following G418 selection of transgenic *An. gambiae* containing a neomycin-resistance gene under *Drosophila hsp70* control was increased by heat shock at 41°C compared with 37°C (Sakai and Miller, 1992).

During transient transformation, non-heat-shocked cells demonstrated a relatively low background level of gene expression from the *Drosophila hsp70* promoter when compared to that of heat-shocked cells (e.g., Durbin and Fallon, 1985). However, in stably transformed C6/36 cells of *Ae. albopictus,* there was no significant difference in transcription from this promoter following heat shock at 37°C, regardless of whether it was driving selected or unselected genes (Monroe *et al.,* 1990; T. J. Monroe, J. O. Carlson, K. E. Olson, D. L. Clemens, and B. J. Beaty, unpublished). Although heat shock of mammalian (BHK-21) cells at 42°C failed to induce the *Drosophila hsp70* promoter, the inducibility of this promoter in mosquito cells was not tested at this temperature. It is unclear whether constitutive expression resulted from the nature of the integration event or whether stress incurred by handling of the cells inadvertently induced a heat-shock response. The level of expression of antibiotic-resistance genes from the *Drosophila hsp70* promoter was sufficient to allow selection of stable transformants of C6/36 cells, in spite of the lack of inducibility (T. J. Monroe, J. O. Carlson, K. E. Olson, D. L. Clemens, and B. J. Beaty, unpublished). However, following heat shock at 37°C, the level of expression of the *Drosophila hsp70*-regulated genes in both cells and transformed embryos has been insufficient to prevent the production of false positives (T. J. Monroe, J. O. Carlson, K. E. Olson, D. L. Clemens, and B. J.

Beaty, unpublished; Miller *et al.*, 1987; McGrane *et al.*, 1988; Morris *et al.*, 1989).

The application of species-specific promoters could provide the benefits of increased levels and control of gene expression. These will soon become available, because numerous groups have cloned a variety of both constitutively expressed and tissue, sex, and stage-specific genes from the mosquitoes *An. gambiae* and *Ae. aegypti,* including those for actin (C. Salazar, D. Mills Hamm, C. B. Beard, and F. Collins, unpublished), vitellogenin (Gemmill *et al.*, 1986; Romans, 1990), a glucosidase, α-amylase, and the female-enriched protein D7 from salivary glands (James *et al.*, 1989, 1991, G. L. Grossman and A. A. James, unpublished), and the *hsp70* and *hsp82* genes of *An. albimanus* (M. Q. Benedict, A. F. Cockburn, and J. A. Seawright, unpublished). These *An. albimanus* heat-shock genes were cloned from a genomic DNA library using a probe containing the N-terminal coding region of the *Drosophila* homologues. Benedict *et al.* found two contiguous *hsp70* genes at two loci on chromosome 2R, and two *hsp82* genes at one locus on chromosome 3L. Each locus contains a pair of genes whose coding regions are divergently oriented. The heat-shock regulatory elements in the promoter regions were identified by sequence comparisons to previously defined elements from other organisms. These putative heat-shock elements from the *An. albimanus* promoters are more numerous and more similar to the consensus sequence than are those from *Drosophila.*

D. REPORTER GENES AND SELECTABLE MARKERS

Transformants may be screened individually, by the expression of an unselected reporter gene that confers a visible phenotype. Such genes are useful for estimating transformation efficiency and also allow the characterization of promoter sequences. Examples of reporter genes include eye color genes such as *rosy* from *Drosophila* or genes from bacteria that encode detectable enzymes such as β-galactosidase (β-gal) or chloramphenicol acetyltransferase (CAT). The genes for both β-gal and CAT have been correctly expressed in cultured mosquito cells (Durbin and Fallon, 1985; Lycett, 1990). CAT activity has been detected in all stages of the mosquito *Ae. aegypti* (embryos, larvae, pupae, and adults) following injection of *hsp70* promoter–CAT gene constructs into embryos, thus CAT is an excellent reporter for hybrid genes in this mosquito (A. Morris and A. James, unpublished). Although β-gal activity has been detected throughout adult *Ae. aegypti* following microinjection of embryos with a *hsp70* promoter–β-gal gene construct, relatively high background activity might obscure analysis of some tissue-specific promoters with this reporter gene (O. Marrinotti and A. James, unpublished).

Transformants may also be mass selected, by the use of a selectable marker gene that confers resistance to various chemical treatments, such as antibiotics. Two widely used selectable markers derived from bacteria, the neomycin phosphotransferase (*neo*) and hygromycin phosphotransferase (*hyg*) genes, confer resistance to the antibiotics neomycin (or its analog G418) and hygromycin, respectively. Like the genes for CAT and β-gal, the *neo* and *hyg* genes have also been expressed in cultured mosquito cells (Lycett, 1990; T. J. Monroe, J. O. Carlson, K. E. Olson, D. L. Clemens, and B. J. Beaty, unpublished). The usefulness of the *hyg* gene in whole animals is unknown (Handler and O'Brochta, 1991), but the *neo* gene has been expressed in the larval stages of *Aedes* and *Anopheles* mosquitoes (Miller *et al.*, 1987; McGrane *et al.*, 1988; Morris *et al.*, 1989). Unfortunately, these mosquito species differed dramatically in their sensitivity to the antibiotic, so that effective concentrations had to be carefully established in each case. Moreover, every transformation experiment with the *neo* gene reported to date has been plagued with leaky selection in which cells or mosquitoes apparently resistant to the antibiotic were not actually transformed. (In these experiments, heat shock of the *Drosophila hsp70* promoter driving *neo* was at 37°C.) Although problems associated with the low activity of the *neo* gene product might be partially overcome with the use of a higher temperature to heat shock, a stronger promoter, or a different *neo* gene, the need for alternate selection procedures based on cloned mosquito genes is evident.

A variety of morphological and biochemical mutant strains of mosquitoes are available for transformation, but, with the exception of a few insecticide-resistance genes, the corresponding genes have yet to be cloned or even mapped in many instances. Clearly much more effort needs to be aimed in the direction of cloning dominant gain-of-function genes.

E. Target Genes and Field Application

Overcoming the obstacles to genetic transformation of the mosquito would provide a powerful tool to study of many aspects of basic mosquito biology. However, as part of a program to control mosquito-borne diseases, genetic transformation is only one of a series of problems requiring solutions, including the identification of relevant target genes (not necessarily from the mosquito genome) and the development of methods to drive the genetic constructs into natural mosquito populations.

Putative mosquito counterparts of genes involved in developmental control and circadian rhythms have been isolated from *Ae. aegypti* (Crampton *et al.*, 1990c; Eggleston, 1990a). These clones were identified by hybridization to synthetic oligonucleotide probes based on conserved sequences of other organisms. The nine different developmental control genes being analyzed hybridized to the *Antennapedia*-like homeobox sequence from *D. melanogaster* (Crampton *et al.*, 1990c). The characterization of these genes and of their expression is valuable, but their fundamental importance to the fitness of the organism implies that any attempts to introduce defective copies into natural populations call for a strongly counterbalancing driving force. A prevalent opinion, shared by the authors of this review, is that new genetic control strategies based on the premise of population reduction or elimination are as unlikely to succeed in the long run as earlier control attempts involving the release of sterilized male mosquitoes or mosquitoes genetically altered to be relatively infertile when mated with the wild population.

More promising in theory is the strategy of replacing vector populations with those of lesser or no capability of transmitting pathogens. Although this could in principle involve factors that modify the overall dynamic interaction between the mosquito and the human population (such as host-seeking behavior), genetic factors that cause minimal changes in the overall ecology of the vector species are likely to be the least problematic. The most promising genetic modification would appear to be one that affects only the ability of the mosquito population to support development of the pathogen it transmits. In some epidemiological settings, even a very small modification in the vectorial capacity of the natural population may be sufficient to achieve complete elimination of the pathogen (Coluzzi, 1992).

As was mentioned in Section IV,C, various groups are involved in trying to understand the basis for vectorial competence with respect to malaria parasites in anophelines and to filarial worms in aedine mosquitoes, and to map and clone the genes involved. Efforts have also been initiated to characterize *Aedes* clones homologous to attacins, cecropins, and diptericins of other insects. Much of the work on genes with tissue-specific expression, particularly tissues such as salivary glands, where arboviruses undergo cycles of intracellular replication, is motivated by the possibility of a genetic control strategy based on an arbovirus-specific antisense RNA produced in the appropriate tissue under the control of a tissue-specific mosquito promoter. Unfortunately, not much is known, especially at the molecular level, in any of these areas and much more basic research is necessary before any of these approaches can be implemented.

The success of any genetic control strategy will ultimately depend on the availability of a mechanism that favors the spread of the introduced genes through natural populations, such as meiotic drive, reduced heterozygote fitness, parasite-mediated cytoplasmic incompatibility, or replicative transposition (see Curtis and Graves, 1988; Crampton *et al.*, 1990b; Eggleston, 1990b). The spread of an intracellular, *Wolbachia*-like parasite causing a unidirectional mating incompatibility has been documented in *D. simulans* (Turelli and Hoffmann, 1991). Because the incompatibility favors infected females, the infection spread rapidly through populations. Desirable traits, such as those controlling vector competence, could potentially be introduced into infected females, who would then drive the trait to fixation along with the infection. It is worth noting that *Wolbachia*-like parasites have been found in many different insects, including various *Culex* species (see Kitzmiller, 1976). Transposable elements, another potential driving mechanism, also behave like intracellular parasites and, given a high enough rate of transposition, can theoretically spread even an unfavorable trait through a population before natural selection has a chance to act on it, in spite of the sometimes deleterious consequences of transposition (J. Ribiero and M. Kidwell, personal communication). Furthermore, transposable elements have shown themselves capable of transfer across species boundaries (Houck *et al.*, 1991; Mizrokhi and Mazo, 1990), a feature that might be exploited in species complexes with multiple, sympatric vector species, such as the *An. gambiae* complex. That transposable elements are capable of rapidly spreading through natural populations has already been demonstrated by the P element, which invaded populations of *D. melanogaster* worldwide within the past 50 years, probably from the distant relative *Drosophila willistoni* (Anxolabhere *et al.*, 1988). These features have fueled the search for mosquito transposable elements that function like the P element, and they also point out the potential importance of identifying elements not only from the vector species but also from related mosquito species that may be uninvolved in pathogen transmission. Though the scenario for using an engineered element introduced across a species boundary to transform effectively an entire wild population, or even species, clearly falls in the realm of scientific speculation, the potential usefulness of transposable element vectors to the study of basic mosquito biology cannot be overstated.

VI. Concluding Remarks

Most molecular research on mosquitoes to date has been undertaken because of the importance of mosquitoes as vectors of human disease. Indeed, the basic vision that underlies much of the molecular work with members of this insect family is the prospect for genetic control. Nevertheless, the long-range significance of molecular research on mosquitoes will clearly extend far beyond this more narrowly defined objective. Several obvious areas of potential significance to basic research questions are already apparent.

The most important immediate contribution will probably be in the areas of molecular systematics and evolutionary biology. In part, this will be due to the obvious applied importance of this knowledge as a prelude to any type of mosquito control effort, including genetic control strategies. Clearly, a great deal of financial and intellectual effort will be directed toward understanding important mosquito vectors such as *An. gambiae* and *Ae. aegypti*, and much of this work will focus on population genetic and evolutionary processes. Such evolutionary work with mosquitoes, particularly studies involving differentiation of populations and the analysis of closely related cryptic species complexes, should be of importance to evolutionary biologists as well as vector biologists.

It is likely that significant basic findings will also emerge from the present concerted effort to develop the tools for transgenic technology, such as transformation vectors, genome maps, and detailed molecular knowledge of specific physiological systems. The mosquitoes *An. gambiae* and *Ae. aegypti*, and to a lesser extent *C. pipiens*, currently stand out as important model organisms for the group as a whole. Though other species will no doubt also be important, either because of their involvement with specific disease systems or their possession of unique biological characteristics, the community of scientists involved with mosquito research will need to develop a consensus on the importance of a small number of model species such as the above three. The physical resources and effort required to maintain mosquito colonies are considerable, especially for *Anopheles* and *Culex* species, and these requirements in combination with possible institutional restrictions on the importation or maintainance of tropical mosquitoes may require an unusual degree of collaboration among research laboratories engaged in mosquito research.

A final point we wish to emphasize is the extraordinarily broad scope of past and present work with mosquitoes, work that has ranged from field ecology through physiology and now to molecular biology. Clearly, molecular genetic analysis of mosquitoes is in its infancy relative to many other eukaryotic organisms, but members of the family Culicidae have long been the subject of detailed field and laboratory study. This rich biological background knowledge will not only prove valuable in current genetic control efforts but may also serve to strengthen the potential importance of mosquitoes as basic biological models.

REFERENCES

Agarwal, M. L., Mouches, C., Beyssat, V., Raymond, M., Magnin, M., Pasteur, N., and Georghiou, G. P. (1986). A specific DNA hybridization assay for detection of a gene responsible for insecticide resistance in mosquitoes. *In* "Contemporary Themes in Biochemistry" (O. L. Kon *et al.*, eds.), Vol. 6, pp. 508–509. Cambridge Univ. Press, New York.

Ananiev, E. V., Gvozdev, V. A., Ilyin, Y. V., Tchurikov, N. A., and Georgiev, G. P. (1978). Reiterated genes with varying location in intercalary heterochromatin of *Drosophila melanogaster* polytene chromosomes. *Chromosoma* **70**, 1–17.

Anxolabehere, D., Kidwell, M. G., and Periquet, G. (1988). Molecular characteristics of diverse populations are consistent with the hypothesis of a recent invasion of *Drosophila melanogaster* by mobile P elements. *Mol. Biol. Evol.* **5**, 252–269.

Ashburner, M. (1980). Some aspects of the structure and function of the polytene chromosomes of the Diptera. *In* "Insect Cytogenetics" (R. L. Blackman, G. M. Hewitt, and M. Ashburner, eds.), Vol. 10, pp. 65–84. Blackwell, Oxford.

Ashburner, M. (1989). "*Drosophila*: A Laboratory Handbook." Cold Spring Harbor Lab. Press, Cold Spring Harbor, New York.

Ballinger-Crabtree, M. E., Black, W. C. IV, and Miller, B. R. (1992). Use of genetic polymorphisms detected by RAPD–PCR for differentiation and identification of *Aedes aegypti* subspecies and populations. *Am. J. Trop. Med. Hyg.*, in press.

Barillas-Mury, C., Graf, R., Hagedorn, H. H., and Wells, M. A. (1991). cDNA and deduced amino acid sequence of a blood meal-induced trypsin from the mosquito, *Aedes aegypti*. *Insect Biochem.* **21**, 825–831.

Bates, M. (1949). "The Natural History of Mosquitoes." Macmillan, New York.

Beach, R. F., Mills, D., and Collins, F. H. (1989). Structure of ribosomal DNA in *Anopheles albimanus* (Diptera: Culicidae). *Ann. Entomol. Soc. Am.* **82**, 641–648.

Beckingham, K. (1982). Insect rDNA. *In* "The Cell Nucleus" (H. Busch and L. Rothblum, eds.), pp. 205–269. Academic Press, New York.

Beerntsen, B. T., and Christensen, B. M. (1990). *Dirofilaria immitis*: Effect on hemolymph polypeptide synthesis in *Aedes aegypti* during melanotic encapsulation reactions against microfilariae. *Exp. Parasitol.* **71**, 406–414.

Benedict, M. Q., Cockburn, A. F., and Seawright, J. A. (1991). Heat-shock mortality and induced thermotolerance in larvae of the mosquito *Anopheles albimanus*. *J. Am. Mosq. Control Assoc.* **7**, 547–550.

Besansky, N. J. (1990a). A retrotransposable element from the mosquito *Anopheles gambiae*. *Mol. Cell. Biol.* **10**, 863–871.

Besansky, N. J. (1990b). Evolution of the T1 retroposon family in the *Anopheles gambiae* complex. *Mol. Biol. Evol.* **7**, 229–246.

Besansky, N. J., and Powell, J. R. (1992). Reassociation kinetics of *Anopheles gambiae* (Diptera: Culicidae) DNA. *J. Med. Entomol.* **29**, 125–128.

Besansky, N. J., Paskewitz, S. M., Mills Hamm, D., and Collins, F. H. (1992). Distinct families of site-specific retrotransposons occupy identical positions in the rRNA genes of *Anopheles gambiae. Mol. Cell. Biol.*, in press.

Beyssat-Arnaouty, V., Mouches, C., Georghiou, G. P., and Pasteur, N. (1989). Detection of organophosphate detoxifying esterases by dot-blot immunoassay in *Culex* mosquitoes. *J. Am. Mosq. Control Assoc.* **5**, 196–200.

Black, W. C., IV, and Rai, K. S. (1988). Genome evolution in mosquitoes: Intraspecific and interspecific variation in repetitive DNA amounts and organization. *Genet. Res.* **51**, 185–196.

Black, W. C., IV, McLain, D. K., and Rai, K. S. (1989). Patterns of variation in the rDNA cistron within and among world populations of a mosquito, *Aedes albopictus* (Skuse). *Genetics* **121**, 539–550.

Boman, H. G., and Hultmark, D. (1987). Cell-free immunity in insects. *Annu. Rev. Microbiol.* **41**, 103–126.

Bonet, R. G., Crampton, J., and Townson, H. (1990). Mosquito P-450 genes and pyrethroid insecticide resistance. *In* "Molecular Insect Science" (H. H. Hagedorn, J. G. Hildebrand, M. G. Kidwell, and J. H. Law, eds.), p. 280. Plenum, New York.

Booth, D. R., Mahon, R. J., and Sriprakash, K. S. (1991). DNA probes to identify members of the *Anopheles farauti* complex. *Med. Vet. Entomol.* **5**, 447–454.

Bouchard, R. A. (1982). Moderately repetitive DNA in evolution. *Int. Rev. Cytol.* **76**, 113–193.

Briegel, H., and Lea, A. (1975). Relationship between protein and proteolytic activity in the midgut of mosquitoes. *J. Insect Physiol.* **21**, 1597–1604.

Britten, R. J., and Kohne, D. E. (1968). Repeated sequences in DNA. *Science* **161**, 529–540.

Brown, A. W. A. (1967). Genetics of insecticide resistance in insect vectors. *In* "Genetics of Insect Vectors of Disease" (J. W. Wright and R. Pal, eds.), pp. 505–552. Elsevier, New York.

Carlson, D. A., and Cockburn, A. F. (1989). Advances in insertion of material into insect eggs via a particle shotgun technique. *In* "Host Regulated Developmental Mechanisms in Vector Arthropods" (D. Borovsky and A. Spielman, eds.), pp. 248–252. Univ. of Florida Press, Vero Beach.

Chaithong, U., and Townson, H. (1990). Immune responses of mosquitoes to filariae and bacteria. *In* "Molecular Insect Science" (H. H. Hagedorn, J. G. Hildebrand, M. G. Kidwell, and J. H. Law, eds.), p. 288. Plenum, New York.

Cho, W.-L., and Raikhel, A. S. (1990). Isolation and characterization of a lysosomal enzyme, cathepsin D, from the mosquito *Aedes aegypti. In* "Molecular Insect Science" (H. H. Hagedorn, J. G. Hildebrand, M. G. Kidwell, and J. H. Law, eds.), p. 290. Plenum, New York.

Christensen, B. M., and Forton, K. F. (1986). Hemocyte-mediated melanization of microfilariae in *Aedes aegypti. J. Parasitol.* **72**, 220–225.

Christensen, B. M., and Nappi, A. J. (1988). Immune responses of arthropods. *ISI Atlas Sci.: Anim. Plant Sci.* pp. 15–19.

Christensen, B. M., and Tracy, J. W. (1989). Arthropod-transmitted parasites: Mechanisms of immune interaction. *Am. Zool.* **29**, 387–398.

Christensen, B. M., Huff, B. M., Miranpuri, G. S., Harris, K. L., and Christensen, L. A. (1989). Hemocyte population changes during the immune response of *Aedes aegypti* to inoculated microfilariae of *Dirofilaria immitis. J. Parasitol.* **75**, 119–123.

Cockburn, A. F. (1990). A simple and rapid technique for identification of large numbers of individual mosquitoes using DNA hybridization. *Arch. Insect Biochem. Physiol.* **14**, 191–199.

Cockburn, A. F., and Mitchell, S. E. (1989). Repetitive DNA interspersion patterns in Diptera. *Arch. Insect Biochem. Physiol.* **10**, 105–113.

Cockburn, A. F., Mitchell, S. E., and Seawright, J. A. (1990). Cloning of the mitochondrial genome of *Anopheles quadrimaculatus*. *Arch. Insect Biochem. Physiol.* **14,** 31–36.

Collins, F. H., Paskewitz, S. M., and Crews-Oyen, A. E. (1991). A genetic study of *Plasmodium* susceptibility in the African malaria vector *Anopheles gambiae. Ann. Belg. Soc. Trop. Med. Hyg.* **71,** 225–232.

Collins, F. H., Sakai, R. K., Vernick, K. D., Paskewitz, S., Seeley, D. C., Miller, L. H., Collins, W. E., Campbell, C. C., and Gwadz, R. W. (1986). Genetic selection of *a* Plasmodium-refractory strain of the malaria vector *Anopheles gamibae. Science* **234,** 607–610.

Collins, F. H., Mendez, M. A., Rasmussen, M. O., Mehaffey, P. C., Besansky, N. J., and Finnerty, V. (1987). A ribosomal RNA gene probe differentiates member species of the *Anopheles gambiae* complex. *Am. J. Trop. Med. Hyg.* **37,** 37–41.

Collins, F. H., Petrarca, V., Mpofu, S., Brandling-Bennett, A. D., Were, J. B. O., Rasmussen, M. O., and Finnerty, V. (1988a). Comparison of DNA probe and cytogenetic methods for identifying field collected *Anopheles gambiae* complex mosquitoes. *Am. J. Trop. Med. Hyg.* **39,** 545–550.

Collins, F. H., Mehaffey, P. C., Rasmussen, M. O., Brandling-Bennett, A. D., Odera, J. S., and Finnerty, V. (1988b). Comparison of DNA-probe and isozyme methods for differentiating *Anopheles gambiae* and *Anopheles arabiensis* (Diptera: Culicidae). *J. Med. Entomol.* **25,** 116–120.

Collins, F. H., Paskewitz, S. M., and Finnerty, V. (1989). Ribosomal RNA genes of the *Anopheles gambiae* species complex. *In* "Advances in Disease Vector Research" (K. F. Harris, ed.), Vol. 6, pp. 1–28. Springer-Verlag, New York.

Collins, F. H., Porter, C. H., and Cope, S. E. (1990). Comparison of rDNA and mtDNA in the sibling species *Anopheles freeborni* and *A. hermsi. Am. J. Trop. Med. Hyg.* **42,** 417–423.

Coluzzi, M. (1984). Heterogeneities of the malaria vectorial system in tropical Africa and their significance in malaria epidemiology and control. *Bull. W. H. O.* **62,** 107–113.

Coluzzi, M. (1992). Malaria vector analysis and control. *Parasitol. Today* **8,** 113–118.

Coluzzi, M., Petrarca, V., and Di Deco, M. A. (1985). Chromosomal inversion intergradation and incipient speciation in *Anopheles gambiae. Boll. Zool.* **52,** 45–63.

Cooper, L., Cooper, R. D., and Burkot, T. R. (1991). The *Anopheles punctulatus* complex: DNA probes for identifying the Australian species using isotopic, chromogenic, and chemiluminescence detection systems. *Exp. Parasitol.* **72,** 27–35.

Crampton, J. M., Morris, A., Lycett, G., Warren, A., and Eggleston, P. (1990a). Molecular characterization and genome manipulation of the mosquito, *Aedes aegypti. In* "Molecular Insect Science" (H. H. Hagedorn, J. G. Hildebrand, M. G. Kidwell, and J. H. Law, eds.), pp. 1–11. Plenum, New York.

Crampton, J. M., Morris, A., Lycett, G., Warren, A., and Eggleston, P. (1990b). Transgenic mosquitoes: A future vector control strategy? *Parasitol. Today* **6,** 31–36.

Crampton, J. M., Cullen, S., and Knapp, T. (1990c). Periodicity genes in the mosquito. *In* "Molecular Insect Science" (H. H. Hagedorn, J. G. Hildebrand, M. G. Kidwell, and J. H. Law, eds.), p. 293–294. Plenum, New York.

Curtis, C. F., and Graves, P. M. (1988). Methods for replacement of malaria vector populations. *J. Trop. Med. Hyg.* **91,** 43–48.

Dary, O., Georghiou, G. P., Parsons, E., and Pasteur, N. (1990). Microplate adaptation of Gomori's assay for quantitative determination of general esterase activity in single insects. *J. Econ. Entomol.* **83,** 2187–2192.

Dary, O., Georghiou, G. P., Parsons, E., and Pasteur, N. (1991). Dot-blot test for identification of insecticide-resistant acetylcholinesterase in single insects. *J. Econ. Entomol.* **84**, 28–33.

DeSalle, R., Freedman, T., Prager, E. M., and Wilson, A. C. (1987). Tempo and mode of sequence evolution in mitochondrial DNA of Hawaiian *Drosophila. J. Mol. Evol.* **26**, 157–164.

Dubin, D. T., HsuChen, C.-C., and Tillotson, L. E. (1986). Mosquito mitochondrial transfer RNAs for valine, glycine and glutamate: RNA and gene sequences and vicinal genome organization. *Curr. Genet.* **10**, 701–707.

Durbin, J. E., and Fallon, A. M. (1985). Transient expression of the chloramphenicol acetyltransferase gene in cultured mosquito cells. *Gene* **36**, 173–178.

Eggleston, P. (1990a). Homeobox genes in the mosquito, *Aedes aegypti. In* "Molecular Insect Science" (H. H. Hagedorn, J. G. Hildebrand, M. G. Kidwell, and J. H. Laws, eds.), p. 300. Plenum, New York.

Eggleston, P. (1990b). The control of insect-borne disease through recombinant DNA technology. *Heredity* **66**, 161–172.

Engels, W. R. (1989). P elements in *Drosophila. In* "Mobile DNA" (D. Berg and M. Howe, eds.), pp. 437–484. Am. Soc. Microbiol., Washington, DC.

Fallon, A. M. (1989). Optimization of gene transfer in cultured insect cells. *J. Tissue Culture Methods* **12**, 1–6.

Faria, F. S., and Leoncini, O. (1988). Evolution of ribosomal genes in the family Culicidae. *Rev. Braz. Genet.* **11**, 275–285.

Feldmann, A. M., and Ponnudurai, T. (1989). Selection of *Anopheles stephensi* for refractoriness and susceptibility to *Plasmodium falciparum. Med. Vet. Entomol.* **3**, 41–52.

Feldmann, A. M., Billingsley, P. F., and Savelkoul, E. (1990). Bloodmeal digestion by strains of *Anopheles stephensi* Liston (Diptera: Culicidae) of differing susceptibility to *Plasmodium falciparum. Parasitology* **101**, 193–200.

Ferrari, J. A., and Rai, K. S. (1989). Phenotypic correlates of genome size variation in *Aedes albopictus. Evolution (Lawrence, Kans.)* **43**, 895–899.

Feyereisen, R., Koener, J. F., Cariño, F. A., and Daggett, A. S. (1990). Biochemistry and molecular biology of insect cytochrome P450. *In* "Molecular Insect Science" (H. H. Hagedorn, J. G. Hildebrand, M. G. Kidwell, and J. H. Law, eds.), pp. 263–272. Plenum, New York.

Finnerty, V., and Collins, F. H. (1988). Ribosomal DNA probes for identification of member species of the *Anopheles gambiae* complex. *Fla. Entomol.* **71**, 288–294.

Fournier, D., Bride, J.-M., Mouches, C., Raymond, M., Magnin, M., Berge, J.-B., Pasteur, N., and Georghiou, G. P. (1987). Biochemical characterization of the esterases A1 and B1 associated with organophosphate resistance in the *Culex pipiens* L. complex. *Pestic., Biochem. Physiol.* **27**, 211–217.

Fritz, B. N., Narang, S. K., Kline, D. L., Seawright, J. A., Washino, R. K., Porter, C. H., and Collins, F. H. (1991). Diagnostic characterization of *Anopheles freeborni* and *An. hermsi* by hybrid crosses, frequencies of polytene X chromosomes and by rDNA restriction enzyme fragments. *J. Am. Mosq. Control Assoc.* **7**, 198–206.

Gale, K. R., and Crampton, J. M. (1987a). DNA probes for species identification of mosquitoes in the *Anopheles gambiae* complex. *Med. Vet. Entomol.* **1**, 127–136.

Gale, K. R., and Crampton, J. M. (1987b). A DNA probe to distinguish the species *Anopheles quadriannulatus* from other species of the *Anopheles gambiae* complex. *Trans. R. Soc. Trop. Med. Hyg.* **81**, 842–846.

Gale, K. R., and Crampton, J. M. (1988). Use of a male-specific DNA probe to distinguish female mosquitoes of the *Anopheles gambiae* species complex. *Med. Vet. Entomol.* **2**, 77–79.

Gale, K. R., and Crampton, J. M. (1989). The ribosomal genes of the mosquito, *Aedes aegypti. Eur. J. Biochem.* **185,** 311–317.

Gemmill, R. M., Hamblin, M., Glaser, R. L., Racioppi, J. V., Marx, J. L., White, B. N., Calvo, J. M., Wolfner, M. F., and Hagedorn, H. H. (1986). Isolation of mosquito vitellogenin genes and induction of expression by 20-hydroxyecdysone. *Insect Biochem.* **16,** 761–774.

Georghiou, G. P. (1986). Insecticide resistance: The Tephritidae next? *In* "Fruit Flies" (A. P. Economopoulos, ed.), pp. 27–40. Elsevier, New York.

Gerbi, S. A. (1985). Evolution of ribosomal DNA. *In* "Molecular Evolutionary Genetics" (R. MacIntyre, ed.), pp. 419–517. Plenum, New York.

Gerenday, A., Park, Y.-J., and Fallon, A. M. (1989). Expression of a heat-inducible gene in transfected mosquito cells. *Insect Biochem.* **19,** 679–686.

Golic, K. G., and Lindquist, S. (1989). The FLP recombinase of yeast catalyzes site-specific recombination in the *Drosophila* genome. *Cell (Cambridge, Mass.)* **59,** 499–509.

Gordon, J. W., and Ruddle, F. H. (1985). DNA-mediated genetic transformation of mouse embryos and bone marrow: A review. *Gene* **33,** 121–136.

Graf, R., and Briegel, H. (1989). The synthetic pathway of trypsin in the mosquito *Aedes aegypti* L. (Diptera: Culicidae) and *in vitro* stimulation in isolated midguts. *Insect Biochem.* **19,** 129–137.

Graves, P. M., and Curtis, C. F. (1982). Susceptibility of *Anopheles gambiae* to *Plasmodium yoelli* nigeriensis and *Plasmodium falciparum. Ann. Trop. Med. Parasitol.* **76,** 633–639.

Graziosi, C., Sakai, R. K., Romans, P., Miller, L. H., and Wellems, T. (1990). Method for *in situ* hybridization to polytene chromosomes from ovarian nurse cells of *Anopheles gambiae* (Diptera: Culicidae). *J. Med. Entomol.* **27,** 905–912.

Grossman, G. L., and James, A. A. (1990). Localized gene expression in the salivary glands of the mosquito, *Aedes aegypti. In* "Molecular Insect Science" (H. H. Hagedorn, J. G. Hildebrand, M. G. Kidwell, and J. H. Law, eds.), p. 307. Plenum, New York.

Gwadz, R. W., Kaslow, D., Lee, J.-Y., Maloy, W. L., Zasloff, M., and Miller, L. H. (1989). Effects of magainins and cecropins on the sporogonic development of malaria parasites in mosquitoes. *Infect. Immun.* **57,** 2628–2633.

Hall, L. M. C., and Malcolm, C. A. (1991). The acetycholinesterase gene of *Anopheles stephensi. Cell. Mol. Neurobiol.* **11,** 131–141.

Handler, A. M., and O'Brochta, D. A. (1991). Prospects for gene transformation in insects. *Annu. Rev. Entomol.* **36,** 159–183.

Harrison, G. (1978). "Mosquitoes, Malaria and Man." E. P. Dutton, New York.

Hays, A. R., and Raikhel, A. (1990). A novel female-specific protein produced by the vitellogenic fat body and accumulated in ovaries in the mosquito *Aedes aegypti. In* "Molecular Insect Science" (H. H. Hagedorn, J. G. Hildebrand, M. G. Kidwell, and J. H. Law, eds.), p. 311. Plenum, New York.

Hill, S. M., Urwin, R., Knapp, T. F., and Crampton, J. M. (1991a). Synthetic DNA probes for the identification of sibling species within the *Anopheles gambiae* complex. *Med. Vet. Entomol.* **5,** 455–463.

Hill, S. M., Urwin, R., and Crampton, J. M. (1991b). A comparison of nonradioactive labeling and detection systems with synthetic oligonucleotide probes for the species identification of mosquitoes in the *Anopheles gambiae* complex. *Am. J. Trop. Med. Hyg.* **44,** 609–622.

Hill, S. M., Urwin, R., and Crampton, J. M. (1992). A simplified, non-radioactive DNA probe protocol for the field identification of insect vector specimens. *Trans. R. Soc. Trop. Med. Hyg.* (in press).

Hotchkin, P. G., and Fallon, A. M. (1987). Ribosome metabolism during the vitellogenic cycle of the mosquito, *Aedes aegypti. Biochim. Biophys. Acta* **924**, 352–359.

Houck, M. A., Clark, J. B., Peterson, K. R., and Kidwell, M. G. (1991). Possible horizontal transfer of *Drosophila* genes by the mite *Proctolaelaps regalis. Science* **253**, 1125–1128.

HsuChen, C.-C., Kotin, R. M., and Dubin, D. T. (1984). Sequences of the coding and flanking regions of the large ribosomal subunit RNA gene of mosquito mitochondria. *Nucleic Acids Res.* **12**, 7771–7785.

Jacob, H. J., Lindpaintner, K., Lincoln, S. E., Kusumi, K., Bunker, R. K., Mao, Y.-P., Ganten, D., Dzau, J. J., and Lander, E. S. (1991). Genetic mapping of a gene causing hypertension in the stroke-prone spontaneously hypertensive rat. *Cell (Cambridge, Mass.)* **67**, 213–224.

Jacobs, E., Dewerchin, M., and Boeke, J. D. (1988). Retrovirus-like vectors for *Saccharomyces cerevisiae*: Integration of foreign genes controlled by efficient promoters into yeast chromosomal DNA. *Gene* **67**, 259–269.

James, A. A., and Rossignol, P. A. (1991). Mosquito salivary glands: parasitological and molecular aspects. *Parasitol. Today* **7**, 267–271.

James, A. A., Blackmer, K., and Racioppi, J. V. (1989). A salivary gland-specific, maltase-like gene of the vector mosquito, *Aedes aegypti. Gene* **75**, 73–83.

James, A. A., Blackmer, K., Marinotti, O., Ghosn, C. R., and Racioppi, J. V. (1991). Isolation and characterization of the gene expressing the major salivary gland protein of the female mosquito, *Aedes aegypti. Mol. Biochem. Parasitol.* **44**, 245–254.

Jost, E., and Mameli, M. (1972). DNA content in nine species of *Nematocera* with special reference to the sibling species of the *Anopheles maculipennis* group and the *Culex pipiens* group. *Chromosoma* **37**, 201–208.

Kaiser, P. E., Seawright, J. A., and Joslyn, D. J. (1979). Cytology of a genetic sexing system in *Anopheles albimanus. Can. J. Genet. Cytol.* **21**, 201–211.

Kambhampati, S., and Rai, K. S. (1991a). Variation in mitochondrial DNA of *Aedes* species (Diptera: Culicidae). *Evolution (Lawrence, Kans.)* **45**, 120–129.

Kambhampati, S., and Rai, K. S. (1991b). Mitochondrial DNA variation within and among populations of the mosquito *Aedes albopictus. Genome* **34**, 288–292.

Kambhampati, S. and Rai, K. S. (1991c). Temporal variation in the ribosomal DNA nontranscribed spacer of *Aedes albopictus* (Diptera: Culicidae). *Genome* **34**, 293–297.

Kambhampati, S., Black, W. C. IV, and Rai, K. S. (1992). RAPD-PCR for identification and differentiation of mosquito species and populations: techniques and statistical analysis. *J. Med. Entomol.*, in press.

Kertbundit, S., Rajkulchai, P., Kashemsanta, A., and Panyim, S. (1986). Molecular cloning in *E. coli* of the DNA fragments specific for *Anopheles dirus* B. In "Contemporary Themes in Biochemistry" (Kon *et al.*, eds.), Vol. 6, pp. 144–145. Cambridge Univ. Press, New York.

Kitzmiller, J. B. (1976). Genetics, cytogenetics, and evolution of mosquitoes. *Adv. Genet.* **18**, 315–433.

Kjer, K. M., and Fallon, A. M. (1991). Efficient transfection of mosquito cells is influenced by the temperature at which DNA-calcium phosphate coprecipitates are prepared. *Arch. Insect Biochem. Physiol.* **16**, 189–200.

Knapp, T., and Crampton, J. (1990). Sequences related to immune proteins in the mosquito, *Aedes aegypti. In* "Molecular Insect Science" (H. H. Hagedorn, J. G. Hildebrand, M. G. Kidwell, and J. H. Law, eds.), pp. 324–325. Plenum, New York.

Kumar, A., and Rai, K. S. (1990a). Chromosomal localization and copy number of 18S + 28S ribosomal RNA genes in evolutionarily diverse mosquitoes (Diptera, Culicidae). *Hereditas* **113**, 277–289.

Kumar, A., and Rai, K. S. (1990b). Intraspecific variation in nuclear DNA content among world populations of a mosquito, *Aedes albopictus* (Skuse). *Theor. Appl. Genet.* **79**, 748–752.

Kumar, A., and Rai, K. S. (1991a). Organization of a cloned repetitive DNA fragment in mosquito genomes (Diptera: Culicidae). *Genome* **34**, 998–1006.

Kumar, A., and Rai, K. S. (1991b). Chromosomal localization and genomic organization of cloned repetitive DNA fragments in mosquitoes (Diptera: Culicidae). *J. Genet.* **70**, 189–202.

Lackie, A. M. (1988). Immune mechanisms in insects. *Parasitol. Today* **4**, 98–105.

Lan, Q., and Fallon, M. (1990). Small heat shock proteins distinguish between two mosquito species and confirm identity of their cell lines. *Am. J. Trop. Med. Hyg.* **43**, 669–676.

Lea, A. O., and Brown, M. R. (1990). Neuropeptides of mosquitoes. *In* "Molecular Insect Science" (H. H. Hagedorn, J. G. Hildebrand, M. G. Kidwell, and J. H. Law, eds.), pp. 181–188. Plenum, New York.

Li, J., Tracy, J. W., and Christensen, B. M. (1989). Hemocyte monophenol oxidase activity in mosquitoes exposed to microfilariae of *Dirofilaria immitis*. *J. Parasitol.* **75**, 1–5.

Li, J., Tracy, J. W., and Christensen, B. M. (1992). Phenol oxidase activity in hemolymph compartments of *Aedes aegypti* during melanotic encapsulation reactions against microfilariae. *Dev. Comp. Immunol.* **16**, 41–48.

Lycett, G. J. (1990). DNA transfection of mosquito cells in culture. *Insect Mol. Genet. Newsl.* **4**, 1–3.

Macdonald, W. W. (1962). The genetic basis of susceptibility to infection with semi-periodic *Brugia malayi* in *Aedes aegypti*. *Ann. Trop. Med. Parasitol.* **56**, 373–382.

Macdonald, W. W., and Ramachandran, C. P. (1965). The influence of the gene *fm* (filarial susceptibility, *Brugia malayi*) on the susceptibility of *Aedes aegypti* to seven strains of *Brugia, Wuchereria* and *Dirofilaria*. *Ann. Trop. Med. Parasitol.* **59**, 64–73.

Malcolm, C. A., and Hall, L. M. C. (1990). Cloning and characterization of a mosquito acetylcholinesterase gene. *In* "Molecular Insect Science" (H. H. Hagedorn, J. G. Hildebrand, M. G. Kidwell, and J. H. Law, eds.), pp. 57–65. Plenum, New York.

Manning, J. E., Schmid, C. W., and Davidson, N. (1975). Interspersion of repetitive and nonrepetitive DNA sequences in the *Drosophila melanogaster* genome. *Cell (Cambridge, Mass.)* **4**, 141–155.

Marinotti, O., and James, A. A. (1990a). The alpha-glucosidase of *Aedes aegypti* salivary glands. *In* "Molecular Insect Science" (H. H. Hagedorn, J. G. Hildebrand, M. G. Kidwell, and J. H. Law, eds.), p. 333. Plenum, New York.

Marinotti, O., and James A. A. (1990b). An alpha-glucosidase in the salivary glands of the vector mosquito, *Aedes aegypti*. *Insect Biochem.* **20**, 619–623.

Marinotti, O., James, A. A., and Ribeiro, J. M. C. (1990). Diet and salivation in female *Aedes aegypti* mosquitoes. *J. Insect Physiol.* **36**, 545–548.

Mattingly, P. F. (1973). Culicidae (Mosquitoes). *In* "Insects and Other Arthropods of Medical Importance" (K. G. V. Smith, ed.), pp. 37–107. Trustees of the British Museum (Natural History), London.

McAlpine, J. F., and Wood, D. M. (1989). "Manual of Neartic Diptera," Vol. 3, Monogr. No. 32. Research Branch Agriculture, Canada.

McGrane, V., Carlson, J. O., Miller, B. R., and Beaty, B. J. (1988). Microinjection of DNA into *Aedes triseriatus* ova and detection of integration. *Am. J. Trop. Med. Hyg.* **39**, 502–510.

McLain, D. K., and Collins, F. H. (1989). Structure of rDNA in the mosquito *Anopheles gambiae* and rDNA sequence variation within and between species of the *A. gambiae* complex. *Heredity* **62**, 233–242.

McLain, D. K., Rai, K. S., and Fraser, M. J. (1986). Interspecific variation in the abundance of highly repeated DNA sequences in the *Aedes scutellaris* (Diptera: Culicidae) subgroup. *Ann. Entomol. Soc. Am.* **79**, 784–791.

McLain, D. K., Rai, K. S., and Fraser, M. J. (1987). Intraspecific and interspecific variation in the sequence and abundance of highly repeated DNA among mosquitoes of the *Aedes albopictus* subgroup. *Heredity* **58**, 373–381.

McLain, D. K., Collins, F. H., Brandling-Bennett, A. D., and Were, J. B. O. (1989). Microgeographic variation in rDNA intergenic spacers of *Anopheles gambiae* in western Kenya. *Heredity* **62**, 257–264.

Meyer-Leon, L., Senecoff, J. F., Bruckner, R. C., and Cox, M. M. (1984). Site-specific genetic recombination promoted by the FLP protein of the yeast 2-micron plasmid *in vitro*. *Cold Spring Harbor Symp. Quant. Biol.* **49**, 797–804.

Miller, B. R., and Mitchell, C. J. (1991). Genetic selection of a flavivirus-refractory strain of the yellow fever mosquito *Aedes aegypti*. *Am. J. Trop. Med. Hyg.* **45**, 399–407.

Miller, L. H., Sakai, R. K., Romans, P., Gwadz, R. W., Kantoff, P., and Coon, H. G. (1987). Stable integration and expression of a bacterial gene in the mosquito *Anopheles gambiae*. *Science* **237**, 779–781.

Mitchell, S. E. (1990). Mitochondrial and ribosomal DNA analysis for identification of sibling species of the mosquito, *Anopheles quadrimaculatus*. Ph.D. Dissertation, University of Florida, Gainesville.

Mizrokhi, L. J., and Mazo, A. M. (1990). Evidence for horizontal transmission of the mobile element *jockey* between distant *Drosophila* species. *Proc. Natl. Acad. Sci. U.S.A.* **87**, 9216–9220.

Monroe, T. J., Carlson, J. O., Clemens, D. L., and Beaty, B. J. (1990). Selectable markers for transformation of mosquito and mammalian cells. *In* "Molecular Insect Science" (H. H. Hagedorn, J. G. Hildebrand, M. G. Kidwell, and J. H. Law, eds.), p. 337. Plenum, New York.

Morris, A. C., Eggleston, P., and Crampton, J. M. (1989). Genetic transformation of the mosquito *Aedes aegypti* by microinjection of DNA. *Med. Vet. Entomol.* **3**, 1–7.

Morris, A. C., Schaub, T. L., and James, A. A. (1991). FLP-mediated recombination in the vector mosquito, *Aedes aegypti*. *Nucleic Acids Res.* **19**, 5895–5900.

Mouches, C., Pasteur, N., Berge, J. B., Hyrien, O., Raymond, M., Robert De Saint Vincent, B., De Silvestri, M., and Georghiou, G. P. (1986). Amplification of an esterase gene is responsible for insecticide resistance in a California *Culex* mosquito. *Science* **233**, 778–780.

Mouches, C., Magnin, M., Bergé, J.-B., De Silvestri, M., Beyssat, V., Pasteur, N., and Georghiou, G. P. (1987). Overproduction of detoxifying esterases in organophosphate-resistant *Culex* mosquitoes and their presence in other insects. *Proc. Natl. Acad. Sci. USA.* **84**, 2113–2116.

Mouches, C., Pauplin, Y., Agarwal, M., Lemieux, L., Herzog, M., Abadon, M., Beyssat-Arnaouty, V., Hyrien, O., Robert De Saint Vincent, B., Georghiou, G. P., and Pasteur, N. (1990). Characterization of amplification core and esterase B1 gene responsible for insecticide resistance in *Culex*. *Proc. Natl. Acad. Sci. U.S.A.* **87**, 2574–2578.

Munstermann, L. E. (1990a). Gene map of the yellow fever mosquito *Aedes aegypti*. *In* "Genetic Maps: Locus Maps of Complex Genomes" (S. J. O'Brien, ed.), 5th ed., pp. III, 179–III, 184. Cold Spring Harbor Lab. Press, Cold Spring Harbor, New York.

Munstermann, L. E. (1990b). Gene map of the eastern North American tree hole mosquito *Aedes triseriatus*. *In* "Genetic Maps: Locus Maps of Complex Genomes" (S. J. O'Brien, ed.), 5th ed., pp. III, 185–III, 188. Cold Spring Harbor Lab. Press, Cold Spring Harbor, New York.

Nance, E., Heyse, D., Britton-Davidian, J., and Pasteur, N. (1990). Chromosomal organization of the amplified esterase B1 gene in organophosphate-resistant *Culex pipiens quinquefasciatus* Say (Diptera, Culicidae). *Genome* **33**, 148–152.

Nappi, A. J., and Christensen, B. M. (1986). Hemocyte cell surface changes in *Aedes aegypti* in response to microfilariae of *Dirofilaria immitis*. *J. Parisitol.* **72**, 875–879.

Narang, S. K., and Seawright, J. A. (1990). Linkage map of the mosquito *Anopheles albimanus*. *In* "Genetic Maps: Locus Maps of Complex Genomes" (S. J. O'Brien, ed.), 5th ed., pp. III, 191–III, 194. Cold Spring Harbor Lab. Press, Cold Spring Harbor, New York.

Narang, S. K., Seawright, J. A., and Mitchell, S. E. (1990). Linkage map of the mosquito *Anopheles qudarimaculatus*. *In* "Genetic Maps: Locus Maps of Complex Genomes" (S. J. O'Brien, ed.), 5th ed., pp. III, 195–III, 197. Cold Spring Harbor Lab. Press, Cold Spring Harbor, New York.

O'Brochta, D. A., and Handler, A. M. (1988). Mobility of P elements in drosophilids and nondrosophilids. *Proc. Natl. Acad. Sci. USA* **85**, 6052–6056.

O'Gorman, S., Fox, D. T., and Wahl, G. M. (1991). Recombinase-mediated gene activation and site-specific integration in mammalian cells. *Science* **251**, 1351–1355.

Panyim, S., Yasothornsrikul, S., Tungpradubkul, S., Baimai, V., Rosenberg, R., Andre, R. F., and Green, C. A. (1988). Identification of isomorphic malaria vectors using a DNA probe. *Am. J. Trop. Med. Hyg.* **38**, 47–49.

Paskewitz, S. M., and Collins, F. H. (1989). Site-specific ribosomal DNA insertion elements in *Anopheles gambiae* and *A. arabiensis*: Nucleotide sequence of gene-element boundaries. *Nucleic Acids Res.* **17**, 8125–8133.

Paskewitz, S. M., and Collins, F. H. (1990). Use of the polymerase chain reaction to identify mosquito species of the *Anopheles gambiae* complex. *Med. Vet. Entomol.* **4**, 367–373.

Paskewitz, S. M., Brown, M. R., Lea, A. O., and Collins, F. H. (1988). Ultrastructure of the encapsulation of *Plasmodium cynomolgi* (B strain) on the midgut of a refractory strain of *Anopheles gambiae*. *J. Parasitol.* **74**, 432–439.

Paskewitz, S. M., Brown, M. R., Collins, F. H., and Lea, A. O. (1989). Ultrastructural localization of phenoloxidase in the midgut of refractory *Anopheles gambiae* and association of the enzyme with encapsulated *Plasmodium cynomolgi*. *J. Parasitol.* **75**, 594–600.

Pasteur, N., Raymond, M., Pauplin, Y., Nance, E., and Heyse, D. (1990). Role of gene amplification in insecticide resistance. *In* "Pesticides and Alternatives" (J. E. Casida, ed.), pp. 439–447. Elsevier, Amsterdam.

Porter, C. H., and Collins, F. H. (1991). Species-diagnostic differences in a ribosomal DNA internal transcribed spacer from the sibling species *Anopheles freeborni* and *Anopheles hermsi* (Diptera: Culicidae). *Am. J. Trop. Med. Hyg.* **45**, 271–279.

Raikhel, A. S., Dhadialla, T. S., Cho, W.-L., Hays, A. R., and Koller, C. N. (1990). Biosynthesis and endocytosis of yolk proteins in the mosquito. *In* "Molecular Insect Science" (H. H. Hagedorn, J. G. Hildebrand, M. G. Kidwell, and J. H. Law, eds.), p. 147–154. Plenum, New York.

Rao, P. N. (1985). Nuclear DNA and chromosomal evolution in mosquitoes. Ph.D. Dissertation, University of Notre Dame, Notre Dame. IN.

Rao, P. N., and Rai, K. S. (1987a). Comparative karyotypes and chromosomal evolution in some genera of nematocerous (Diptera: Nematocera) families. *Ann. Entomol. Soc. Am.* **80**, 321–332.

Rao, P. N., and Rai, K. S. (1987b). Inter and intraspecific variation in nuclear DNA content in *Aedes* mosquitoes. *Heredity* **59**, 253–258.

Rao, P. N., and Rai, K. S. (1990). Genome evolution in the mosquitoes and other closely related members of the superfamily Culicoidea. *Hereditas* **113**, 139–144.

Raymond, M., and Pasteur, N. (1989). The amplification of B1 esterase gene in the mosquito *Culex pipiens* is present in gametes. *Nucleic Acids Res.* 17, 7116.

Raymond, M., Beyssat-Arnaouty, V., Sivasubramanian, N., Mouches, C., Georghiou, G. P., and Pasteur, N. (1989). Amplification of various esterase B's responsible for organophosphate resistance in *Culex* mosquitoes. *Biochem. Genet.* 27, 417–423.

Raymond, M., Callaghan, A., Fort, P., and Pasteur, N. (1991). Worldwide migration of amplified insecticide resistance genes in mosquitoes. *Nature (London)* 350, 151–153.

Reiss, R. A. (1990). A method of *in situ* hybridization to the polytene chromosomes of *Anopheles gambiae*. *Insect Mol. Genet. Newsl.* 4, 4–5.

Ribeiro, J. M. C. (1987). Role of saliva in blood-feeding by arthropods. *Annu. Rev. Entomol.* 32, 463–478.

Ribeiro, J. M. C., Sarkis, J. J. F., Rossignol, P. A., and Spielman, A. (1984). Salivary apyrase of *Aedes aegypti*: Characterization and secretory fate. *Comp. Biochem. Physiol. B* 79B, 81–86.

Romans, P. (1990). The vitellogenin genes of the malaria vector *Anopheles gambiae*. In "Molecular Insect Science" (H. H. Hagedorn, J. G. Hildebrand, M. G. Kidwell, and J. H. Law, eds.), p. 353. Plenum, New York.

Romans, P., Seeley, D. C., Kew, Y., and Gwadz, R. W. (1991). Use of a restriction fragment length polymorphism (RFLP) as a genetic marker in crosses of *Anopheles gambiae* (Diptera: Culicidae): independent assortment of a diphenol oxidase RFLP and an esterase locus. *J. Med. Entomol.* 28, 147–151.

Rossignol, P. A., and Lueders, A. M. (1986). Bacteriolytic factor in the salivary glands of *Aedes aegypti*. *Comp. Biochem. Physiol. B* 83B, 819–822.

Sakai, R. K., and Miller, L. H. (1992). Effects of heat shock on the survival of transgenic *Anopheles gambiae* (Diptera: Culicidae) under antibiotic selection. *J. Med. Entomol.* 29, 374–376.

Saunders, R. D. C., Glover, D. M., Ashburner, M., Sidén-Kiamos, I., Louis, C., Monastirioti, M., Savakis, C., and Kafatos, F. C. (1989). PCR amplification of DNA microdissected from a single polytene chromosome band: A comparison with conventional microcloning. *Nucleic Acids Res.* 17, 9027–9037.

Shotkoski, F. A., and Fallon, A. M. (1990). Genetic changes in methotrexate-resistant mosquito cells. *Arch. Insect Biochem. Phys.* 15, 79–92.

Shotkoski, F. A., and Fallon, A. M. (1991). An amplified insect dihydrofolate reductase gene contains a single intron. *Eur. J. Biochem.* 201, 157–160.

Sidén-Kiamos, I., Saunders, R. D. C., Spanos, L., Majerus, T. Treanear, J. Savakis, C., Louis, C., Glover, D. M., Ashburner, M., and Kafatos, F. C. (1990). Towards a physical map of the *Drosophila melanogaster* genome: Mapping of cosmid clones within defined genomic divisions. *Nucleic Acids Res.* 18, 6261–6270.

Spray, F. J., and Christensen, B. M. (1991). *Aedes aegypti*: Characterization of hemocyte polypeptide synthesis during an immune response to microfilariae. *Exp. Parasitol.* 73, 481–488.

Stark, G. R., and Wahl, G. M. (1984). Gene amplification. *Annu. Rev. Biochem.* 53, 447–491.

Steiner, W. W. M., Tabachnick, W. J., Rai, K. S., and Narang, S., eds. (1982). "Recent Developments in the Genetics of Insect Disease Vectors." Stipes Publishing Co., Champaign, IL.

Sun, S.-C., Lindstrom, I., Boman, H. G., Faye, I., and Schmidt, O. (1990). Hemolin: An insect-immune protein belonging to the immunoglobulin superfamily. *Science* 250, 1729–1732.

Swerdel, M. R., and Fallon, A. M. (1987). Phosphoribosylation of xanthine by extracts from insect cells. *Insect Biochem.* **17,** 1181–1186.

Turelli, M., and Hoffmann, A. A. (1991). Rapid spread of an inherited incompatability factor in California *Drosophila. Nature (London)* **353,** 440–442.

Vernick, K. D., and Collins, F. H. (1989). Association of a *Plasmodium*-refractory phentype with an esterase locus in *Anopheles gambiae. Am. J. Trop. Med. Hyg.* **40,** 593–597.

Vernick, K. D., Collins, F. H., and Gwadz, R. W. (1989). A general system of resistance to malaria infection in *Anopheles gambiae* controlled by two main genetic loci. *Am. J. Trop. Med. Hyg.* **40,** 585–592.

Warburg, A., and Miller, L. H. (1991). Critical stages in the development of *Plasmodium* in mosquitoes. *Parasitol. Today* **7,** 179–181.

Warren, A. M., and Crampton, J. M. (1991). The *Aedes aegypti* genome: Complexity and organization. *Genet. Res.* **58,** 225–232.

Warren, M., and Collins, W. E. (1981). Vector–parasite interactions and the epidemiology of malaria. *In* "Parasitological Topics" (E. U. Canning, ed.), pp. 266–274. London Society for Protozoology, London.

Wattam, A. R., and Christensen, B. M. (1992). Induced polypeptides associated with filarial worm refractoriness in *Aedes aegypti* mosquitoes. *Proc. Natl. Acad. Sci. U.S.A.* (in press).

White, G. B. (1980). Academic and applied aspects of mosquito cytogenetics. *In* "Insect Cytogenetics" (R. L. Blackman, G. M. Hewitt, and M. Ashburner, eds.), pp. 245–274. Blackwell, Oxford.

Whyard, S., Tittiger, C., Downe, A. E. R., and Walker, V. K. (1990). A malathion degrading enzyme in the mosquito *Culex tarsalis. In* "Molecular Insect Science" (H. H. Hagedorn, J. G. Hildebrand, M. G. Kidwell, and J. H. Law, eds.), p. 382. Plenum, New York.

Williams, J. G. K., Kubelik, A. R., Livak, K. J., Rafalski, J. A., and Tingey, S. V. (1991). DNA polymorphisms amplified by arbitrary primers are useful as genetic markers. *Nucleic Acids Res.* **18,** 6531–6536.

World Health Organization (1991a). *TDR News* **35.**

World Health Organization (1991b). "Prospects for Malaria Control by Genetic Manipulation of its Vectors," TDR/BCV/MAL-ENT/91.3. WHO, Geneva.

Wright, J. W., and Pal, R., eds. (1967). "Genetics of Insect Vectors of Disease." Elsevier, New York.

Xiong, Y., and Eickbush, T. H. (1988). Similarity of reverse transcriptase-like sequences of viruses, transposable elements, and mitochondrial introns. *Mol. Biol. Evol.* **5,** 675–690.

Xiong, Y., and Eickbush, T. H. (1990). Origin and evolution of retroelements based upon their reverse transcriptase sequences. *EMBO J.* **9,** 3353–3362.

Yasothornsrikul, S., Tungpradabkul, S., and Panyim, S. (1986). Species specific DNA probe for identification of *Anopheles dirus* sibling species D. *In* "Contemporary Themes in Biochemistry" (Kon *et al.,* eds.), Vol. 6, pp. 142–143. Cambridge Univ. Press, New York.

Yasothornsrikul, S., Panyim, S., and Rosenberg, R. (1988). Diagnostic restriction fragment patterns of DNA from the four isomorphic species of *Anopheles dirus. Southeast Asian J. Trop. Med. Public Health* **19,** 703–708.

Zheng, L., Saunders, R. D. C., Fortini, D., dellaTorre, A., Coluzzi, M., Glover, D., and Kafatos, F. C. (1991). Low resolution genome map of the malaria mosquito, *Anopheles gambiae. Proc. Natl. Acad. Sci. U.S.A.* **88,** 11187–11191.

GENETICS AND MOLECULAR BIOLOGY OF TELOMERES

Harald Biessmann* and James M. Mason †

*Developmental Biology Center, University of California, Irvine,
Irvine, California 92717
† Experimental Carcinogenesis and Mutagenesis Branch, National Institute of
Environmental Health Sciences, Research Triangle Park, North Carolina 27709

"When *I* use a word," Humpty Dumpty said . . . , "it means just what I choose it to mean—neither more nor less. . . . The question is, . . . which is to be the master—that's all."

Lewis Carroll, *Through the Looking Glass* (1896)

I. Introduction and Premise

Over the years the term "telomere" has been used by different groups to mean different things. To Muller (1938, 1940), it was a hypothetical genetic entity responsible for maintaining chromosome stability. To Westergaard and von Wettstein (1972; von Wettstein *et al.*, 1984), it is a cytologically visible chromosome marker identified as a thickening of the chromosome end associated with the nuclear envelope at leptotene. To Moyzis (1991), it is a DNA sequence generated by a template-containing reverse transcriptase. Are all of these telomeres the same? Or is the term a convenience, and are there several struc-

185

Copyright © 1992 by Academic Press, Inc.
All rights of reproduction in any form reserved.

tural elements near each other, each with its own function? It may be possible to answer these questions within the next few years. Our guess, and it is only a guess at this point, is that the latter is the case; a number of structural elements are associated with chromosome ends. Bearing in mind Dobzhansky's (1972) statement that "nothing in biology makes sense except in light of evolution," it becomes reasonable to ask what are these structures and what functions they have.

For simplicity, we define telomeres as complex structures at the ends of linear chromosomes that perform several cellular functions. Among the functions proposed for telomeres at least two are vital: protection of the chromosome end from degradation and fusion (termed "capping") and complete replication of DNA sequences at chromosome ends. Telomeres are also involved in associations with other telomeres and structures within the nucleus, but the functions of these associations are not clearly understood. Although the concept of the telomere was developed, and the term was coined, more than 50 years ago (Muller, 1938), the structure of the telomere and its functions are still not fully understood.

II. Structure of Telomeres

Telomeres consist of a number of DNA sequences, including a simple repeat that is G rich on one strand at the extreme terminus of the chromosome (termed "telomeric repeat") and more complex repetitive sequences proximal to it (termed "telomere-associated DNA" or "subtelomeric repeats"). Proteins have been identified that bind preferentially to these sequences. Except for evidence that the G-rich telomere repeat, combined with a telomere-specific reverse transcriptase (telomerase), is important for the replication of chromosome ends, it is not clear which components within the telomere perform which functions.

A. DNA COMPONENTS

1. Terminal Repeats

In a variety of eukaryotic organisms specific DNA sequences are present at the termini of linear chromosomes consisting of tandemly repeated short, simple sequences with a G-rich strand oriented in the 5′ to 3′ direction toward the end of the chromosome (reviewed by Zakian, 1989). The telomeric repeat is highly conserved, with similar sequences found in protozoa, nematodes, lower and higher plants, and vertebrates. Species specificity of the telomeric repeat sequence is determined in an elongation process catalyzed by a telomerase, which is a reverse transcriptase containing an RNA molecule that functions as template for the telomeric repeat (reviewed by Blackburn, 1991). The unit length of the telomeric repeat is generally in the range of 6–10 base pairs (bp), but the number of copies may vary greatly, so that the overall length of the repeat motif may be as little as 20 bp in *Oxytricha* or as long as 20 kilobase pairs (kbp) in humans (de Lange *et al.*, 1990; Hastie *et al.*, 1990) and up to 150 kbp in mice (Kipling and Cooke, 1990; Starling *et al.*, 1990). Within one organism the amount of telomeric DNA may vary among chromosomes, among cell types, and according to age. Though in many species the repeating unit is simple, in *Saccharomyces, Dictyostelium,* and possibly other fungi, slime molds and cellular slime molds, there may not be a simple repeat per se, although there is a repeating motif that is similar to the telomeric repeats of other species.

The guanine-rich strand of the terminal repeat forms a 3′ overhang of 12–16 bases in the few species analyzed to date (Klobutcher *et al.*, 1981; Pluta *et al.*, 1982). The guanine residues in the single-stranded overhang may hydrogen bond with each other to form an unusual, non-Watson–Crick, four-stranded structure termed a "G-quartet" (Fig. 1) or "G-DNA" (Cech, 1988; Henderson *et al.*, 1987; Williamson *et al.*, 1989). A G-quartet consists of four guanines Hoogsteen-paired in a planar, symmetric array and stabilized by a central monovalent cation. Quadruplex structures formed by stacked arrays of these G–G base pairs have been implicated in the formation of unique intrastrand foldback structures as well as telomere–telomere associations (Sen and Gilbert, 1988; Sundquist and Klug, 1989; Acevedo *et al.*, 1991). Conformational changes between the proposed intra- and interstrand quadruplex structures (Fig. 2) may be mediated by local relative concentrations of K^+ and Na^+ (Sen and Gilbert, 1990; Hardin *et al.*, 1991). The G-quartet, however, is not required for the replication of the telomere, because the G-rich strand can be extended by telomerase even when the G-quartet structure is disrupted by methylation of the dG residues or substitution of deoxyinosine for dG (Henderson *et al.*, 1990). The opposite effect is more likely because K^+-stabilized G-quartet structures will inhibit telomere elongation by telomerase *in vitro* (Hardin *et al.*, 1991; Zahler *et al.*, 1991).

In the animal and plant kingdoms several phyla are notably absent from the list of organisms shown to have telomeric repeats, and the question remains whether echinoderms, mollusks, and arthropods also contain similar telomeric repeats that have not been discovered, or whether they have evolved different telomere structures. Because the

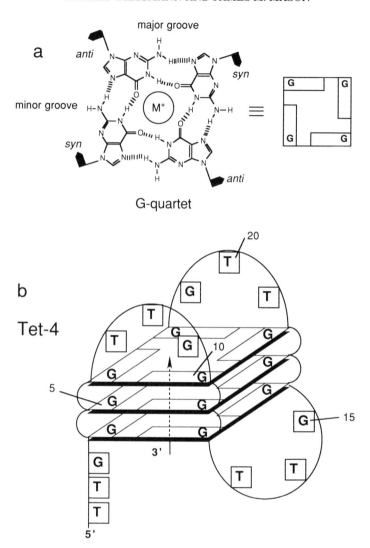

FIG. 1. Models for the structure of telomeric oligonucleotides. (a) The G-quartet. Four guanines are hydrogen bonded in a square-planar symmetric array. Each base is both the donor and acceptor in a Hoogsteen base pair. The cavity at the center is a possible binding site for a monovalent cation, M^+. The glycosidic torsions alternate between the syn and anti conformations, which creates major and minor grooves when several quartets are stacked. (b) Model for *Tetrahymena* fold-back quadruplex. Three G-quartets are stacked over each other and are connected at adjacent corners by loops of d(GTT). Reproduced with permission from Williamson *et al.* (1989).

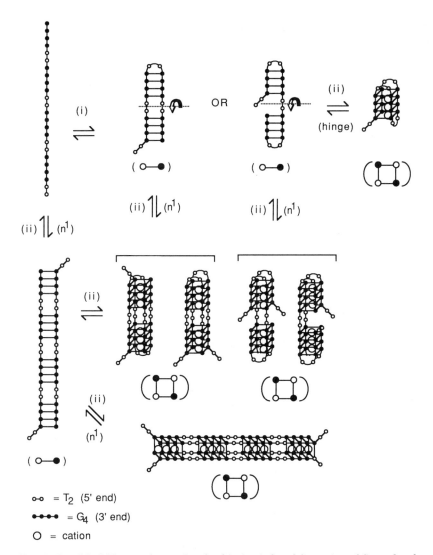

FIG. 2. Possible folding pathways involved in ion-induced formation of G-quadruplex conformations in *Tetrahymena* telomeric DNA. Individual folding steps involve one of two mechanisms: (i) pairing of individual strands to form non-Watson–Crick duplex (hairpin) structures driven by counterion condensation and hydrophobic effects or (ii) pairing of non-Watson–Crick duplexes to form quadruplexes driven by formation of octahedral coordination complexes involving a monovalent cation "caged" within two stacked G-quartets. Reactions are zero order with respect to DNA concentration in the intramolecular folding steps and first order (n^1) in the intermolecular folding steps. Intramolecular quadruplex formation is identified by the term "hinge." Symbols: (○) thymidine residues; (●) deoxyguanine residues. The strand polarities are indicated below the structures. Open circles represent 5′-to-3′ strand directionality going into the page; filled circles represent the opposite strand direction. Note that only antiparallel conformations are shown. Analogous parallel-stranded structures may also be possible. Also note that one of the bimolecular structures formed by the condensation of two hairpins cannot form via the hinge mechanism because the loops are at opposite ends. Reproduced with permission from Hardin *et al.* (1991).

telomeric repeats are paramount for the telomerase-catalyzed elongation mechanism, the absence of simple telomeric repeats would also indicate the possibility of a different telomere elongation mechanism. A few attempts have been made to detect telomeric repeats with suitable oligonucleotide probes in a variety of invertebrate organisms. These may clarify the distribution of terminal repeats among organisms. Cross-reactions with terminal repeats of other organisms or with various similar oligonucleotide probes have failed to provide evidence for telomeric repeats in *Drosophila* (Richards and Ausubel, 1988; Meyne *et al.*, 1989). Though it is conceivable that the difference between *Drosophila melanogaster* and other species is quantitative rather than qualitative and that *Drosophila* simply has fewer or more heterogeneous G-rich repeats, the alternative is that *Drosophila* may have a very different telomere structure. The absence of simple repeats would imply that the formation and replication of telomeres would also differ.

2. Telomere-Associated Sequences

The telomeres of many eukaryotic organisms contain extensive regions of repetitive sequences that are proximal to the simple telomeric repeats. These regions vary in length from chromosome to chromosome, e.g., occupying some 100 kbp in *Plasmodium* (Foote and Kemp, 1989) and up to 2600 kbp in humans (Ellis and Goodfellow, 1989), and may account for about 12–18% of total DNA in *Secale cereale* (Bedbrook *et al.*, 1980a). Two major types of repetitive sequences have been found in telomeric regions: satellite-like, short, tandemly reiterated sequences and more complex, moderately repeated sequences.

a. Complex Moderately Repetitive Sequences. Moderately repetitive sequences, ranging from 350 bp to several thousand base pairs, have been described in *Saccharomyces cerevisiae* (Chan and Tye, 1983a,b), *Plasmodium* (Pace *et al.*, 1987; Dore *et al.*, 1990), *D. melanogaster* (Rubin, 1978; Young *et al.*, 1983), rye (Bedbrook *et al.*, 1980b), and humans (Rouyer *et al.*, 1990).

i. Yeast. Saccharomyces cerevisiae contains two types of subtelomeric repeats, X and Y′, that are located immediately proximal to the terminal $C_{1-3}A$ repeats. Yeast telomeres can be divided into two classes. Most chromosomes carry one to four Y′ elements and an X element (Chan and Tye, 1983a,b), but some lack the Y′ sequences (Horowitz *et al.*, 1984; Walmsley *et al.*, 1984; Walmsley and Petes, 1985;

Jäger and Philippsen, 1989a). $C_{1-3}A$ repeats are found between Y' elements and separating Y' from X. Though Y' elements are quite homogeneous in size, the X repeats vary in length from 0.3 to 3.75 kbp (Chan and Tye, 1983a,b), which seems to be due to length variations in the proximal direction. The organization of a minimal telomere from the left arm of chromosome III, which lacks Y' sequences, has been elucidated (Button and Astell, 1986). In this telomere, the X repeat is 1.2 kbp in length and is located directly proximal to the terminal $C_{1-3}A$ repeats. Its nucleotide sequence, together with partial sequences from two other X elements that originated from a Y'–X telomere (Walmsley *et al.*, 1984), reveals 80–90% sequence similarity in the distal region among the three members of the family. It is worth noting that the distal end of the X repeat abutting the $C_{1-3}A$ repeats in both telomeric situations, either with or without Y' present, is almost identical in the three copies sequenced to date: TAA-Py-ACA/TA-Py-C-Py. There is considerable variation among strains in the distribution of Y' and X sequences (Zakian and Blanton, 1988; Louis and Haber, 1990a). It is interesting that *Saccharomyces carlsbergensis*, a hybrid between *S. cerevisiae* and another (unknown) species, has no Y' sequences, even though it has several *S. cerevisiae* chromosomes (*S. carlsbergensis* is listed as *Saccharomyces uvarum* by Jäger and Philippsen, 1989a).

The organization of Y'-containing telomeres has been studied in great detail (Chan and Tye, 1983a,b; Shampay *et al.*, 1984; Walmsley *et al.*, 1984). In their natural telomeric positions, Y' sequences are distally and in most cases also proximally flanked by $C_{1-3}A$ repeats, as if they are embedded in the distal telomeric $C_{1-3}A$ repeats of chromosome ends. Frequent recombination among the Y' repeats (Louis and Haber, 1990b) is very likely to be responsible for the high degree of polymorphism in the telomeric region (Horowitz *et al.*, 1984). Y' sequences can also be transferred by recombination between linear plasmid ends and chromosomal telomeres (Dunn *et al.*, 1984).

Y' elements are quite homogeneous in size. They occur in two variants, a long form of 6.7 kbp and a short form of 5.2 kbp, which is derived from the long form by an internal deletion (Louis and Haber, 1990a,b). Partial sequences of Y' have been published (Horowitz and Haber, 1984; Shampay *et al.*, 1984) but a complete sequence of Y' elements has only recently become available (Louis and Haber, 1992). Y' elements appear to contain two overlapping, out-of-frame open reading frames (ORFs) that potentially encode polypeptides with similarities to some DNA and RNA helicases as well as to retroviral proteins, respectively. RNA homologous to each ORF was detectable. No similarity to reverse transcriptases, as encoded by retrotransposable elements, could be found, nor do Y′ elements contain long terminal repeats (LTRs). Although the sequence similarities are rather weak, they raise the possibility that Y′ sequences may be related to DNA viruses or transposable elements. Although Y′ elements have been found in autonomously replicating extrachromosomal circles (Horowitz and Haber, 1984), it is unclear whether they are capable of transposing, because the vast majority of movements of marked Y′ elements to new chromosome ends can be accounted for by recombination events (Louis and Haber, 1990b). Thus, Louis and Haber (1992) suggest that the present dispersal of Y′ sequences to a variety of different chromosome ends has resulted from one fortuitous insertion of a single Y′ element into the telomeric region of an ancestral *S. cerevisiae* strain, which then became dispersed by recombination between the subtelomeric sequences. An example of a very recent integration/recombination-mediated transfer of a mitochondrial intron sequence into the telomeric X–Y′ junctions in *S. cerevisiae* has been reported (Louis and Haber, 1991), suggesting that fortuitous and rare integration events may be conserved by recombination between subtelomeric repeats without detrimental effect or functional gain for the organism. Such a mechanism has also been proposed for the subtelomeric dispersal of the *SUC* and *MAL* loci discussed below (Carlson *et al.*, 1985; Charron *et al.*, 1989). The possibility remains, however, that Y′ elements might be stimulated to transpose under unusual conditions, termed "genomic stress" (McClintock, 1984), as has been suggested for the yeast Ty1 transposable element (Morawetz, 1987; Boeke and Corces, 1989; Bradshaw and McEntee, 1989; Morawetz and Hagen, 1990).

The question arises whether X and Y′ repeats are a general feature of telomeric structure. The absence of a Y′ repeat on chromosome III L (Button and Astell, 1986) clearly indicates that at least Y′ sequences are not required for stability and maintenance of this particular telomere. Moreover, hybridization of chromosomes separated by pulsed-field electrophoresis with Y′ and X probes indicated the absence of Y′ from some chromosomes in some strains (Zakian and Blanton, 1988; Jäger and Philippsen, 1989a). The evidence for X is much less clear. Although in the same experiment chromosome I was found to lack X sequences, this result is less conclusive because X repeats can be quite short and may be significantly diverged from the X sequence used as probe. Furthermore, the experiments of Murray and Szostak (1986), in which the natural yeast chromosome end was substituted with a telomere from *Tetrahymena,* also seem to support the idea that Y' and X sequences are not essential for mitotic stability of chromosome III. However, the yeast chromosome III deletion derivative used in this study carried a terminal fragment of *Tetrahymena* ribosomal DNA, which may have substituted for Y' and X element functions.

ii. Drosophila. The subtelomeric regions of *D. melanogaster* are not well characterized to date, and we do not know how many different components they contain. However, one family of telomere-associated DNA sequences has been defined by a set of DNA fragments cloned in the phage λT-A (Young *et al.,* 1983). The cytological position of the original λT-A clone is not known, but available evidence suggests that it originated from the X chromosome telomere (Valgeirsdottir *et al.,* 1990). Subfragments of this recombinant phage hybridize exclusively with the telomeric region of each of the polytene chromosomes and with the pericentric β-heterochromatin, hence they were termed the HeT family of sequences. A similar cytological distribution has been found for the terminal telomeric repeats in vertebrate (Meyne *et al.,* 1990) and plant (Richards *et al.,* 1991) genomes, where homologous and slightly degenerate telomeric sequences are also found in the centromeric heterochromatin. Why these repeats appear in centromeric locations is unclear, but they might be remnants of chromosome rearrangements that occurred during genome evolution, or they might have been carried along by transposable elements and expanded in their new location by unequal recombination.

HeT DNA appears to be a major family of telomere-associated DNA in *Drosophila.* Due to underrepresentation of telomeric and pericentric DNA sequences in polytene chromosomes, the intensity of hybridization of HeT sequences to polytene chromosomes does not reflect their relative abundance in diploid chromosomes. *In situ* hybridization to metaphase chromosomes reveals that HeT sequences are most abundant in the heterochromatic Y chromosome, followed by the pericentric regions of the large autosomes, and of the X chromosome (Traverse and Pardue, 1989). Judged from these results, the amount of HeT sequences in telomeric regions must be comparatively low. A member of the HeT sequence family has been isolated that clearly originated from the Y chromosome (Danilevskaya *et al.,* 1991).

Other clones that have been reported to hybridize to telomeric and pericentric regions (Rubin, 1978; Renkawitz-Pohl and Bialojan, 1984; Danilevskaya *et al.,* 1991) all have homology to λT-A, which apparently represents a mosaic of several kinds of repeated elements (Young *et al.,* 1983; Valgeirsdottir *et al.,* 1990) that may occur in different combinations at other locations. Related sequences also occur in arrays of tandem repeats of a 3-kbp element (Rubin, 1978; Young *et al.*, 1983) that are located on the Y chromosome. Interestingly, similar tandem repeats of sequences with identical telomeric and pericentric cytological distribution have been isolated from *Drosophila miranda* (Steinemann, 1984; Steinemann and Nauber, 1986). However, the repeat length is 4.4 kbp, and a possible sequence similarity to the 3-kbp *D. melanogaster* repeats has not been established, although HeT-A sequences do cross-react at reduced stringency with *D. miranda* genomic DNA (Traverse and Pardue, 1989).

The most abundant sequence component in λT-A is called HeT-A. Sequence comparisons of several isolates (Biessmann *et al.*, 1990b, 1992a) indicate that the HeT-A element represents a novel family of *Drosophila* retroposons, a class of nonviral, non-LTR-containing, oligoadenylated, transposable elements that also include vertebrate long interspersed elements (LINES) (Berg and Howe, 1989). HeT-A elements appear to have a role in the healing of broken chromosomes (Biessmann *et al.*, 1990b). At least 23 new transpositions of HeT-A elements to broken chromosome ends have been detected, and some of them have been sequenced (Biessmann *et al.*, 1992b). The junction of the HeT-A elements with each broken chromosome end always contains an oligo(A) region at the proximal, 3' end of the element, suggesting that transposition has occurred via an RNA intermediate. A polyadenylation signal precedes the oligo(A) tail by 53 bp. Newly transposed HeT-A elements often appear to be truncated at their distal, 5' ends, which is also characteristic of retroposons. This makes it difficult to assess the size of the complete element, but the longest HeT-A element that has been observed to transpose is at least 12 kbp in length (Biessmann *et al.*, 1992b). This is considerably longer than the other *Drosophila* non-LTR, oligo(A) retroposons, which range from 3.5 to 5.4 kbp. However, unlike other LINE-like retroposons that contain open reading frames encoding proteins involved in their transposition (Berg and Howe, 1989), no ORFs have been detected in the 3'-most 3 kbp of any of the HeT-A elements sequenced to date (Biessmann *et al.*, 1992a). It is quite possible that such ORFs lie still more 5' in the element, or there are only a few intact HeT-A elements in the genome that supply the transposition functions, or these functions are encoded elsewhere in the genome.

iii. Human. The human pseudoautosomal region (PAR) is a 2600-kbp-long subtelomeric region that is shared between the X and Y chromosomes (reviewed by Ellis and Goodfellow, 1989). The PAR is part of the region of pairing between the sex chromosomes during pachytene and is thus believed to be required for correct sex chromosome segregation in males. This region has been studied extensively in recent years and substantial progress has been made in mapping and characterizing it genetically and molecularly. The results obtained may give an indication of the complexity and organization of the subtelomeric regions of higher eukaryotes. One major characteristic of the PAR is its high rate of recombination during male meiosis (Rouyer *et al.*, 1986). The boundary between the PAR and Y-specific sequences has been defined genetically. Molecular cloning of DNA from humans (Ellis *et al.*, 1989) and apes and monkeys (Ellis *et al.*, 1990) has revealed that an *Alu* retrotransposable element is inserted at the boundary in humans and apes but not in Old World monkeys. Because the position of the boundary is the same in all species studied, the *Alu* element probably inserted after the determination of the boundary has occurred, suggesting that the element does not define the boundary.

Using several DNA probes and recombination mapping, a long-range linkage map of the PAR has been established (Brown, 1988; Petit *et al.*, 1988). Although some loci have been mapped by restriction site polymorphism, it is not known how many different families of sequences are obligatory residents of the PAR and which sequences make up the major portion of the PAR. Islands of CpG-rich repeats are spread throughout the PAR and are clustered at the distal end. It is still an open question whether any genes reside within the PAR. In analogy with yeast, in which the *SUC* genes are embedded between the X and Y′ subtelomeric repeats (Carlson *et al.*, 1985), genes might well be interspersed in the subtelomeric regions of human chromosomes. So far only one gene, *MIC2*, encoding a cell surface antigen, has been identified at the proximal boundary of the PAR (Goodfellow *et al.*, 1986).

Subtelomeric regions of human autosomes are much shorter. Maps of terminal regions have been established of the short arm for chromosome 4 (Allitto *et al.*, 1991; Bucan *et al.*, 1990), the long arm of chromosome 7 (Dietz-Band *et al.*, 1990), the short arm of chromosome 16 (Wilkie *et al.*, 1991), and the long arm of chromosome 21 (Burmeister *et al.*, 1991). In two cases, the distal-most single-copy genes have also been mapped in the process, thus these maps provide a good estimate of the lengths of the subtelomeric regions. The gene encoding the b subunit of the neuronal calcium-binding protein S100 maps between 50 and 200 kbp from the telomere at 21q (Burmeister *et al.*, 1991), and the α-globin genes reside at either 170, 350, or 430 kbp proximal to the telomere on 16p (Wilkie *et al.*, 1991). This study also demonstrates clearly the high degree of stable length polymorphism at this telomere.

Although most of the molecular probes used for mapping the PAR recognize highly polymorphic, short minisatellites, a more complex subtelomeric interspersed repeat (STIR) has been characterized (Simmler *et al.*, 1985; Petit *et al.*, 1990; Rouyer *et al.*, 1990). Copies of the repeat are about 350 bp long with a 270-bp core of 80% homology among the pseudoautosomal (and 55–70% homology among the autosomal) members of the family. STIRs occur on a subset of telomeres, where they are interspersed either in small clusters or singly within the subtelomeric regions. They seem to be located at least 20 kbp proximal to the telomeric repeats and have sometimes been found in the noncoding regions of genes located near telomeres. The function of STIRs is unknown, but it has been proposed that they may play a role in the increased recombination frequency of the PAR and between homologous autosomal telomeric regions, perhaps by promoting initiation of pairing at meiosis (Rouyer *et al.*, 1990). A less well-characterized repeat (20–25 copies per genome) has been found to be predominantly associated with the telomeric region of 4p, but also occurs near the end of chromosomes 14 and 21 (Altherr *et al.*, 1989). This repeat may be part of a larger array of yet uncharacterized subtelomeric repeats.

b. Satellite-like Repeats. Besides the more complex moderately repeated sequences, subtelomeric regions of many species contain satellite-like tandem arrays of 21- to 120-bp repeats. Such repeats have been described in *Plasmodium* (Oquendo *et al.*, 1986; Corcoran *et al.*, 1988), *Dictyostelium* (Emery and Weiner, 1981), *Physarum* (Johnson, 1980; Bergold *et al.*, 1983), *Drosophila* (Bachmann *et al.*, 1990), *Chironomus* (Carmona *et al.*, 1985; Saiga and Edström, 1985; Nielsen *et al.*, 1990; Cohn and Edström, 1991), rye (Bedbrook *et al.*, 1980a), and humans (Simmler *et al.*, 1985, 1987; Brown *et al.*, 1990; de Lange *et al.*, 1990). It appears that these repeats rearrange frequently to produce the high degree of polymorphism at the chromosome ends that has been found in several species and is most clearly demonstrated in *Plasmodium* (Foote and Kemp, 1989) and humans (Cooke *et al.*, 1985; Simmler *et al.*, 1985).

In *Plasmodium*, the genes encoding surface antigens are located within the polymorphic telomeric region. Because some of the genes are composed of reiterated sequences, antigenic polymorphism probably arises by intragenic recombination between alleles (Pologe and Ravetch, 1986), enhanced by a high degree of recombination between the satellite-like telomeric repeats (Corcoran *et al.*, 1988). However, this may also lead to occasional partial or complete deletions of these genes from their telomeric locations (Cappai *et al.*, 1989; Foote and Kemp, 1989).

   c. Distribution of Subtelomeric Repeats in Human. Most of the molecular probes that have been used to map the human PAR, e.g., those defining the loci *DXYS 14, 15, 17,* and *20,* hybridize to highly polymorphic minisatellites (Cooke *et al.*, 1985; Simmler *et al.*, 1985); two of them have been sequenced (Simmler *et al.*, 1987). Despite these short-range polymorphisms, there is surprisingly little long-range polymorphism in the PAR, making it possible to establish a linkage map (Brown, 1988; Petit *et al.*, 1988).

   A detailed molecular analysis of the most distal part of human telomeres has been made possible by cloning of human chromosome ends on linear yeast artificial chromosomes, or YACs (Brown, 1989; Cheng *et al.*, 1989; Cross *et al.*, 1989; Riethman *et al.*, 1989; Bates *et al.*, 1990; Cheng and Smith, 1990; Guerrini *et al.*, 1990). This technique relies on the observation that telomeric repeats from a number of organisms can provide telomere function in yeast (Pluta *et al.*, 1984; Shampay *et al.*, 1984), and together with other cloning techniques has made it possible to study the most distal portion of the complex subtelomeric repeat region.

   By cloning and sequencing several kilobase pairs from three different human telomeres, Brown *et al.* (1990) found several families of short, tandemly repeated minisatellites adjacent to the telomeric (TTAGGG)$_n$ repeat. Two of these telomeres contained the same set of minisatellites in the same proximal-to-distal arrangement, but a different number of repeats in each cluster, and the third telomere contained a different set of repeats. A 29-bp hypervariable, GC-rich minisatellite, also described by de Lange *et al.* (1990), may be responsible for CpG islands at the distal end of the PAR and might correspond to some of the GC-rich repeats defined by Cheng *et al.* (1989) using hybridization. Brown *et al.* (1990) also found that DNA segments located 2 to 5 kbp proximal to the telomeric repeat on different chromosomes hybridize to different subsets of telomeres and that the subsets vary among individuals. Such a polymorphic pattern is reminiscent of the distribution of the Y' repeats in yeast. One subtelomeric probe was specific for the long arm of chromosome 7, which was cloned independently by Riethman *et al.* (1989). One of the subtelomeric probes also hybridized reproducibly to an interstitial site on chromosome 2. It is not clear what type of repetitive element these probes represent, but the sequence of one of the subtelomeric regions indicates that it is not internally repetitive like a satellite. Sequences of the other probes are as yet undetermined. It is intriguing that each subset of chromosomes that hybridizes to a particular subtelomeric probe exhibits an almost identical restriction map in the 10 kbp adjacent to the telomeric repeats. Thus, the subtelomeric probes that identify a given subset of telomeres seem to be embedded in the same molecular environment at the same distance from the telomeric repeat. Proximally, the restriction patterns diverge.

Taken together, the mapping data suggest that the human subtelomeric region is organized as a series of repeated sequences (Cheng *et al.*, 1989; Riethman *et al.*, 1989; Brown *et al.*, 1990; de Lange *et al.*, 1990; Rouyer *et al.*, 1990). Only a small number of subtelomeric repeats have been identified to date, some of which are organized as short minisatellites, whereas others are longer repeats that are interspersed into the subtelomeric regions. Although some members seem to be restricted to one telomere, most are present at a subset of telomeres, suggesting a polymorphic distribution varying from individual to individual, making it unlikely that each telomere is defined by a combination of such subtelomeric repeats. Moreover, no subtelomeric element has been defined that is present at detectable levels at all of the telomeres, making it an obligatory component of the telomeric structure. Yet much of the subtelomeric region remains to be analyzed and it is possible that such an element will be found. Alternatively, such a conserved subtelomeric element may not exist, and the only sequence that all telomeres of one species have in common would be the terminal telomeric repeat. In that case, the function of the subtelomeric region may be to allow frequent recombination, which can be achieved by a variety of different sequences, or act as a buffer zone between the chromosome end and the most distal gene.

B. PROTEINS

Gene-sized DNA fragments in the *Oxytricha* macronucleus, containing C_4A_4 repeats at their ends, have been used to study protein/DNA interactions and nucleosome spacing. It has been shown that the termini are tightly complexed with proteins and are, thus, protected from micrococcal nuclease and MPE-Fe II attack (Gottschling and Cech, 1984). Two basic proteins, 56 and 41 kDa in size, have been identified that bind strongly to the terminal regions, protecting them from exonuclease Bal 31 digestion (Gottschling and Zakian, 1986) and from methylation of certain guanine residues (Price and Cech, 1987). The proteins seem to form a heterodimer (Price and Cech, 1989) that is tenaciously bound to DNA, making the complex stable in 2 M NaCl (Gottschling and Zakian, 1986; Price and Cech, 1987, 1989). In binding experiments *in vitro,* the proteins show a strong preference for the natural terminus containing a 3′ single-stranded overhang over random single-stranded tails or altered subterminal duplex sequences (Gottschling and Zakian, 1986; Raghuraman *et al.,* 1989), but do not bind to folded termini in the G-quartet configuration (Raghuraman and Cech, 1990). Different types of complexes can be formed *in vitro* and *in vivo* between the *Oxytricha* telomere-binding protein and the telomeric DNA. These indicate a dynamic state of the nucleoprotein complex and have been implicated in mediating telomere–telomere associations (Raghuraman and Cech, 1989).

The gene for the 41-kDa protein has been cloned and shown to encode a protein with similarity to histone H1 and to contain a hydrophobic domain that may function in protein–protein interactions (Hicke *et al.,* 1990). Cloning of the gene encoding the 56-kDa protein has recently been reported (Gray *et al.,* 1991). *In vitro* binding experiments of these two proteins to macronuclear DNA showed that the large subunit binds tightly to 3′ single-stranded, telomeric overhangs and that the small subunit is required to generate a nucleoprotein complex that resembles the *in vivo* situation. The homologous genes have also been cloned from *Stylonychia* (Fang and Cech, 1991). The genes for these telomere-binding proteins are sufficiently conserved to exhibit cross-reactivity with *Saccharomyces, Trypanosoma, Caenorhabditis,* and chicken genomic DNA, indicating the possibility that comparable proteins performing similar functions might be present in these organisms (Fang and Cech, 1991).

A protein with similar features has recently been characterized in another hypotrichous ciliate, *Euplotes* (Price, 1990). Tightly binding proteins and altered chromatin structure have also been reported at the termini of extrachromosomal ribosomal genes of *Tetrahymena* (Blackburn and Chiou, 1981; Budarf and Blackburn, 1986) and *Physarum* (Cheung *et al.,* 1981; Lucchini *et al.,* 1987; Coren *et al.,* 1991).

In yeast, a telomere-binding protein has been described with high *in vitro* binding specificity for the native yeast telomeric sequence (Berman *et al.,* 1986; Buchman *et al.,* 1988a,b). It has recently been demonstrated that this protein is the transcriptional regulator RAP1 (Longtine *et al.,* 1989; Conrad *et al.,* 1990; Lustig *et al.,* 1990). However, several lines of evidence suggest (Blackburn, 1991) that a different group of proteins might bind the single-stranded overhangs at the very ends of termini. These are distinct from the RAP1-type proteins that bind to the internal double-stranded regions of the telomeric repeat sequence. Overexpression of RAP1 causes chromosome instability and loss, and it has thus been implicated in telomeric interaction with the nuclear lamina or matrix (Conrad *et al.*, 1990). Indeed, RAP1 can mediate DNA loop formation (Hofmann *et al.*, 1989) and fractionates with the nuclear scaffold (Cardenas *et al.*, 1990).

Two other telomere-binding proteins, TBFα and TBFβ, with properties distinct from RAP1, have recently been isolated from yeast. They bind to the telomeric repeat and to the junction of the subtelomeric X element and the adjacent $C_{1-3}A$ tracts (Liu and Tye, 1991). The large number of telomere-binding proteins suggests that complex interactions are probably involved in the telomeric functions.

The introduction of extra telomeric DNA sequences on high-copy, linear plasmids causes telomere elongation, leading Runge and Zakian (1989) to postulate the existence of a titratable protein that normally prevents telomere elongation. It is not known whether this hypothetical protein is related to the other telomere-associated proteins.

Bearing in mind the recovery of chromosomes (described below) that are deleted for terminal regions, another protein, identified in humans, should be noted. This protein, known alternatively as Ku (Mimori and Hardin, 1986; Mimori *et al.*, 1986) or nuclear factor IV (NFIV) (de Vries *et al.*, 1989; Stuiver *et al.*, 1990), is a heterodimer that recognizes and binds to the ends of double-stranded DNA, but not a specific DNA sequence, and may slide along the DNA until movement is blocked by the presence of NFI or NFIII. It has been suggested that NFIV plays a role in DNA replication (Stuiver *et al.*, 1991), but a protein with similar binding properties has been postulated as a telomere-capping telomere in *Drosophila* (Biessmann *et al.*, 1990a).

Naturally occurring linear plasmids as well as poxviruses and parvoviruses have been identified in a number of eukaryotic organisms, including fungi (Kikuchi *et al.*, 1984; Meinhardt *et al.*, 1986), plants (Kemble and Thompson, 1982; Erickson *et al.*, 1985), and animals (Rekosh *et al.*, 1977). Many of these have common structural features, including inverted terminal repeats and a protein covalently bound to the 5' terminus at each end. Linkage of the initial 3' nucleotide to the covalently bound protein is thought to act as a primer for DNA synthesis.

C. Conclusion

Despite the fact that the molecular analysis of telomeres from a variety of organisms has made much progress in the past few years, it is far from complete. Our knowledge of the obligatory components of a telomere is still limited because it is difficult to distinguish between them and fortuitous residents of the telomeric region. As we pointed out earlier, there seem to be only very few obligatory telomeric structural components: some tightly binding proteins, telomeric repeats in organisms that depend on the telomerase elongation mechanism, and perhaps the X repeats in yeast and the previously described HeT-A retroposon in *Drosophila*. The latter two are the only DNA sequences besides telomeric repeats that appear to be consistently found at the telomeres of all the chromosomes in *S. cerevisiae* or *Drosophila*, respectively.

III. Function of Telomeres

At least four functions have been attributed to telomeres. Two of these, protection of the chromosome ends from degradation and fusion (capping), and replication, are essential for the long-term survival of the cell. The other two, control of gene expression and nuclear organization, are more problematic. It is not clear whether the telomere is involved or a closely linked chromosomal domain, such as heterochromatin, is involved. Nor is it clear whether these functions are important in themselves or are the consequences of another activity or association.

A. CHROMOSOME CAPPING

A predominant feature of the telomere is its capping function, as was recognized by Muller (1940; Muller and Herskowitz, 1954). Caps appear to be necessary for chromosome stability because nontelomeric ends of broken chromosomes have the tendency to fuse with other broken ends. Such a process ligates unrelated DNA ends together and has been investigated extensively by monitoring the recovery of radiation-induced chromosome rearrangements in the germ line of *Drosophila*. Here we will summarize only a few pertinent results from a vast literature. There seems to be no DNA repair mechanism in spermiogenesis, and lesions induced in chromosomes of mature sperm are repaired in the zygote (Muller, 1940; Maddern and Leigh, 1976) under the genetic control of the mother (Graf *et al.*, 1979). Radiation-induced lesions are actively repaired and rejoined within 15 minutes in stage 7 oocytes (Parker and Hammond, 1958) but remain open in stage 14 oocytes for several hours (Würgler and Matter, 1968), often until after fertilization. In the zygote, DNA breaks are repaired after fertilization in a 15- to 20-minute temporal window (Muller, 1940; Sobels, 1974; Maddern and Leigh, 1976; Leigh, 1978) before the formation of the gonomeric spindle of the first zygotic mitosis. Thus, the capping function of telomeres is required throughout these times of active DNA ligation in order to prevent fusion of chromosome ends. A single broken chromosome end derived from a meiotic anaphase II dicentric chromatid bridge in *Drosophila* appears to be a dominant embryonic lethal (Sturtevant and Beadle, 1936; Novitski, 1952, 1955), although the reason is not known.

When proliferating mammalian cells are exposed to DNA-damaging agents, such as chemicals that cause DNA cross-links (Konopa, 1988), X irradiation (Kimler *et al.*, 1981), and topoisomerase inhibitors (Barlogie *et al.*, 1976; Drewinko and Barlogie, 1976), they arrest in the G_2 phase of the cell cycle. It is thought that a signal is released in response to DNA damage, including broken chromosome ends, to arrest the cell cycle, and that the G_2 arrest allows time to repair the damage before mitosis. Treatment of cells with the topoisomerase inhibitor etoposide or with γ irradiation causes a rapid inhibition of the p34^{cdc2} kinase activity (Lock and Ross, 1990). This kinase is active primarily at the G_2–M boundary and is thought to play a role in the control of the transition into mitosis. Similarly, in yeast, chromosome breaks cause a delay in G_2 until repair can be effected. This delay is under the genetic control of the *RAD9* gene (Weinert and Hartwell, 1988). A chromosome break-induced cell cycle delay also has been proposed for *Drosophila* (Smith *et al.*, 1985; Baker *et al.*, 1987).

We know very little about the molecular nature of the telomeric chromosome cap. It is conceivable that capping is achieved by an unusual secondary structure of the DNA molecule which would prevent ligation. A possible candidate is the G-quartet structure (Henderson *et al.*, 1987; Williamson *et al.*, 1989; Hardin *et al.*, 1991) formed by non-Watson–Crick base pairing between the terminal guanine residues in the single-stranded 3' overhang of the telomeric G-rich repeats, as described for *Tetrahymena* and *Oxytricha*. However, no evidence exists on whether these secondary structures actually form *in vivo* at the physical ends of chromosomes. If they do, they might interact with proteins, but *in vitro* binding studies with the telomere-binding protein from *Oxytricha* showed no binding to the four-stranded G-quartet structure (Raghuraman and Cech, 1990). It is also possible that proteins that preferentially bind to the telomeric repeats may protect the ends from degradation and fusion (Gottschling and Cech, 1984). Several telomere-binding proteins have been isolated from a variety of organisms (described earlier), and a protective function may be achieved by the interaction of several protein components with each other and the DNA termini. However, as we show later, neither of these hypotheses adequately explains the recovery of chromosomes in *Drosophila* with broken and receding ends (Biessmann and Mason, 1988; Levis, 1989; Biessmann *et al.*, 1990a).

B. REPLICATION AND ELONGATION

1. Incomplete Replication of Chromosome Ends

As all known DNA polymerases require a primer with a free 3′ hydroxyl group, and synthesize in the 5′ to 3′ direction, telomere replication poses special problems for the linear chromosomes of eukaryotes (Watson, 1972). Removal of the RNA primer of the lagging strand will result in a 3′ terminal overhang in one of the daughter strands and thus the DNA molecule will lose sequences from the end with each round of replication. Therefore, eukaryotic organisms and their viruses must possess a mechanism to avoid or counterbalance this inevitable loss of DNA in order to prevent loss of important genetic information.

One method to avoid the loss of telomeric sequence from a linear chromosome has been adopted by the poxviruses. These are linear, cytoplasmic DNA viruses containing closed terminal loops (Baroudy *et al.*, 1982a) such that the double-stranded DNA molecule contains a single continuous polynucleotide chain. The telomere regions are AT rich, incompletely base paired, and exist in two inverted and complementary forms (reviewed by DeLange and McFadden, 1990). Replication around the terminal hairpin loop produces concatemers (Moyer and Graves, 1981; Baroudy *et al.*, 1982b), and unit length chromosomes are resolved by a process that is poorly understood.

Though some DNA viruses avoid the problem of incomplete replication using terminal hairpin structures, most eukaryotic organisms counterbalance loss of material from chromosome termini by elongating telomere sequences. Therefore, it should be possible to observe loss of DNA from chromosomal termini when the counterbalancing mechanism is disrupted. DNA loss has been observed in *Drosophila* at terminal deletions of the X chromosome (Biessmann and Mason, 1988; Biessmann *et al.*, 1990a, 1992b) and at the end of the right arm of chromosome 3 carrying a *white* (*w*) gene in a P element inserted into the subtelomeric region (Levis, 1989). In both cases a continuous loss of nucleotides occurs from the chromosome end at a rate of 70–80 bp per fly generation. As the chromosome end recedes, sequences are lost from the *yellow* (*y*) gene at the tip of the X chromosome (Biessmann and Mason, 1988) or from the *w* gene at the tip of 3R (Levis, 1989). As neither gene is vital, and the X chromosome deletions are kept in males carrying a genetically marked duplication, the receding chromosome end causes visible changes toward yellow body color or white eye pigmentation. The observed rate of loss per generation can be correlated with the average number of mitoses in the male germ line (Levis, 1989; Biessmann *et al.*, 1990a). Given a *Drosophila* eight-nucleotide RNA primer (Kitani *et al.*, 1984), replication of a blunt ended chromosome is expected to produce one daughter molecule with an eight-nucleotide, 3' single-stranded overhang and one full-sized daughter. Assuming that the RNA primer is removed and the DNA single-stranded overhang is not, the next round of replication will yield one daughter of original length, two with an eight-nucleotide overhang, and one with a blunt end shortened by 8 bp. Thus, in the absence of DNase activity, the average shortening of a linear molecule would be 2 bp per round of replication. Some fraction of the resulting chromosomes is expected to shorten as much as twice as fast, and some not at all. With a minimum of 18 and a maximum of approximately 48 mitoses in the germ line during the life span of a male (Lindsley and Tokuyasu, 1980), 36 to 96 bp would be lost per generation, which agrees quite well with the observed figure of 70–80 bp.

Similar cases of telomere shortening have been reported in other organisms. Under normal conditions, terminal nucleotide loss caused by incomplete DNA replication is balanced by one of several mechanisms. If, however, a compensating mechanism is impaired by a mutation in an essential gene of the telomere extension pathway, terminal loss of nucleotides can be observed. This is the case in *S. cerevisiae* carrying the mutant allele *est1* (Lundblad and Szostak, 1989). The *EST1* gene encodes a protein with similarities to reverse transcriptases (Lundblad and Blackburn, 1990). Telomerase is also believed to a specialized reverse transcriptase (for a review, see Blackburn, 1990a). Telomeres in the haploid yeast *est1* strain shorten at a rate of 4 bp per cell cycle, and the cells eventually senesce and die.

Telomere shortening that is correlated with cell proliferation has also been observed in human cells. Telomere length of nontransformed human fibroblasts decreases steadily with continuous passage in culture (Harley *et al.*, 1990). Because the total amount of telomeric DNA decreases, and internal repetitive sequences do not seem to be rearranged, the observation can most easily be explained by terminal loss of nucleotides due to incomplete replication of the termini, although mechanisms involving unequal recombination cannot be ruled out. The decrease in mean telomere length is about 50 bp per population doubling. The loss of telomeric DNA is greater than would be expected from the simple loss of the RNA primer, and is consistent with the loss of an Okazaki fragment, although the mechanism of loss is not understood. If, indeed, terminal loss is caused by incomplete replication, it is implied, but not proved, that the telomere elongation mechanism is not functional in these fibroblasts. This question has recently been addressed by Counter $et\ al.$ (1992). Human embryonic kidney cells, which can be cultured for only 16 population doublings, show extended life span $in\ vitro$ after transfection with SV40 tumor antigens. During extended culture, telomere repeat lengths decrease at a rate of about 65 bp/generation until they reach an average length of 1.5 kbp, at which time the cell line dies. These cells do not contain detectable telomerase activity, which would explain why telomeres become shorter. However, in one case an immortal cell line was derived in which telomere lengths stabilized and telomerase activity was detectable, suggesting a correlation.

Harley $et\ al.$ (1990) found that telomere length in fibroblasts declines with the age of the donor, indicating that telomere shortening also occurs $in\ vivo$ and suggesting that the length of telomeric repeats may be inversely correlated with the number of DNA replications the cells have undergone. Analyzing telomere lengths in various human tissues, it was found that this correlation generally holds (Cooke and Smith, 1986; Allshire $et\ al.$, 1988, 1989; de Lange $et\ al.$, 1990; Hastie $et\ al.$, 1990). Telomere repeats are longest in sperm [10–14 kbp (de Lange $et\ al.$, 1990) to 20 kbp (Hastie $et\ al.$, 1990)] and placenta, and they are 1–2 kbp shorter in fetal liver and kidney, 12 kbp shorter in intestinal mucosa, and on average 14 kbp shorter in adult blood. Generally there is also a significant reduction of telomere length in a given tissue as a function of age (Hastie $et\ al.$, 1990), again supporting the notion that telomeric length decreases with the number of DNA replications. Thus, the hypothesis has been put forward that telomerase may be active only in the germ line, where the telomere length is "reset" during gametogenesis (Cooke and Smith, 1986). At the calculated rate of telomere repeat loss in blood cells of 33 bp per year, however, telomeres will not be "used up" during the lifetime of an individual.

A similar correlation has been found in tumors compared with their tissues of origin. For instance, colorectal carcinomas generally exhibit considerably shorter telomeres than the mucosa from which they are derived, and most of the telomere shortening occurs in the adenoma stage when proliferation is highest (Hastie *et al.*, 1990). Similar results were obtained in Wilm's tumors, a childhood kidney neoplasia, and mammary carcinomas (de Lange *et al.*, 1990). Age- and tissue-dependent telomere shortening has not been detected in mice; this may be due to the inability to resolve relatively small differences in length in mouse telomeres, which are up to 150 kbp long (Kipling and Cooke, 1990; Starling *et al.*, 1990).

Although these data all show the same trend, a strict cause-and-effect relation between cell proliferation *in vivo* and telomere length reduction has not been established. Because not all mechanisms that control telomere length are fully understood, other factors, such as exonuclease degradation or unequal recombination, may also play a role.

2. Telomere Shortening and Cellular Senescence

Shortened telomere repeats have a severe effect on the cell or organism. As observed by Lundblad and Szostak (1989), *est1* yeast cells exhibit a senescence phenotype associated with the progressive decrease in telomere length, which is characterized by a gradual reduction in viability and growth rate and an increase in cell death. Moreover, *est1* cells show increased loss of linear plasmid molecules. To explain this phenotype it was hypothesized that a decreased telomere repeat may cause chromosome rearrangement and eventually chromosome loss. There are at least two possibilities for how a receding telomere may cause chromosome loss. First, the loss of a telomere may expose nontelomeric sequences that are not capped, and are therefore sticky, initiating a bridge–breakage–fusion cycle. Support for this hypothesis comes from the observation that senescent fibroblasts in culture that are subject to telomere shortening (Harley *et al.*, 1990) exhibit a high frequency of chromosome fusion (Benn, 1976; Sherwood *et al.*, 1988). A high number of dicentric chromosomes also occurred in human embryonic kidney cells with shortened telomeric repeat lengths after prolonged *in vitro* cultivation (Counter *et al.*, 1992). Second, the interaction of normal-length telomeres with each other, and with nuclear components such as the matrix and the lamina, may be required for proper chromosome segregation. If telomeres are too short these interactions may be severely affected, perhaps because binding sites for important proteins are diminished or lost, or because telomeric hybridization, recombination, or other DNA-specific interactions, such as quadruplex formation, are reduced due to the loss of telomeric repetitive DNA sequences. Using a value of telomere shortening of 50 bp/

generation in human fibroblasts in culture, Levy et al. (1992) estimated that the mean telomere repeat length at senescence would be about 2 kbp. With the length variations of telomeres at different chromosomes, some chromosomes may have little, if any, telomeric repeats left at this point. Moreover, this rate of shortening is consistent with the cell cycle kinetics of fibroblasts in culture, suggesting a releationship between telomere shortening and the limited life span of cultured fibroblasts, as proposed earlier by Olovnikov (1973).

Another case of altered telomeric structure that is correlated with cellular senescence has recently been reported (Yu et al., 1990). When Tetrahymena is transformed with a high-copy-number plasmid overexpressing a mutated sequence of the RNA component of telomerase, severe phenotypic effects are observed. For two altered sequences, introduction of the mutated gene resulted in the in vivo synthesis of the new telomeric repeat sequences, and the telomere repeats were longer than in control cells. In the case of a third altered sequence, telomere shortening was observed, possibly due to an interference of the mutated RNA template with telomerase activity. However, all these mutations caused a senescence phenotype, accompanied by impaired nuclear and cellular divisions and reduced viability. Before the cells died, the shorter telomere transformants, which exhibited the most severe phenotype, were characterized by much larger than normal cells with enlarged and irregularly shaped nuclei.

The observation that receding, as well as longer, telomeres with altered nucleotide sequence may cause very similar morphological changes, described as senescence, suggests that quality as well as quantity of telomeric repeats is important for their role in nuclear and cellular division. The function of telomeres in maintaining nuclear morphology and their involvement in chromosome segregation are not well understood. It is, however, conceivable that the senescence phenotype is due to an impaired interaction of the altered or shortened telomere repeats with specific telomere-binding proteins (described earlier in this review). In fact, overexpression of the telomere-binding protein RAP1 in yeast causes telomere lengthening and a concomitant decrease in chromosome stability (Conrad et al., 1990), suggesting that the interaction of telomere-binding proteins with telomere repeats and possibly other nuclear structures is important for proper chromosome segregation. Moreover, telomere repeats complexed with specific proteins may associate with the nuclear lamina (Agard and Sedat, 1983; Hochstrasser et al., 1986; Hochstrasser and Sedat, 1987; Shoeman et al., 1988; Hiraoka et al., 1990; Paddy et al., 1990; Shoeman and Traub,

1990), or nuclear matrix (de Lange, 1992), and thus be involved in maintaining nuclear morphology and allow faithful chromosome segregation. Further studies of these interactions will help to elucidate the various aspects of cellular senescence and the role telomeres play in this complex phenotype.

In several studies discussed above, telomere shortening was observed in tumors when compared to their tissue of origin (Hastie *et al.*, 1990; de Lange *et al.*, 1990). Other studies have shown that neoplastic cells often contain chromosomes that are fused at their telomeres (Fitzgerald and Morris, 1984; Mandahl *et al.*, 1985, 1988; Morgan *et al.*, 1986; Dewald *et al.*, 1987; Kovacs *et al.*, 1987; Pathak *et al.*, 1988). It is, therefore, conceivable that telomere shortening and increased levels of telomere–telomere fusion observed in tumors and in senescent fibroblasts may be related (Benn, 1976; Harley *et al.*, 1990). However, it remains to be demonstrated whether telomere losses play a role in causing malignancy or are a consequence of tumorigenesis. As discussed by Hastie and Allshire (1989) and by Hastie *et al.* (1990), telomere fusions may arise because chromosome ends with reduced telomeric repeat lengths expose nontelomeric sequences and thus become fusigenic, or alternatively, end-to-end fusions might occur normally as part of the telomere interactions and be resolved more slowly in the tumor cells. Regardless of the mechanism, this telomeric fusion may be responsible in part for the loss of alleles that often occurs in tumors (Knudson, 1986; Stanbridge, 1990). The predominant mechanism causing inactivation of tumor suppressor genes in tumors is abnormal chromosome segregation. Nondisjunction leading to allele loss and hemizygosity (or after reduplication of the remaining chromosome, to homozygosity) of the mutant allele has been shown to occur in retinoblastomas (Cavenee *et al.*, 1983, 1985), in Wilm's tumors (Fearon *et al.*, 1984; Koufos *et al.*, 1984; Orkin *et al.*, 1984), and in colorectal carcinomas (Vogelstein *et al.*, 1989). In addition, mitotic recombination (Cavenee *et al.*, 1983) and somatically occurring internal deletions (Fung *et al.*, 1987) are also involved in the somatic inactivation of the retinoblastoma gene.

3. Telomere Elongation Mechanisms

In order to avoid loss of genetic information, the continuous loss of nucleotides due to incomplete replication of DNA ends must be avoided or counterbalanced. Though there is some evidence suggesting that telomere elongation may be restricted to the germ line, at least in mammals, this needs to be substantiated. In yeast, elongation of plasmids occurs in vegetative cells. Telomere elongation can be achieved by several mechanisms with very specific requirements for each one. Because the two most well-studied elongation mechanisms, telomerase and recombination (Fig. 3), have recently been reviewed extensively (Zakian, 1989; Zakian *et al.*, 1990; Greider, 1990; Blackburn, 1990a,b, 1991), we will discuss here only a few recent findings and then concentrate on the discussion of a possible third mechanism, namely, transposition.

a. Telomerase. This elongation mechanism requires as one component the simple telomeric repeats that have been found at the end of chromosomes in protozoa, vertebrates, and higher plants. A second requirement is the enzyme telomerase, which recognizes these repeats and elongates them with an intrinsic RNA molecule as template. An altered sequence of this RNA template results in correspondingly altered telomeric repeats (Yu *et al.*, 1990). A reverse transcriptase activity that would be required for the activity of this enzyme has been identified tentatively by the *est1* mutation in yeast (Lundblad and Szostak, 1989). However, a functional telomerase activity has not yet been demonstrated in yeast by using *in vitro* assay conditions that successfully identify telomerase activity in *Tetrahymena* (Greider and Blackburn, 1985, 1987, 1989) and HeLa cells (Morin, 1989). Telomerases have a range of primer specificity and will recognize slightly variant telomeric repeats, but can only add repeats encoded by the intrinsic RNA template (Greider and Blackburn, 1987; Blackburn *et al.*, 1989; Morin, 1989; Shippen-Lentz and Blackburn, 1989). Telomerase activity is probably also responsible for developmentally programmed attachment of telomeric repeats to fragmented chromosomes in *Tetrahymena* and *Ascaris,* and for "healing" of broken chromosomes in some organisms by adding telomeric repeats at nontelomeric sites.

Such healing by telomerase has been observed in a human α-thalassemia case, in which telomeric repeats have been added to the end of a truncated chromosome 16 (Wilkie *et al.*, 1990a,b, 1991). Sequence analysis of the junction indicated similarities to human telomeric repeats that may have been recognized by telomerase *in vivo. In vitro* assays using oligonucleotides from the distal-most part of the junction confirmed that human telomerase recognizes the healing site via limited base pairing with the template and is capable of adding telomeric repeats (Morin, 1991). *In vitro* studies of *Tetrahymena* telomerase have shown recently that telomerase can also elongate DNA oligonucleotides that are not complementary to the telomerase RNA template

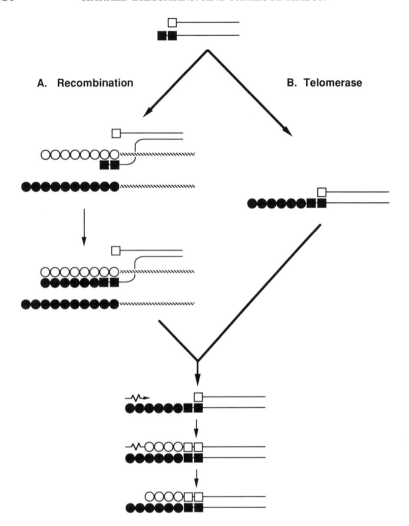

FIG. 3. Two possible pathways for formation of new telomeres in yeast. Different telomeric repeat sequences (represented by squares) can promote telomere formation in
yeast by serving as substrates for the addition of yeast $C_{1-3}A$ repeats (represented by
circles). (In both cases, the C-rich strand is represented by an open symbol, and the G-
rich strand is represented by the solid symbol.) After introduction of the plasmid into
yeast, replication occurs. After removal of the RNA primer, the G-rich strand is longer
than the C-rich strand (represented by an unpaired solid block). Telomere formation can
occur by either of two pathways: gene conversion (recombination) with another telomere,
or telomerase. Both pathways utilize the 3'-OH of the single-strand G-rich tail. (A) Recombination: the G-rich single-strand tail invades another telomere (donor) (the donor
chromosomal DNA is represented by hatched lines). At the donor end, most (all?) recombination events are initiated at the junction between telomeric DNA and unique DNA.

(Fig. 4). However, two telomeric repeats are required, which can be up to 36 bp away from the 3' end that is elongated (Harrington and Greider, 1991). Even though the specificity of telomerase *in vivo* may be more restricted than *in vitro*, the two results indicate that the enzyme may recognize telomeric repeats in the vicinity of a broken DNA end and elongate its 3' end by adding telomeric repeats directly onto non-telomeric sequences. This type of healing may also have occurred at a broken chromosome end in yeast, where a tract of telomeric repeats is found about 100 bp away from the break (Murray *et al.*, 1988). However, with lack of evidence for a telomerase activity in yeast, it has also been proposed tht the telomeric repeats could have been added by homologous recombination between the $C_{1-3}A$ telomeric repeats and the internal telomere-like stretch (Zakian *et al.*, 1990).

b. Recombination. A number of models have been proposed that invoke recombination as a mechanism for telomere elongation. These have been reviewed extensively (Dancis and Holmquist, 1979; Zakian, 1989). We will, therefore, concentrate on recent data that bear on the relevance of recombination for telomere elongation and that help to distinguish among the proposed mechanisms.

In some cellular slime molds and yeasts, the telomeric DNA sequence is more complex than would be predicted based on the simple telomerase model described above (Emery and Weiner, 1981; Shampay *et al.*, 1984). Although the telomeric DNA has a G-rich strand and may form tertiary structures similar to those described for telomeric repeats, the runs of guanines are irregular in length, and there is no obvious repeat unit. The best evidence in favor of a recombinational mechanism for telomere elongation comes from a series of experiments

---

The invading G-rich strand is extended about 300 bases by replication using the donor telomere as a template. After dissociation, the terminus carries a 300-base $G_{1-3}T$ single-strand tail. (B) Telomerase: a hypothetical yeast telomerase adds $G_{1-3}T$ repeats to the G-rich single-strand left after RNA primer removal. The end product of gene conversion and telomerase is the same, a telomere with an extended 3' single-strand tail. The extended strand can serve as a template for primase and conventional DNA polymerase-mediated replication of the complementary strand. Subsequent removal of the RNA primer (wavy line) would still leave a gap at the 5' end of the newly replicated strand, but no sequence information would be lost. (Alternatively, the extended G-rich strand could fold back on itself to form a terminal hairpin and provide the primer for replication of the C-rich strand.) Either gene conversion or telomerase (or both) could also mediate telomere replication in each cell cycle. In this case, the number of $G_{1-3}T$ repeats added might be fewer than during telomere formation. Reproduced with permission from Zakian *et al.* (1990).

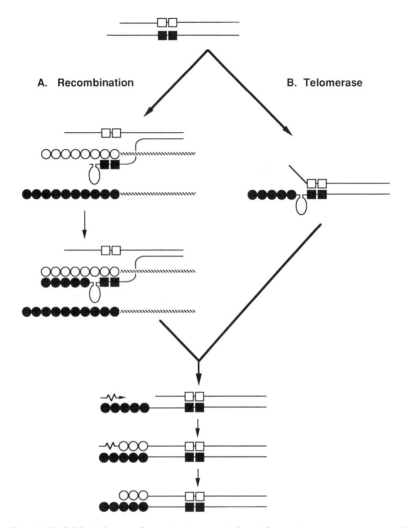

Fig. 4. Model for telomere formation in yeast when telomeric sequences are near, but not at the terminus. Sequence-specific associations may form between the telomeric repeats on the recipient stand and those on the donor. These associations might be based on Watson–Crick base complementarity or on non-Watson–Crick (quadruplex?) associations. (A) To be recognized and extended by DNA polymerase, the base pairing at the terminus of the recipient may be as short as 5–8 bp and may be facilitated by the complexity of yeast telomeric DNA. (B) Telomerase does not require that telomeric sequences be at the terminus. It may extend a single strand when telomeric sequences are within 30–40 bases of the end.

on linear plasmids carrying terminal *Tetrahymena* $(C_4A_2)_n$ or *Oxytricha* $(C_4A_4)_n$ repeats introduced into *S. cerevisiae*. When positioned in the natural orientation with respect to the terminus, the sequences at the ends of these plasmids are recognized as telomeric by the transformed cell and *Saccharomyces* telomeric sequences are added to the ends (Szostak and Blackburn, 1982; Dani and Zakian, 1983; Pluta *et al.*, 1984). When the transforming plasmid has *Tetrahymena* repeats at one end and *Oxytricha* repeats at the other, sequence information is transferred from one end to the other by recombination (Pluta and Zakian, 1989). Sequence data on the recombination products (Wang and Zakian, 1990b) reveal that long stretches of C_4A_2 repeats can be transferred from one end of the molecule to the other by a nonreciprocal, gene conversion mechanism without loss of C_4A_2 repeats from their original position (Fig. 3). Reciprocal recombination between the ends of two different plasmids can be ruled out because the observed addition occurs without the loss of C_4A_4 repeats from the recipient end that would be predicted by reciprocal recombination.

Elaboration of yeast sequences onto protozoan telomeric repeats is independent of *RAD52* function (Dunn *et al.*, 1984; Zakian *et al.*, 1985) and may occur even when short intervening nontelomeric sequences are at the terminus (Murray *et al.*, 1988; Wang and Zakian, 1990b). These short nontelomeric sequences are not lost, rather the new telomeric sequences are added to their distal ends. This has been interpreted as evidence against recombination, because recombination is expected to occur between homologous sequences (Murray *et al.*, 1988). As pointed out by Wang and Zakian (1990b), however, strand extension by a gene conversion mechanism may require very limited homology. In fact, slippage by DNA polymerase has been shown to account for frameshift mutations (Streisinger *et al.*, 1966; Ripley, 1990). Slippage during DNA synthesis can account for frameshift mutations of hundreds of base pairs and requires homology at the $3'$ end of the primer of as little as 5 bp. Thus, given the sequence similarity between yeast telomeric DNA and the terminal repeats on the plasmids, slippage can account easily for the addition of yeast telomeric DNA onto the ends of linear plasmids when protozoan telomeric repeats are nearby (Fig. 4), and recombination can account entirely for the elongation and maintenance of yeast telomeres.

There may be two domains of telomeric DNA at the ends of yeast chromosomes (Wang and Zakian, 1990a). Sequence data extending virtually to the chromosome terminus reveals a modified version of the yeast consensus sequence in the distal region of the telomeric DNA. Furthermore, the sequence of the distal region varies among descendants of a single transformant, whereas the sequence in the proximal 120- to 150-bp region varies little, if at all. It has been suggested that the distal domain participates in recombination and elongation, but that the proximal region is protected from these processes (Wang and Zakian, 1990a). Whether the protection of the telomeric domain is related to the hypothetical factor that limits telomere growth (Runge and Zakian, 1989) remains to be seen.

Recombination may also be used to maintain telomeres in the mitochrondria of *Tetrahymena* (Morin and Cech, 1986, 1988) and the extrachromosomal rDNA molecule of *Physarum* (Bergold *et al.*, 1983). The mitochondrial telomeres in *Tetrahymena* are composed of tandem repeats, but these repeats differ from the repeats on nuclear chromosomes in sequence and in size. Moreover, mitochondrial telomeres in different *Tetrahymena* species differ in sequence and length, with sequences from six species ranging from 31 to 53 bp in length. Unequal homologous recombination accompanied by expansion of molecules with many copies and loss of molecules with no copies of the telomeric repeat can account for the maintenance of telomere length, the homogeneity of the repeat sequence within a species, and its rapid evolution (Morin and Cech, 1986, 1988).

c. Transposition. It is conceivable that telomere elongation may be achieved by the addition of transposable elements to the end of chromosomes. This mechanism is particularly attractive for those organisms that appear to lack simple telomeric repeats required for elongation by telomerase or the more complex sequences that appear to play a role in recombination. As was pointed out earlier, no simple telomere repeats have yet been found in a variety of animal phyla, including arthropods, although negative evidence is not conclusive. The inability of several laboratories to detect any "typical" G-rich telomeric repeat in *Drosophila* with oligonucleotide probes (Richards and Ausubel, 1988; Meyne *et al.*, 1989) further suggests that *Drosophila* either has fewer and/or more degenerate telomeric repeats, or has a very different telomere elongation mechanism that does not require such repeats. Although the possibility exists that X repeats in yeast and HeT-A repeats in *Drosophila* share sequence similarities, as has been noted by low-stringency cross-hybridization (Chan and Tye, 1983b), this needs to be substantiated by more sequence analyses, and its significance, if any, needs to be elucidated. We raise here the evidence that telomere elongation in *Drosophila* might be achieved by DNA transposition.

If telomere elongation in *Drosophila* is based on transposition, the transposable element involved must have some unusual characteristic, such as a specificity for chromosome ends, DNA breaks, or a specific chromatin structure, and it must have a sufficiently high transposition rate and a relatively large size to ensure sufficient DNA transfer in a single transposition event. The HeT-A retroposon possesses some of these important features. It is found at all telomeres in *Drosophila* (Young *et al.*, 1983); it transposes preferentially, if not exclusively, to DNA ends (Biessmann *et al.*, 1990b, 1992b), and its transposition frequency and size would be capable of counterbalancing terminal nucleotide loss of 70–80 bp per generation incurred by incomplete replication of DNA ends (Biessmann and Mason, 1988; Levis, 1989; Biessmann *et al.*, 1990a). In order to assess transposition frequency and specificity of the HeT-A retroposon and to investigate events that occur at the end of a terminally deleted X chromosome, populations originating from single males were monitored for 17 generations (Biessmann *et al.*, 1992b). Daughters of single males from the population were tested by genomic Southern blot hybridization with a single-copy DNA probe homologous to a region adjacent to the terminus. In the majority of cases, DNA is lost from the distal end at 75 bp per generation, confirming earlier results on broken chromosome ends. However, 23 cases were discovered in which the terminal DNA fragment had suddenly increased in size by 3 to 12 kbp. DNA from six such elongated termini was sequenced, and all were found to be additions of HeT-A retroposons with their oligo(A) tail attached to the previously receding chromosome end. From these data, Biessmann *et al.* (1992b) determined a transposition rate for HeT-A elements of 0.7–1.0% per X chromosome per generation. Because the average estimated size of the often truncated HeT-A retroposon is 7 kbp, the net addition of sequences per X chromosome end would be 50–70 bp in a stochastic process considering all X chromosomes in the population. Theoretically, this lengthening caused by HeT-A transposition to the ends would be sufficient to counterbalance the progressive loss due to incomplete DNA replication of the chromosome ends.

Although transposition of this HeT-A retroposon has so far only been observed to occur to broken chromosome ends, where it may be a step in generating a new telomere, it may also occur regularly at natural chromosome ends. However, due to the repetitive nature of DNA sequences in the subtelomeric region, such events would be difficult to detect. If true, this would represent a telomere elongation mechanism that is drastically different from one based on the addition of telomeric repeats by telomerase. A long enough zone of middle repetitive sequences in the subtelomeric region would provide a "buffer" that would prevent loss of single-copy genes from the receding ends before elongation occurs again by transposition of a new HeT-A element to the end. Such a model makes several testable predictions. First, *Drosophila* chromosome ends would not be defined by short tandem telomeric repeats but different sequences are exposed at any given time at different chromosome ends due to terminal loss of nucleotides. Second, the vast majority of terminal fragments would originate from HeT-A elements, i.e., HeT-A sequences from any region within the possibly complete 12-kbp element would almost always be at the very end of a natural chromosome. Third, transposition of new HeT-A elements to old terminal, and thus receding, HeT-A elements would result in tandem arrays of these elements. Finally, the proximal end of each element in an array would be the same, but the distal end of each element would differ, depending on how long the element was when it transposed, and how long it had been exposed on the terminus before a new HeT-A element replaced it.

d. Conclusions. The mechanism of elongation by telomerase is currently quite well understood, and details of the mechanism, such as recognition requirements of the enzyme, will be worked out soon. However, though the RNA components of several telomerases are well characterized at the primary (Greider and Blackburn, 1989; Shippen-Lentz and Blackburn, 1990) and secondary structure level (Romero and Blackburn, 1991), almost nothing is known about the protein components, due to the fact that the enzyme has not yet been purified to homogeneity. Only in yeast, for which no telomerase activity has yet been demonstrated in extracts, has a mutation been identified (*est1*) that causes telomere shortening (Lundblad and Szostak, 1989), and because the sequence of the gene suggests homology to reverse transcriptases (Lundblad and Blackburn, 1990), it makes a good candidate for the still hypothetical yeast telomerase.

Transposition of HeT-A elements to broken chromosomes has been demonstrated in several cases, but it has yet to be shown that it occurs at natural telomeres in *Drosophila* or other arthropods. Moreover, much remains to be learned about this special transposition mechanism and its components, especially the necessary reverse transcriptase moiety. All known retroposons in *Drosophila,* as well as most of the human LINE elements, encode a gag-like protein and their own reverse transcriptase (Berg and Howe, 1989). Sequence analysis of some 5 kbp adjacent to the oligo(A) tail of the HeT-A element has provided evidence for a 2.8 kbp open reading frame encoding a 918 amino acid polypeptide with three zinc finger nucleic acid binding motifs and other similarities to gag-like proteins. No reverse transcriptase-encoding reading frame has yet been found (Biessmann *et al.*, 1992b). There are at least two possibilities of the source of reverse transcriptase. Either there are only very few intact "master" elements in the genome that provide this function, or the reverse transcriptase used for HeT-A transposition is encoded elsewhere in the genome, not on a HeT-A element, similar to the situation of the much shorter human SINE elements (Berg and Howe, 1989).

An intriguing aspect of telomerase has been discussed by Weiner (1988). He hypothesized that telomerase with its intrinsic RNA template and the required reverse transcriptase activity may be an evolutionarily very ancient enzyme. It may have evolved from a catalytic RNA to an RNP, which adds nucleotides to the termini of RNA molecules, and later must have acquired its reverse transcriptase activity when DNA became the molecule for storage of genetic information. Interestingly, the hypothetical telomere elongation mechanism by transposition of a retroposon also involves a reverse transcriptase activity and may have evolved independently.

4. Control of Telomere Length

Several studies have shown that telomere repeat length in a given organism can vary substantially under a variety of conditions. The mechanisms that reduce telomere length and those that cause its increase have to be balanced in order to maintain the size of a telomere. However, not much is known about the mechanisms of how this balance is maintained. As we discussed above, it is obvious that a progressively shortened telomere has severe effects on chromosome segregation and cell viability.

Telomeric length variations have been studied in trypanosomes, *Plasmodium, Tetrahymena,* and yeast. In continuous log-phase cultures of trypanosomes (Bernards *et al.*, 1983), the terminal restriction fragments of selected telomeres gradually increase in size, varying among chromosome ends. Telomere elongation occurs at a rate of 6–10 bp per generation, simultaneously causing an increase in overall length heterogeneity in the population (Van der Ploeg *et al.*, 1984). However, the chromosome ends do not continue to grow to infinite length. Two out of four random subclones from the population exhibit drastically shortened terminal fragments that were probably caused by large deletions in the telomeric region. It is unlikely that they were present in low abundance in the original population, rather it seems possible that the deletions were caused by the cloning procedure, which exposes the trypanosomes to heat shock. Longer telomeres seem to be more susceptible to deletions than do shorter ones. Although the molecular causes of gradual elongation and the sudden drop in telomere length have not been elucidated, these results are consistent with continuous lengthening of telomeric repeats during telomere replication or by the action of a telomerase, followed by sudden internal deletions that might occur in the subtelomeric region.

The telomeric regions of *Plasmodium* species exhibit considerable size variations that can mainly be attributed to frequent rearrangements within the subtelomeric repeats (Foote and Kemp, 1989). This size variation can be due to both the expansion or contraction of a hypervariable minisatellite (Åslund *et al.*, 1985; Oquendo *et al.*, 1986), or to changes in the copy number of a subtelomeric, internally repetitive 2.3-kbp DNA element (Pace *et al.*, 1987; Dore *et al.*, 1990). Amplification of this tandemly organized element can add up to 400 kbp to the size of a chromosome, and seems to account for most of the observed size polymorphisms during mitotic growth (Pace *et al.*, 1990; Ponzi *et al.*, 1990). However, the mechanisms that regulate amplification and loss of these subtelomeric repeats are not understood.

In vegetatively growing *Tetrahymena* cells, telomeric sequences of ribosomal and nonribosomal DNA in the macronucleus initially lengthen coordinately (Larson *et al.*, 1987). The rate during the first 300 log-phase divisions is 3–10 bp per generation, after which the length of the telomeric restriction fragments reaches a maximum that varies from strain to strain. Then, in contrast to the situation in trypanosomes, subpopulations with shorter telomeres gradually took over in several cultures, perhaps due to a slight growth advantage. In stationary cultures, telomeric repeat length is regulated at about 50–70 repeats.

The range of the telomere repeat tracts in *S. cerevisiae* is between 350 and 710 bp (Wang and Zakian, 1990a) and varies between strains by up to twofold (Walmsley and Petes, 1985). Individual telomeres in yeast show independence in length variation in a clonal population (Shampay and Blackburn, 1988). This clonal variation is independent of the major recombination pathway and was also observed in a strain with *rad52*.

It is not known how telomeric repeat copy number is regulated, but metabolic effects or growth conditions may be one factor controlling overall telomere length, e.g., causing the drastic reduction in trypanosome telomeres. Regulation may also occur at the level of a population, in which individuals with shorter telomeres may have a growth advantage. Alternatively, length may be regulated at the level of each telomere, as suggested by the genetic analysis of the process in yeast.

Two genes, *TEL1* and *TEL2*, that have been identified in yeast influence telomere length (Lustig and Petes, 1986). Mutations in these genes result in reduced but stable telomere length but do not show the progressive telomere reduction and senescence phenotype of the *est1* mutation. An increase in telomere length is observed when temperature-sensitive mutants of *cdc17*, encoding the catalytic subunit of DNA polymerase I, are grown at the permissive or semipermissive temperature (Carson and Hartwell, 1985). Interestingly, in both situations, the adjustment to the new, increased or decreased telomere length occurs gradually over many generations. This indicates that the balance existing under normal conditions between growing and receding at the chromosome ends is disturbed in these mutant strains. The rates by which the new telomere lengths are achieved may give an indication of the mechanisms involved. In *tel1* mutants, 150 divisions are required for a 300-bp telomere reduction. The same rate of reduction of 2 bp per generation was observed when long telomeres produced in *cdc17* mutants at elevated temperature gradually became shorter after a wild-type *CDC17* gene was crossed back into the cells. This rate is very similar to the rate of reduction observed in the *est1* mutant (4 bp/generation) and in terminally deficient chromosomes of *Drosophila* (2 bp/generation). This raises the possibility that a highly conserved mechanism, such as DNA replication, may be responsible for the very similar rates of telomere shortening in both organisms. As we have discussed, removal of the last RNA primer from the lagging strand, which would result in an 8-bp gap in *Drosophila* (Kitani *et al.*, 1984) and in an 11-bp gap in *S. cerevisiae* (Singh and Dumas, 1984), could explain the observed rates of telomere shortening if the elongation mechanism were fully inhibited. When *cdc17* cells are shifted from lower, permissive to higher, sublethal temperatures, telomeres grow. When shifted to 28° C they grow during the first 30 generations at 11 bp/generation, afterwards slowing to 5 bp/generation. When shifted to 32° C, telomeres initially grow at a rate of 18 bp/generation, slowing to 6 bp/generation after 30 generations. Thus, the telomere elongation mechanism has the capacity to modulate telomere length in response to changes in growth conditions, but eventually slows down to the rate required to balance the 2- to 4-bp loss that normally occurs at every round of replication. However, it is not yet understood how the telomere elongation mechanisms are regulated.

Insights into these mechanisms may be obtained from some recent experiments that were aimed at disturbing the balance between telomere elongation and shortening. All results indicate the involvement of proteins interacting with DNA to regular telomere length. Introducing extra telomeric DNA sequences via high-copy-number plasmids into *S. cerevisiae* cells causes a gradual increase in the length of chromosomal telomeric repeats that is proportional to the plasmid copy number (Runge and Zakian, 1989). The longer telomeres appear to be stable for many generations, indicating that a new equilibrium length has been reached. Comparing circular and linear plasmids containing telomeric repeats, it was concluded that the total increase in telomere length, not the increase in the number of DNA termini, was responsible for the observed chromosomal telomere lengthening. The results have been interpreted by postulating that a protein factor that normally binds to long $C_{1-3}A$ telomeric repeats is competed away from the chromosomal termini, thus allowing easier access of the elongation machinery, resulting in telomeric growth.

Another candidate for controlling telomere length is the RAP1 protein. Temperature-sensitive *rap1* mutants exhibit reduced telomere tract lengths under semipermissive temperature conditions (Conrad *et al.*, 1990; Lustig *et al.*, 1990), effects that are opposite those observed by Runge and Zakian (1989). Obviously, the interactions between proteins and telomeric DNA are more complex, and other protein factors are very likely to be involved. Overexpression of *RAP1* does not simply increase telomeric length but often causes an increased heterogeneity that may be accompanied by DNA rearrangements (Conrad *et al.*, 1990). When only the C-terminal half of *RAP1* is overexpressed, substantial telomere lengthening is observed. These results are difficult to interpret at this time, especially given that the RAP1 protein binds to many sites on yeast chromosomes (Buchman *et al.*, 1988b), but it has been hypothesized that the *RAP1*-encoded product may mediate the interactions between chromosomes and the nuclear scaffold or lamina. *RAP1* overexpression results in an elevated rate of chromosome loss and reduced viability that is reminiscent of *est1* cells, suggesting that these interactions may be important for proper chromosome segregation.

Quite complex effects on telomere length have also been observed in *Tetrahymena* that were transformed with plasmids overexpressing mutated telomerase RNA genes (Yu *et al.*, 1990). These experiments demonstrate the coding capacity of telomerase RNA, because two of the mutated RNA templates gave rise to altered telomeric DNA sequences corresponding to the mutation in the RNA template. Moreover, in these two transformed strains telomeric repeats were longer than in wild type. If one assumes a dynamic equilibrium of telomere length as seen in other organisms, which is negatively influenced by telomere-binding protein factors and positively influenced by the activity of telomerase, these results might be interpreted as follows. A lower affinity of the negatively regulating protein for the mutated telomeric repeat sequence might allow easier access of the telomerase. In the case of a third mutant telomerase RNA gene, the transformed telomerase RNA led to telomere shortening, perhaps by reducing telomerase affinity or activity on the mutant telomere. Mutated telomeres (either longer, or shorter) also led to a cellular senescence phenotype in *Tetrahymena* similar to that described in yeast, supporting the idea that the cause may be an impaired interaction with telomere-binding proteins. Some telomere-binding proteins, similar to those described in other organisms, or the G-quartet structure, possibly complexed with proteins, may act as a negative regulator of telomere elongation (Zahler *et al.*, 1991).

5. *Control and Timing of Telomere Replication*

Telomeric regions are usually heterochromatic and are replicated late in S phase (Berendes and Meyer, 1968; Lima-de-Faria and Jaworska, 1968; Crossen *et al.*, 1975). Recent analyses in yeast have established that the region of late replication includes not only the telomere-associated Y' sequences but also the proximally adjacent single-copy regions (McCarroll and Fangman, 1988). Both classes of yeast subtelomeric repeated elements, X and Y', contain autonomously replicating sequences (ARSs) that might function as origins of replication (Chan and Tye, 1983b; Walmsley *et al.*, 1984). However, different results have been obtained from different telomeric regions. Kinetic data on the replication of the left arm of chromosome III, which contains an X element but lacks a Y' element (Button and Astell, 1986), suggest that late replication of this telomeric region may not be due to a late activation of the ARS contained on the X element. Late replication is more likely to be caused by the delayed arrival of a replication fork that was initiated at a more proximal, earlier activated region (Reynolds *et al.*, 1989). The situation is quite different at the telomere of the right arm of chromosome V, which contains a single X and a single Y' element. Here, late replication is achieved by forks that initiate at origins that lie within the late-replicating telomeric region and are activated late in S phase (Ferguson *et al.*, 1991). Though one late-activated ARS element is not contained in either the X or Y' elements, other late-activated ARSs are probably located in one of these elements. In fact, when all Y' elements were examined together, the results suggest that most Y' ARSs are used as active replication origins in their telomeric chromosomal positions. These findings indicate that Y' elements may be used but are not necessary to ensure late replication of telomeric regions. It is not clear how late activation is achieved.

The possibility exists that either the affinity of these ARSs for the replication machinery and/or the chromosomomal environment of these telomeric ARS elements determines the late onset of replication of the telomeric region. Moreover, in view of results with human telomeric sequences, which appear to be replicated throughout S phase (Ten-Hagen et al., 1990), it is not clear that late replication of telomeric regions is essential for their proper function. Late replication may simply be a consequence of some structural feature of the region or of interactions with other nuclear components.

C. EFFECT ON GENE EXPRESSION

In general, telomeres exhibit heterochromatic characteristics such as condensed chromatin structure, late replication, and a high amount of repetitive DNA sequences. It has been well documented that the pericentric heterochromatin exhibits a negative effect on the level of expression of adjacent genes, resulting in position-effect variegation, the mechanism of which is not fully understood (Spofford, 1976; Eissenberg, 1989; Spradling and Karpen, 1990; Tartof et al., 1989). Similar effects on genes that integrated near the telomere have only recently been observed in Drosophila (Gehring et al., 1984; Hazelrigg et al., 1984; Levis et al., 1985) and yeast (Gottschling et al., 1990).

It appears, at least in yeast, that the physical distance from the chromosome end and not the vicinity of $C_{1-3}A$ tracts may be responsible for the telomeric position-effect variegation that acts at the transcriptional level (Gottschling et al., 1990). It has also been estimated that the position effect may extend 6 kbp but not as far as 20 kbp from the chromosome end. If the $C_{1-3}A$ repeats in yeast are not directly involved in mediating suppression of adjacent genes, proteins bound to the ends of chromosomes may be responsible for exerting the negative effect on expression of nearby genes. The nature of these proteins remains to be determined, and it is not clear whether they are the same as the ones that negatively control telomere repeat length, or are involved in chromosome capping, or mediate the interaction between telomeres and nuclear structures. Conceivably, the same group of proteins, perhaps interacting cooperatively, may perform all of these functions.

This idea is substantiated by the observation that products of several genes (SIR2, SIR3, SIR4, NAT1, ARD1, and HHF1) are required for transcriptional silencing of telomere-adjacent genes (Aparicio et al., 1991). The same genes also control repression of the two silent mating loci, which are located 12 and 25 kbp away from the termini of chromosome III. The products of these genes may be involved in chromatin

conformation at the telomeric regions, converting the chromosome termini into heterchromatic structures. Although the structure of these
proteins and their possible interactions are not fully understood, the
involvement of *HHF1* (encoding histone H4), *NAT1* and *ARD1* (encoding subunits of N-terminal acetyltransferase, which acetylates histone
H2B), and *SIR4* (which shares sequence similarities with the coiled-
coil domains of nuclear lamins that might interact with telomeres),
supports this notion.

Telomeric position effects have also been observed in *Drosophila*
(Gehring *et al.*, 1984; Hazelrigg *et al.*, 1984; Levis *et al.*, 1985). A *w*
gene exhibited variegated expression after P-element mediated integration at the tip of chromosome 3R. However, the variegation is partially relieved in "revertants" that exhibit a higher degree of eye pigmentation. Molecular analyses demonstrated that such revertants are
associated with terminal deletions that have lost distal DNA sequences
(Levis, 1989). Because the deletions place the chromosome end closer
to the telomeric *w* gene, it seems unlikely that the vicinity of the chromosome end per se is responsible for the position effect. Similarly,
when the end of a terminally deleted RT chromosome is located 0.3 to
2.5 kbp upstream of the start site for transcription of the *y* gene (Biessmann and Mason, 1988), no variegated expression of *y* was observed.
The RT chromosome ends do not carry any repetitive DNA sequences
(Biessmann *et al.*, 1990a). Although it is not clear whether any proteins are associated with the RT chromosome end, such an association
has been inferred from the observation that this broken end behaves
as if it were "capped" (Biessmann and Mason, 1988). Both observations
are consistent with the hypothesis that specific DNA sequences, normally present in the telomeric region, are causing the variegation, and
that variegation is relieved when these sequences are removed by a
terminal deficiency. The nature of such DNA sequences and the proteins that bind to them remain to be determined, but possible candidates include minisatellites that have been found in telomeric regions
of many organisms, including *Drosophila*. The HeT-A retroposon does
not seem to exert a position effect on nearby genes because the *y* gene
in the RT394 chromosome that carries a HeT-A retroposon adjacent to
y does not show variegated expression (Biessmann *et al.*, 1990b).

Genes that lie beyond the region influenced by the position effect are
expressed normally but are subject to rearrangements and translocations. For example, several members of the *SUC* gene family (structural genes for invertase) of *S. cerevisiae* are located in the telomeric
region of different chromosomes in different strains. Most strains do
not carry *SUC* alleles at all six loci. Molecular analysis of this gene

family has shown (Carlson *et al.*, 1985) that at least some members are embedded in the telomeric region between the X and the Y' repeats and that telomeric suc^0 loci lack *SUC* genes together with portions of the conserved flanking sequence. These findings suggest that the *SUC* genes might be dispersed by rearrangements of subtelomeric regions. Because the *SUC* genes are expressed normally in their telomeric positions, the vicinity of X and Y' elements has apparently no effect on their activity. No attempt has been made to determine the actual distance of the *SUC* genes from the chromosome ends. The *MAL* multigene family in yeast is also closely associated with telomeres. It has also been suggested (Charron *et al.*, 1989) that during the evolution of this gene family, the *MAL* loci have been translocated to different chromosome ends by rearrangements of the subtelomeric regions. Similar observations have been made with the telomeric acid phosphatase genes (Venter and Hörz, 1989).

Activation of telomeric expression sites of trypanosome surface antigen genes can be achieved by several mechanisms, including *in situ* activation without DNA rearrangement, total or partial gene conversion, and reciprocal telomeric recombination (reviewed by Pays and Steinert, 1988). The telomere location of the expression sites and the array of 70-bp repeats located at the proximal end of the surface antigen genes provide favorable conditions for the antigen switch by gene conversion and recombination. Moreover, interaction between telomeres seems to be important for antigen gene recombination (Van der Werf *et al.*, 1990). Much less is known about the *in situ* activation that may be achieved by yet unknown regulatory proteins, perhaps similar to the ones involved in *S. cerevisiae* transcriptional silencing of telomere-adjacent genes (Aparicio *et al.*, 1991). A possible analogy of the *in situ* switching mechanism with the yeast telomeric position effect has been discussed by Gottschling *et al.* (1990). In yeast, a gene next to a telomere can be either active or repressed due to the influence of the telomere. It was found that the transcriptional state of such a gene is reversible but is inherited in a semistable fashion and maintained in its original state for at least 15–20 generations before switching occurs.

D. NUCLEAR ORGANIZATION

Telomeres may play a role in establishing the architecture of the interphase and/or the meiotic prophase nucleus. During interphase in some cell types the chromosomes are arranged in the nucleus with the centromeres close together and near each other at one side of the nuclear envelope and the telomeres on the opposite side. This has been termed the "Rabl" orientation because of early descriptions of this arrangement by Rabl (1885). It has been suggested that this orientation occurs as a natural consequence of the decondensation of chromosomes after anaphase. This would predict that telomeres in sister cells would be oriented toward each other. In fact, in the *Drosophila* blastoderm, in which the mitotic spindle is parallel to the surface of the animal, cells in interphase have their centromeres pointed toward the outside of the animal and their telomeres pointed inward (Foe and Alberts, 1985). In *Vicia faba,* telomeres are clustered in the opposite direction from that predicted (Rawlins *et al.,* 1991), and in *Drosophila* salivary glands, in which the polytene chromosomes assumed a Rabl-like arrangement, there is no preferential alignment of the chromocenter or any other chromosome region (Hochstrasser *et al.,* 1986; Mathog and Sedat, 1989). Nuclear rotation has been proposed to account for these discrepancies, but little is known about this phenomenon (Billia and De Boni, 1991, and references therein; Rawlins *et al.,* 1991).

In many species telomeres may be found associated with each other or the nuclear envelope, but these associations are not universal. Localization of telomeres within root tip nuclei of two plant species by *in situ* hybridization and three-dimensional confocal microscopy showed that most telomeres were adjacent to the nuclear envelope, except for a few next to the nucleolus (Rawlins *et al.,* 1991). In *V. faba,* however, the telomeres were clustered together, whereas in *Pisum sativum* telomeres were generally observed at one side of the nucleus, although there were always a few elsewhere. In *Trypanosoma brucei,* on the other hand, telomeres were clustered together, but were not always located at the nuclear periphery (Chung *et al.,* 1990). Instead, many cells contained what appeared to be a network of hybridizing material running through the nuclear interior. In cultured mammalian cells telomeres occurred in pairs distributed through the nucleus (Billia and De Boni, 1991). Moreover, though mammalian chromosomes appear to occupy specific regions of the nucleus, a Rabl orientation is not evident in these cells. One exception to this in human cells is the Barr body, which forms a loop with the two telomeres closely associated with each other. In 75–80% of cells the two telomeres are adjacent to the nuclear envelope (Walker *et al.,* 1991).

Chromosomes of *Drosophila* also occupy specific regions of the nucleus and are attached, or at least consistently apposed to the nuclear envelope, at several sites along each arm. This appears to be true for diploid (Foe and Alberts, 1985) as well as polytene cells (Mathog *et al.*, 1984). Oxygen deprivation of diploid cells in living embryos causes interphase chromosomes to condense and adhere to the inner surface of the nuclear envelope along their entire lengths. Polytene chromosomes are closely associated with the nuclear envelope at a number of sites (Hochstrasser and Sedat, 1987). These sites tend to be heterochromatic, including telomeres as well as sites of intercalary heterochromatin, and are generally the same in different cell types, although few, if any, sites are always apposed to the envelope. Given the general similarity of the nuclear architecture among cells with different transcriptional activities and the lack of absolute reproducibility of apposition sites in cells with similar activities, Hochstrasser and Sedat (1987) have argued against the hypothesis that placement of a gene within the nucleus controls its activity.

Chromosomes in meiotic prophase are often oriented with centromeres to one side of the nucleus and telomeres to the other, in a "bouquet" arrangement. Although it is tempting to speculate that the apposition of homologous telomeres helps to align chromosomes in preparation for synapsis (reviewed by Dancis and Holmquist, 1979; Gillies, 1975; Rouyer *et al.*, 1990), the genetic effects of chromosomal rearrangements argue against such a mechanism. Thus, though meiotic exchange is reduced, it occurs at a fairly high frequency within a heterozygous inversion, but is also reduced between the inversion breakpoint and the telomere (Sturtevant and Beadle, 1936). Based on similar observations in a number of species, Maguire (1984) has argued that the arrangement of telomeres in the bouquet is neither necessary nor sufficient for synapsis and recombination of homologues. Thus, for cells in meiotic prophase as well as in mitotic interphase, telomeres tend to be associated with each other and with the nuclear envelope. It remains to be seen whether any of these associations have a functional significance.

IV. De Novo Telomere Formation

Most chromosomes, fragments, or linear plasmids without telomeres are lost rapidly, though some are recovered. When examined closely, most of those recovered have acquired a telomeric DNA sequence. A few, however, have not (Biessmann and Mason, 1988; Biessmann *et al.*, 1990a; Levis, 1989), or at least have not been shown to have acquired such a DNA sequence (McClintock, 1941). Because these chromosomes with broken ends are recoverable and are genetically stable over several generations, they are believed to have acquired new telomeres. The physical nature of these new telomeres is not known.

A. DEVELOPMENTALLY REGULATED TELOMERE ADDITION

Ciliated protozoa provide the best examples of developmentally programmed formation of new telomeres. During macronucleus formation, the normal-sized chromosomes of the micronucleus are broken up, DNA is eliminated, and, after DNA replication, abundant small minichromosomes are generated with new telomeres (reviewed by Yao, 1989). Sequence comparison of macronuclear DNA fragments and their micronuclear origins have shown that the newly acquired telomeric repeats at the ends of macronuclear DNA are not present at the equivalent positions in micronuclear DNA, arguing strongly for a *de novo* generation of telomeric repeats. Breakage of micronuclear chromosomes occurs at so-called fragmentation sites, which are very well defined in *Tetrahymena* (Godiska and Yao, 1990) and less well in *Euplotes* (Baird and Klobutcher, 1989). In contrast, DNA breakage in *Oxytricha* and *Paramecium* can occur over a wider range of sites. Sequence analysis of multiple independent clones of particular macronuclear DNA has shown that addition of telomeric repeats occurs within a 9-bp interval in *Oxytricha nova* (Baird and Klobutcher, 1989), within a 33-bp interval in *Oxytricha fallax* (Herrick *et al.*, 1987), and within hundreds of base pairs in *Paramecium* (Baroin *et al.*, 1987; Forney and Blackburn, 1988). Because no sequence similarities were found at the different points of telomere addition in *Paramecium,* it is unlikely that a specific cis-acting signal is required for telomere formation, unless a secondary structure is recognized. Regardless of the variability of sites of telomere addition in macronuclear DNA, species-specific repeats are directly added to the broken micronuclear DNA ends without any intervening telomere-associated sequences (Yokoyama and Yao, 1986; Baroin *et al.*, 1987; Herrick *et al.*, 1987; Forney and Blackburn, 1988; Baird and Klobutcher, 1989), although some subtelomeric rearrangement may occur before addition of telomeric repeats (Forney and Blackburn, 1988). By using transformants with a telomerase containing a mutated RNA template, it could be shown that, during macronuclear formation in *Tetrahymena,* telomerase is capable of adding telomeric repeats directly onto broken DNA ends that lack preexisting telomeric repeats (Yu and Blackburn, 1991).

A classical example of chromosome fragmentation and diminution occurs in the nematode *Ascaris,* where germ-line-specific DNA is eliminated in somatic cells during embryogenesis. During this process, telomeric $(TTAGGC)_n$ repeats become attached directly to the chromosomal fragmentation sites (Müller *et al.*, 1991), supporting the notion that the postulated *Ascaris* telomerase is capable of adding telomeric repeats directly onto nontelomeric ends. It is also conceivable that in both processes the chromosomal fragmentation complex that remains after cleavage may play a role in "anchoring" telomerase.

B. Artificially Induced Telomere Addition

1. Injection or Transformation with Linear DNA

If linear DNA, even from prokaryotic sources, is injected into the macronucleus of *Paramecium,* telomeric repeats are added, and the molecules are maintained in high copy number in the transformants (Gilley *et al.*, 1988; Bourgain and Katinka, 1991). This is consistent with the apparent absence of any sequence requirement in macronuclear DNA for the direct attachment of telomeric repeats (Forney and Blackburn, 1988), and although some limited pairing to the template may be required, it suggests that high levels of telomerase, possibly with the aid of accessory enzymes, can add telomeric repeats indiscriminately to any DNA end.

The situation is quite different in *Saccharomyces*. Linear molecules, even if they contain origins of replication, are not maintained as linear fragments when introduced into yeast by transformation. The ends do not acquire telomeric repeats but are highly recombinogenic, resulting in integration, fusion, and circularization (Orr-Weaver *et al.*, 1981; Szostak and Blackburn, 1982). Linear molecules are only maintained, and yeast telomeric DNA is only added when molecules contain sequences that resemble yeast telomeres at their ends (Pluta *et al.*, 1984; Shampay *et al.*, 1984; Zakian, 1989) or within about 100 bp from the end (Murray *et al.*, 1988). Whether this addition is due to a telomerase or to recombination is still an open question (Wang and Zakian, 1990b; Zakian *et al.*, 1990).

Recently, linear DNA fragments carrying human telomeric repeats, together with some subtelomeric repetitive sequences, have been introduced into a hamster/human hybrid cell line by electroporation (Farr *et al.*, 1991). In the majority of cases integration occurred at interstitial sites, but in 22% of cases integration occurred near telomeres, suggesting a mechanism that involves random integration followed by chromosome breakage and formation of a new telomere.

2. Healing of Broken Chromosome Ends

The term "healing" used in reference to the stabilization of broken chromosome ends has two distinct definitions. McClintock (1939, 1941) used the term originally to describe the newly acquired stability of broken chromosomes in the maize embryo. McClintock (1938) examined the behavior of chromosomes broken at meiotic anaphase in the microsporocyte. She found that a single broken chromosome end in the resulting microspore underwent a round of replication, then the two sister broken chromatids fused to produce a dicentric chromosome. At mitosis the dicentric chromosome formed an anaphase bridge and broke. This breakage–fusion–bridge cycle repeated at each succeeding cell cycle during the development of the pollen (McClintock, 1938) and after fertilization of the endosperm (McClintock, 1939). Upon fertilization of the embryo, however, the broken chromosome became stable and the breakage–fusion–bridge cycle stopped (McClintock, 1939, 1941). These newly stable chromosomes were said to be healed, although they have not been characterized molecularly.

Haber and Thorburn (1984) also broke chromosomes using anaphase bridges. They found that, in yeast, a small proportion of recovered stable derivatives had acquired new telomeric repeats directly onto the broken end, while other derivatives became stable via recombination or rearrangement with other chromosomes. The derivatives that carried the telomeric repeats were said to have healed. Others since then have used the term to refer to the addition of telomeric repeats onto broken chromosome ends. According to this definition the stable chromosomes recovered by McClintock cannot be described as healed, because it has not been shown that they have acquired telomeric repeats at their termini.

As will be described below, the stable terminal deletions recovered in *Drosophila* by Mason *et al.* (1984) have healed in the sense that McClintock (1939, 1941) used the term, but have not acquired telomeric repeats. Some of the deleted *Drosophila* chromosomes eventually acquire telomere-associated HeT-A sequences, but this step is subsequent to the initial stabilization. We will use the term to imply the addition of telomeric repeats onto the new chromosome ends.

In *S. cerevisiae* the majority of broken chromosome ends generated mitotically from dicentric chromosomes are healed by homologous recombination. In some cases, however, genetic and molecular analyses suggest the formation of a new telomere at the broken chromosome end (Haber and Thorburn, 1984; Haber *et al.*, 1984), and it could be shown after pulsed-field gel electrophoresis that the stable chromosome fragments had acquired new Y' sequences (Jäger and Philippsen, 1989b). In *Schizosaccharomyces pombe* γ-ray-induced minichromosomes seem to have acquired new telomeres by direct addition of telomeric repeats to the broken ends without any intervening telomere-associated sequences (Matsumoto *et al.*, 1987). Terminal chromosome deletions causing loss of distal genes have been described in *Plasmodium falci-parum* (Cappai *et al.*, 1989; Pologe and Ravetch, 1988) and in *Giardia* (Le Blancq *et al.*, 1991; Adam *et al.*, 1991), where telomeric repeats are attached directly to the broken ends within the known sequence of the affected gene.

An interesting case of human chromosome truncation has been analyzed at the molecular level (Wilkie *et al.*, 1990a,b, 1991). An α-thalassemic patient carried a terminal deficiency of the short arm of chromosome 16, deleting all DNA in the region starting at 50 kbp distal to the α globin genes and extending to the end of the chromosome. The α-globin cluster resides within 400 kbp of the terminus (Harris *et al.*, 1990), and deletion of the region distal to the α-globin genes has no apparent phenotypic effects in a heterozygous individual. The breakpoint of the deletion occurs within a pair of 56-bp tandem repeats, and human telomeric repeats $(CCCTAA)_n$ have been added directly to the broken chromosome end without any intervening subtelomeric repeats, suggesting that, at least for the short period of observation, the telomeric repeat is sufficient for stabilizing the broken chromosome end. The mechanism of this healing event is unknown, but the sequence at the breakpoint may have provided limited pairing with the RNA template in the human telomerase (Morin, 1991).

Most attempts to recover terminal deficiencies in *Drosophila* have been unsuccessful (e.g., Muller and Herskowitz, 1954; Roberts, 1975). However, in the presence of a homozygous mutation of the *mu2* locus, terminal deficiencies of each chromosome are recovered at a high frequency after irradiating females with low doses of X-rays (Mason *et al.*, 1984). Deficiencies that removed the tip of the X chromosome together with several genes residing in the most distal euchromatic region were selected by screening for the loss of the *y* gene, indicated by the complete or partial absence of black melanin pigmentation of the adult cuticle. The deficiencies were termed "Df(1)RT," abbreviated RT. Positions of the distal breakpoints of the RT deficiencies within the *y* gene region were initially determined by genomic Southern blots (Biessmann and Mason, 1988). The terminal restriction fragments of all RT deficiencies appeared "fuzzy" on the genomic Southern blots, indicating length heterogeneity, and they were sensitive to the exonuclease Bal 31 (Biessmann *et al.*, 1990a), suggesting that the breakpoints represent the structural end of the chromosome. Wild-type DNA fragments located more distally than the indicated breakpoints, as well as probes from the telomere-associated HeT-A repeat, failed to hybridize to the deficient chromosomes. This collection of RT-terminal deficiency stocks has yielded two X chromosomes, which appear to be healed. Molecular analyses of these two chromosomes (RT394 and RT473) revealed that, in each case, healing had involved the addition of a 3-kbp DNA fragment at the end of the chromosome. This newly acquired DNA belongs to the *Drosophila* subtelomeric HeT-A retroposon family, which had been added to different positions in the *y* gene with its oligo(A) tail directed proximally. No detectable sequence similarities have been found in the vicinity of the attachment sites (Biessmann *et al.*, 1990b). Thus, in contrast to healing mechanisms in other organisms, the observed healing event at the terminally deficient X-chromosome in *Drosophila* was apparently initiated by a transposition event of a HeT-A retroposon to the broken end, which may represent a step in the process of generating a new telomere. Since then, several other cases of HeT-A retropositions to broken chromosome ends have been observed and analyzed (Biessmann *et al.*, 1992b).

Another example of healed broken chromosome ends in *Drosophila* has recently been described. A spontaneously opened ring X chromosome has been identified that does not give rise to acentric fragments and apparently does not cause detectable loss of genetic information. The two newly generated ends have acquired telomere-associated HeT-A sequences together with apparently fully functional telomeres (Traverse and Pardue, 1988). Although the healed breaks have not been sequenced, the opened ring chromosome suggests that these elements can move from their original positions to "empty" targets, i.e., chromosome breaks.

In light of the apparent absence of telomeric repeats in *Drosophila*, the observations that HeT-A retroposons do indeed move to the ends of terminally deleted chromosomes suggest that *Drosophila* may have a telomere formation and maintenance mechanism that is different from those observed in other organisms. It further indicates an important role of HeT-A retroposons in *de novo* telomere formation.

In summary, it appears that terminal chromosome deficiencies may occur by chromosome breakage or aberrant subtelomeric recombination. Healing is then achieved by different mechanims, involving subsequent attachment of new telomeric components that may eventually rebuild a complete telomeric structure. Thus, the distal-most genes in many organisms are at risk of truncation and loss. In *Drosophila*, the most distal gene on the left arm of chromosome 2 is the gene *lethal(2)giant larvae*, or *l(2)gl*, which controls the growth of imaginal discs. Studying *Drosophila* populations in the Soviet Union, Golubovsky (1978) and Golubovsky and Sokolova (1973) found that 1–2% of all second chromosomes carried heterozygous *l(2)gl* mutations. These findings were substantiated by results from wild populations in the

United States (Green and Shepherd, 1979). The *l(2)gl* locus is also highly susceptible to P-element (Green and Shepherd, 1979) and chemically induced mutagenesis (Mechler *et al.*, 1985). Cloning and molecular analysis of the locus revealed that the vast majority of *l(2)gl* mutations are terminal chromosome deficiencies resulting in truncations of the gene (Mechler *et al.*, 1985). This situation is reminiscent of the terminal deficiencies of the X chromosome and of chromosome 3R that were previously described.

V. Conclusion

It has been agreed generally that telomeres perform at least two essential functions. They protect the ends of linear chromosomes from degradation, recombination, and ligation to other chromosome ends, and they compensate for the inability of DNA polymerase to replicate completely the ends of linear DNA molecules. In a recent review, Zakian (1989) pursued the generally accepted hypothesis that telomeres consist of tandem repeats of a specific DNA sequence and, with evidence for terminal deletions in *Drosophila*, raised the question, "Are telomeres really necessary?" We prefer to start with Muller's (1938) functional definition, that a telomere is a structure that performs vital telomeric functions; assume that, in the absence of redundancy, any structure that performs an essential function is essential; then ask what structures perform the functions ascribed to telomeres. Obviously, given the stability of chromosomes in *Drosophila* without telomeric DNA sequences on one end, both definitions cannot be correct. Zakian concluded that the *Drosophila* chromosomes with tip deletions (Mason *et al.*, 1984; Biessmann and Mason, 1988; Biessmann *et al.*, 1990a) lacked telomeres, in spite of the demonstration by Muller (1940; Muller and Herskowitz, 1954) and others (e.g., Roberts, 1975) that *Drosophila* will not tolerate chromosomes lacking telomeres. We propose that these chromosomes are stable as a result of the nature of the new chromosome ends, and that whatever structure is found to be responsible for stability is a necessary component of the telomere. Similarly, if a DNA sequence is the definition of a telomere, telomerase can be interpreted as an enzyme that uses telomeres as substrate; but if the telomere is responsible for maintaining the stability of the chromosome terminus, telomerase can be interpreted as a protein component of the telomere that, in concert with a specific DNA sequence, maintains chromosome length.

Before issues such as these can be resolved, agreement must be reached on the definitions of the terms "telomere" and "healing." If the latter is defined as the addition of a telomere, its definition would be tied to that of the former. Assuming this approach is adopted and assuming "telomere" refers to a specific DNA sequence, we are faced with evidence suggesting that Muller (1938) was not dealing with telomeres when he coined the term, and we have no evidence that McClintock (1939) was investigating chromosome healing when she coined that term. We prefer to use Muller's functional definition that a telomere is what a telomere does. Though the alternative, that a telomere is a specific DNA sequence, is certainly a testable hypothesis, and there is evidence that the "telomeric DNA" repeats play a role in chromosome replication, at least in some organisms, there is no evidence that the telomeric DNA performs all functions ascribed to telomeres in any one organism. And there is reason to question whether telomeric DNA repeats play the same role in telomere replication in all organisms. Regardless of the outcome of the semantic disagreement, resolution is important because confusion can arise when the alternative definitions are used interchangably.

Acknowledgments

We thank Drs. C. C. Hardin, E. Perkins, and G. B. Morin for critically reading the manuscript; C. B. Bennett and M. E. Dresser for helpful discussions; and our many friends and colleagues, who generously shared their unpublished results.

References

Acevedo, O., Dickinson, L. A., Macke, T. J., and Thomas, C. A., Jr. (1991). The coherence of synthetic telomeres. *Nucleic Acids Res.* **19**, 3409–3419.

Adam, R. D., Nash, T. E., and Wellems, T. E. (1991). Telomeric location of *Giardia* rDNA genes. *Mol. Cell. Biol.* **11**, 3326–3330.

Agard, D. A., and Sedat, J. W. (1983). Three-dimensional architecture of a polytene nucleus. *Nature (London)* **302**, 676–681.

Allitto, B. A., MacDonald, M. E., Bucan, M., Richards, J., Romano, D., Whaley, W. L., Falcone, B., Ianazzi, J., Wexler, N. S., Wasmuth, J. J., Collins, F. S., Lehrach, H., Haines, J. L., and Gusella, J. F. (1991). Increased recombination adjacent to the Huntington disease-linked D4S10 marker. *Genomics* **9**, 104–112.

Allshire, R. C., Gosden, J. R., Cross, S. H., Cranston, G., Rout, D., Sugawara, N., Szostak, J. W., Fantes, P. A., and Hastie, N. D. (1988). Telomeric repeat from *T. thermophila* cross hybridizes with human telomeres. *Nature (London)* **332**, 656–659.

Allshire, R. C., Dempster, M., and Hastie, N. D. (1989). Human telomeres contain at least three types of G-rich repeat distributed non-randomly. *Nucleic Acids Res.* **17**, 4611–4627.

Altherr, M. R., Smith, B., MacDonald, M. E., Hall, L., and Wasmuth, J. J. (1989). Isolation of a novel mildly repetitive DNA sequence that is predominantly located at the terminus of the short arm of chromosome 4 near the Huntington disease gene. *Genomics* **5**, 581–588.

Aparicio, O. M., Billington, B. L., and Gottschling, D. E. (1991). Modifiers of position effect are shared between telomeric and silent mating-type loci in *S. cerevisiae*. *Cell (Cambridge, Mass.)* **66**, 1279–1287.

Åslund, L., Franzen, L., Westin, G., Persson, T., Wigzell, H., and Petterson, U. (1985). Highly reiterated non-coding sequence in the genome of *P. falciparum* is composed of 21 base-pair tandem repeats. *J. Mol. Biol.* **185**, 509–516.

Bachmann, L., Raab, M., and Sperlich, D. (1990). Evolution of a telomere associated satellite DNA sequence in the genome of *Drosophila tristis* and related species. *Genetica* **83**, 9–16.

Baird, S. E., and Klobutcher, L. A. (1989). Characterization of chromosome fragmentation in two protozoans and identification of a candidate fragmentation sequence in *Euplotes crassus*. *Genes Dev.* **3**, 585–597.

Baker, B. S., Carpenter, A. T. C., and Gatti, M. (1987). On the biological effects of mutants producing aneuploidy in *Drosophila*. *In* "Aneuploidy, Part A: Incidence and Etiology" (B. K. Vig and A. A. Sandberg, eds.), pp. 273–296. Liss, New York.

Barlogie, B., Drewinko, B., Johnston, D. A., and Freireich, E. J. (1976). The effect of adriamycin on the cell cycle traverse of a human lymphoid cell line. *Cancer Res.* **36**, 1975–1979.

Baroin, A., Prat, A., and Caron, F. (1987). Telomeric site position heterogeneity in macronuclear DNA of *Paramecium primaurelia*. *Nucleic Acids Res.* **15**, 1717–1728.

Baroudy, B. M., Venkatesan, S., and Moss, B. (1982a). Incompletely base-paired flip-flop terminal loops link the two DNA strands of the vaccinia virus genome into one uninterrupted polynucleotide chain. *Cell (Cambridge, Mass.)* **28**, 315–324.

Baroudy, B. M., Venkatesan, S., and Moss, B. (1982b). Structure and replication of vaccinia virus telomeres. *Cold Spring Harb. Symp. Quant. Biol.* **47**, 723–729.

Bates, G. P., MacDonald, M. E., Baxendale, S., Sedlacek, Z., Youngman, S., Romano, D., Whaley, W. L., Allitto, B. A., Poustka, A., Gusella, J. F., and Lehrach, H. (1990). A yeast artificial chromosome telomere clone spanning a possible location of the Huntington disease gene. *Am. J. Hum. Genet.* **46**, 762–775.

Bedbrook, J. R., Jones, J., O'Dell, M., Thompson, R. D., and Flavell, R. B. (1980a). A molecular description of telomeric heterochromatin in *Secale* species. *Cell (Cambridge, Mass.)* **19**, 545–560.

Bedbrook, J. R., O'Dell, M., and Flavell, R. B. (1980b). Amplification of rearranged repeated DNA sequences in cereal plants. *Nature (London)* **288**, 133–137.

Benn, P. A. (1976). Specific chromosome aberrations in senescent fibroblast cell lines derived from human embryos. *Am. J. Hum. Genet.* **28**, 465–473.

Berendes, H. D., and Meyer, G. F. (1968). A specific chromosome element, the telomere of *Drosophila* polytene chromosomes. *Chromosoma* **25**, 184–197.

Berg, D. E., and Howe, M. M., eds. (1989). "Mobile DNA." Am. Soc. Microbiol., Washington, DC.

Bergold, P. J., Campbell, G. R., Littau, V. C., and Johnson, E. M. (1983). Sequence and hairpin structure of an inverted repeat series at termini of the *Physarum* extrachromosomal rDNA molecule. *Cell (Cambridge, Mass.)* **32**, 1287–1299.

Berman, J., Tachibana, C. Y., and Tye, B. K. (1986). Identification of a telomere-binding activity from yeast. *Proc. Natl. Acad. Sci. U.S.A.* **83**, 3713–3717.

Bernards, A., Michels, P. A., Lincke, C. R., and Borst, P. (1983). Growth of chromosome ends in multiplying trypanosomes. *Nature (London)* **303**, 592–597.

Biessmann, H., and Mason, J. M. (1988). Progressive loss of DNA sequences from terminal chromosome deficiencies in *Drosophila melanogaster*. *EMBO J.* **7**, 1081–1086.

Biessmann, H., Carter, S. B., and Mason, J. M. (1990a). Chromosome ends in *Drosophila* without telomeric DNA sequences. *Proc. Natl. Acad. Sci. U.S.A.* **87**, 1758–1761.

Biessmann, H., Mason, J. M., Ferry, K., d'Hulst, M., Valgeirsdottir, K., Traverse, K. L., and Pardue, M. L. (1990b). Addition of telomere-associated HeT DNA sequences "heals" broken chromosome ends in *Drosophila*. *Cell (Cambridge, Mass.)* **61**, 663–673.

Biessmann, H., Valgeirsdottir, K., Lofsky, A., Chin, C., Ginther, B., Levis, R. W., and Pardue, M. L. (1992a). HeT-A, a transposable element specifically involved in "healing" broken chromosome ends in *Drosophila*. *Mol. Cell. Biol.* (in press).

Biessmann, H., Champion, L. E., O'Hair, M., Ikenaga, K., Kasravi, B., and Mason, J. M. (1992b). Frequent transposition of *Drosophila melanogaster* HeT-A retroposons to receding chromosome ends. *EMBO J.* (in press).

Billia, F., and De Boni, U. (1991). Localization of centromeric satellite and telomeric DNA sequences in dorsal root ganglion neurons, *in vitro*. *J. Cell Sci.* **100**, 219–226.

Blackburn, E. H. (1990a). Telomeres and their synthesis. *Science* **249**, 489–490.

Blackburn, E. H. (1990b). Telomeres: Structure and synthesis. *J. Biol. Chem.* **265**, 5919–5921.

Blackburn, E. H. (1991). Structure and function of telomeres. *Nature (London)* **350**, 569–573.

Blackburn, E. H., and Chiou, S. S. (1981). Non-nucleosomal packaging of a tandemly repeated DNA sequence at termini of extrachromosomal DNA coding for rRNA in *Tetrahymena*. *Proc. Natl. Acad. Sci. U.S.A.* **78**, 2263–2267.

Blackburn, E. H., Greider, C. W., Henderson, E., Lee, M. S., Shampay, J., and Shippen-Lentz, D. (1989). Recognition and elongation of telomeres by telomerase. *Genome* **31**, 553–560.

Boeke, J. D., and Corces, V. G. (1989). Transcription and reverse transcription of retrotransposons. *Annu. Rev. Microbiol.* **43**, 403–434.

Bourgain, F. M., and Katinka, M. D. (1991). Telomeres inhibit end to end fusion and enhance maintenance of linear DNA molecules injected into the *Paramecium primaurelia* macronucleus. *Nucleic Acids Res.* **19**, 1541–1547.

Bradshaw, V. A., and McEntee, K. (1989). DNA damage activates transcription and transposition of yeast Ty retrotransposons. *Mol. Gen. Genet.* **218**, 465–474.

Brown, W. R. (1988). A physical map of the human pseudoautosomal region. *EMBO J.* **7**, 2377–2385.

Brown, W. R. (1989). Molecular cloning of human telomeres in yeast. *Nature (London)* **338**, 774–776.

Brown, W. R., MacKinnon, P. J., Villasante, A., Spurr, N., Buckle, V. J., and Dobson, M. J. (1990). Structure and polymorphism of human telomere-associated DNA. *Cell (Cambridge, Mass.)* **63**, 119–132.

Bucan, M., Zimmer, M., Whaley, W. L., Poustka, A., Youngman, S., Allitto, B. A., Ormondroyd, E., Smith, B., Pohl, T. M., MacDonald, M., Bates, G. P., Richards, J., Volinia, S., Gilliam, T. C., Sedlacek, Z., Collins, F. S., Wasmuth, J. J., Shaw, D. J., Gusella, J. F., Frischauf, A.-M., and Lehrach, H. (1990). Physical maps of 4p16.3, the area expected to contain the Huntington disease mutation. *Genomics* **6**, 1–15.

Buchman, A. R., Kimmerly, W. J., Rine, J., and Kornberg, R. D. (1988a). Two DNA-binding factors recognize specific sequences at silencers, upstream activating sequences, autonomously replicating sequences, and telomeres in *Saccharomyces cerevisiae*. *Mol. Cell. Biol.* **8**, 210–225.

236   HARALD BIESSMANN AND JAMES M. MASON

Buchman, A. R., Lue, N. F., and Kornberg, R. D. (1988b). Connections between transcriptional activators, silencers, and telomeres as revealed by functional analysis of
a yeast DNA-binding protein. *Mol. Cell. Biol.* **8,** 5086–5099.
Budarf, M. L., and Blackburn, E. H. (1986). Chromatin structure of the telomeric region
and 3'-nontranscribed spacer of *Tetrahymena* ribosomal RNA genes. *J. Biol. Chem.*
261, 363–369.
Burmeister, M., Kim, S., Price, E. R., de Lange, T., Tantravahi, U., Myers, R. M., and
Cox, D. R. (1991). A map of the distal region of the long arm of human chromosome
21 constructed by radiation hybrid mapping and pulsed-field gel electrophoresis.
Genomics **9,** 19–30.
Button, L. L., and Astell, C. R. (1986). The *Saccharomyces cerevisiae* chromosome III left
telomere has a type X, but not a type Y', ARS region. *Mol. Cell. Biol.* **6,**
1352–1356.
Cappai, R., van Schravendijk, M. R., Anders, R. F., Peterson, M. G., Thomas, L. M.,
Cowman, A. F., and Kemp, D. J. (1989). Expression of the RESA gene in *Plasmo-
dium falciparum* isolate FCR3 is prevented by a subtelomeric deletion. *Mol. Cell.
Biol.* **9,** 3584–3587.
Cardenas, M. E., Laroche, T., and Gasser, S. M. (1990). The composition and morphology
of yeast nuclear scaffolds. *J. Cell Sci.* **96,** 439–450.
Carlson, M., Celenza, J. L., and Eng, F. J. (1985). Evolution of the dispersed SUC gene
family of *Saccharomyces* by rearrangements of chromosome telomeres. *Mol. Cell.
Biol.* **5,** 2894–2902.
Carmona, M. J., Morcillio, G., Galler, R., Martinez-Salas, E., de la Campa, A. G., and
Edström, J. E. (1985). Cloning and molecular characterization of a telomeric se-
quence from a temperature-induced Balbiani ring. *Chromosoma* **92,** 108–115.
Carson, M. J., and Hartwell, L. (1985). *CDC17:* An essential gene that prevents telomere
elongation in yeast. *Cell (Cambridge, Mass.)* **42,** 249–257.
Cavenee, W. K., Dryja, T. P., Phillips, R. A., Benedict, W. F., Godbout, R., Gallie, B. L.,
Murphree, A. L., Strong, L. C., and White, R. L. (1983). Expression of recessive
alleles by chromosomal mechanisms in retinoblastoma. *Nature (London)* **305,**
779–784.
Cavenee, W. K., Hansen, M. F., Nordenskjold, M., Kock, E., Maumenee, I., Squire, J. A.,
Phillips, R. A., and Gallie, B. L. (1985). Genetic origin of mutations predisposing to
retinoblastoma. *Science* **228,** 501–503.
Cech, T. R. (1988). G-strings at chromosome ends. *Nature (London)* **332,** 777–778.
Chan, C. S., and Tye, B. K. (1983a). A family of *Saccharomyces cerevisiae* repetitive
autonomously replicating sequences that have very similar genomic environments.
J. Mol. Biol. **168,** 505–523.
Chan, C. S., and Tye, B. K. (1983b). Organization of DNA sequences and replication
origins at yeast telomeres. *Cell (Cambridge, Mass.)* **33,** 563–573.
Charron, M. J., Read, E., Haut, S. R., and Michels, C. A. (1989). Molecular evolution of
the telomere-associated MAL loci of *Saccharomyces*. *Genetics* **122,** 307–316.
Cheng, J. F., and Smith, C. L. (1990). YAC cloning of telomeres. *Genet. Anal. Tech. Appl.*
7, 119–125.
Cheng, J. F., Smith, C. L., and Cantor, C. R. (1989). Isolation and characterization of a
human telomere. *Nucleic Acids Res.* **17,** 6109–6127.
Cheung, M. K., Drivas, D. T., Littau, V. C., and Johnson, E. M. (1981). Protein tightly
bound near the termini of the *Physarum* extrachromosomal rDNA palindrome. *J.
Cell Biol.* **91,** 309–314.
Chung, H. M., Shea, C., Fields, S., Taub, R. N., Van der Ploeg, L. H., and Tse, D. B.

(1990). Architectural organization in the interphase nucleus of the protozoan *Trypanosoma brucei:* Location of telomeres and mini-chromosomes. *EMBO J.* **9,** 2611–2619.

Cohn, M., and Edström, J. E. (1991). Evolutionary relations between subtypes of telomere-associated repeats in *Chironomus. J. Mol. Evol.* **32,** 463–468.

Conrad, M. N., Wright, J. H., Wolf, A. J., and Zakian, V. A. (1990). RAP1 protein interacts with yeast telomeres *in vivo:* Overproduction alters telomere structure and decreases chromosome stability. *Cell (Cambridge, Mass.)* **63,** 739–750.

Cooke, H. J., and Smith, B. A. (1986). Variability at the telomeres of the human X/Y pseudoautosomal region. *Cold Spring Harbor Symp. Quant. Biol.* **51,** 213–219.

Cooke, H. J., Brown, W. R., and Rappold, G. A. (1985). Hypervariable telomeric sequences from the human sex chromosomes are pseudoautosomal. *Nature (London)* **317,** 687–692.

Corcoran, L. M., Thompson, J. K., Walliker, D., and Kemp, D. J. (1988). Homologous recombination within subtelomeric repeat sequences generates chromosome size polymorphisms in *P. falciparum. Cell (Cambridge, Mass.)* **53,** 807–813.

Coren, J. S., Epstein, E. M., and Vogt, V. M. (1991). Characterization of a telomere-binding protein from *Physarum polycephalum. Mol. Cell. Biol.* **11,** 2282–2290.

Counter, C. M., Avilion, A. A., Le Feuvre, C. E., Steward, N. G., Greider, C. W., Harley, C. B., and Bacchetti, S. (1992). Telomere shortening associated with chromosome instability is arrested in immortal cells which express telomerase activity. *EMBO J.* **11,** 1921–1929.

Cross, S. H., Allshire, R. C., McKay, S. J., McGill, N. I., and Cooke, H. J. (1989). Cloning of human telomeres by complementation in yeast. *Nature (London)* **338,** 771–774.

Crossen, P. E., Pathak, S., and Arrighi, F. E. (1975). A high resolution study of DNA replication patterns of Chinese hamster chromosomes using sister chromatid differential staining technique. *Chromosoma* **52,** 339–347.

Dancis, B. M., and Holmquist, G. P. (1979). Telomere replication and fusion in eukaryotes. *J. Theor. Biol.* **78,** 211–224.

Dani, G. M., and Zakian, V. A. (1983). Mitotic and meiotic stability of linear plasmids in yeast. *Proc. Natl. Acad. Sci. U.S.A.* **80,** 3406–3410.

Danilevskaya, O. N., Kurenova, E. V., Pavlova, M. N., Bebehov, D. V., Link, A. J., Koga, A., Vellek, A., and Hartl, D. L. (1991). He-T family DNA sequences in the Y chromosome of *Drosophila melanogaster* share homology with the X-linked stellate genes. *Chromosoma* **100,** 118–124.

de Lange, A. M., and McFadden, G. (1990). The role of telomeres in poxvirus DNA replication. *Curr. Top. Microbiol. Immunol.* **163,** 71–92.

de Lange, T. (1992). Human telomeres are attached to the nuclear matrix. *EMBO J.* **11,** 717–724.

de Lange, T., Shiue, L., Myers, R. M., Cox, D. R., Naylor, S. L., Killery, A. M., and Varmus, H. E. (1990). Structure and variability of human chromosome ends. *Mol. Cell. Biol.* **10,** 518–527.

de Vries, E., van Driel, W., Bergsma, W. G., Arnberg, A. C., and van der Vliet, P. C. (1989). HeLa nuclear protein recognizing DNA termini and translocating on DNA forming a regular DNA-multimeric protein complex. *J. Mol. Biol.* **208,** 65–78.

Dewald, G. W., Dahl, R. J., Spurbeck, J. L., Carney, J. A., and Gordon, H. (1987). Chromosomally abnormal clones and nonrandom telomeric translocations in cardiac myxomas. *Mayo Clin. Proc.* **62,** 558–567.

Dietz-Band, J., Riethman, H., Hildebrand, C. E., and Moyzis, R. (1990). Characterization of polymorphic loci on a telomeric fragment of DNA from the long arm of human chromosome 7. *Genomics* **8**, 168–170.

Dobzhansky, T. (1972). Nothing in biology makes sense except in the light of evolution. *Am. Biol. Teacher* **35**, 125–129.

Dore, E., Pace, T., Ponzi, M., Picci, L., and Fontali, C. (1990). Organization of subtelomeric repeats in *Plasmodium berghei*. *Mol. Cell. Biol.* **10**, 2423–2427.

Drewinko, B., and Barlogie, B. (1976). Survival and cycle-progression delay of human lymphoma cells *in vitro* exposed to VP-16-213. *Cancer Treat. Rep.* **60**, 1295–1306.

Dunn, B. Szauter. P., Pardue, M. L., and Szostak, J. W. (1984). Transfer of yeast telomeres to linear plasmids by recombination. *Cell (Cambridge, Mass.)* **39**, 191–201.

Eissenberg, J. C. (1989). Position-effect variegation in *Drosophila:* Towards a genetics of chromatin assembly. *BioEssays* **11**, 14–17.

Ellis, N. A., and Goodfellow, P. N. (1989). The mammalian pseudoautosomal region. *Trends Genet.* **5**, 406–410.

Ellis, N. A., Goodfellow, P. J., Pym, B., Smith, M., Palmer, M., Frischauf, A. M., and Goodfellow, P. N. (1989). The pseudoautosomal boundary in man is defined by an Alu repeat sequence inserted on the Y chromosome. *Nature (London)* **337**, 81–84.

Ellis, N. A., Yen, P., Neiswanger, K., Shapiro, L. J., and Goodfellow, P. N. (1990). Evolution of the pseudoautosomal boundary in Old World monkeys and great apes. *Cell (Cambridge, Mass.)* **63**, 977–986.

Emery, H. S., and Weiner, A. M. (1981). An irregular satellite sequence is found at the termini of the linear extrachromosomal rDNA in *Dictyostelium discoideum*. *Cell (Cambridge, Mass.)* **26**, 411–419.

Erickson, L., Beversdorf, W. D., and Pauls, K. R. (1985). Linear mitochondrial plasmid in *Brassica* has terminal protein. *Curr. Genet.* **9**, 679–682.

Fang, G., and Cech, T. R. (1991). Molecular cloning of telomere-binding protein genes from *Stylonychia mytilis*. *Nucleic Acids Res.* **19**, 5515–5518.

Farr, C., Fantes, J., Goodfellow, P. N., and Cooke, H. (1991). Functional reintroduction of human telomeres into mammalian cells. *Proc. Natl. Acad. Sci. U.S.A.* **88**, 7006–7010.

Fearon, E. R., Vogelstein, B., and Feinberg, A. P. (1984). Somatic deletion and duplication of genes on chromosome 11 in Wilms' tumours. *Nature (London)* **309**, 176–178.

Ferguson, B. M., Brewer, B. J., Reynolds, A. E., and Fangman, W. L. (1991). A yeast origin of replication is activated late in S phase. *Cell (Cambridge, Mass.)* **65**, 507–515.

Fitzgerald, P. H., and Morris, C. M. (1984). Telomeric association of chromosomes in B-cell lymphoid leukemia. *Hum. Genet.* **67**, 385–390.

Foe, V. E., and Alberts, B. M. (1985). Reversible chromosome condensation induced in *Drosophila* embryos by anoxia: Visualization of interphase nuclear organization. *J. Cell Biol.* **100**, 1623–1636.

Foote, S. J., and Kemp, D. J. (1989). Chromosomes of malaria parasites. *Trends Genet.* **5**, 337–342.

Forney, J. D., and Blackburn, E. H. (1988). Developmentally controlled telomere addition in wild-type and mutant paramecia. *Mol. Cell. Biol.* **8**, 251–258.

Fung, Y. K., Murphree, A. L., T'Ang, A., Qian, J., Hinrichs, S. H., and Benedict, W. F. (1987). Structural evidence for the authenticity of the human retinoblastoma gene. *Science* **236**, 1657–1661.

Gehring, W. J., Klemenz, R., Weber, U., and Kloter, U. (1984). Functional analysis of the *white*⁺ gene in *Drosophila* by P-factor-mediated transformation. *EMBO J.* **3**, 2077–2085.

Gilley, D., Preer, J. J., Aufderheide, K. J., and Polisky, B. (1988). Autonomous replication and addition of telomerelike sequences to DNA microinjected into *Paramecium tetraurelia* macronuclei. *Mol. Cell. Biol.* **8**, 4765–4772.

Gillies, C. B. (1975). Synaptonemal complex and chromosome structure. *Annu. Rev. Genet.* **9**, 91–109.

Godiska, R., and Yao, M. C. (1990). A programmed site-specific DNA rearrangement in *Tetrahymena thermophila* requires flanking polypurine tracts. *Cell (Cambridge, Mass.)* **61**, 1237–1246.

Golubovsky, M. D. (1978). The "lethal giant larvae"—The most frequent second chromosome lethal in natural populations of *D. melanogaster. Drosophila Inf. Serv.* **53**, 179.

Golubovsky, M. D., and Sokolova, K. B. (1973). The expression and interaction of different alleles at the *l(2)gl* locus. *Drosophila Inf. Serv.* **50**, 124.

Goodfellow, P. J., Darling, S. M., Thomas, N. S., and Goodfellow, P. N. (1986). A pseudoautosomal gene in man. *Science* **234**, 740–743.

Gottschling, D. E., and Cech, T. R. (1984). Chromatin structure of the molecular ends of *Oxytricha* macronuclear DNA: Phased nucleosomes and a telomeric complex. *Cell (Cambridge, Mass.)* **38**, 501–510.

Gottschling, D. E., and Zakian, V. A. (1986). Telomere proteins: Specific recognition and protection of the natural termini of *Oxytricha* macronuclear DNA. *Cell (Cambridge, Mass.)* **47**, 195–205.

Gottschling, D. E., Aparicio, O. M., Billington, B. L., and Zakian, V. A. (1990). Position effect at *S. cerevisiae* telomeres: Reversible repression of Pol II transcription. *Cell (Cambridge, Mass.)* **63**, 751–762.

Graf, U., Green, M. M., and Würgler, F. E. (1979). Mutagen-sensitive mutants in *Drosophila melanogaster:* Effects on premutational damage. *Mutat. Res.* **63**, 101–112.

Gray, J. T., Celander, D. W., Price, C. M., and Cech, T. R. (1991). Cloning and expression of genes for the *Oxytricha* telomere-binding protein: Specific subunit interactions in the telomeric complex. *Cell (Cambridge, Mass.)* **67**, 807–814.

Green, M. M., and Shepherd, S. H. (1979). Genetic instability in *Drosophila melanogaster:* The induction of specific chromosome 2 deletions by MR elements. *Genetics* **92**, 823–832.

Greider, C. W. (1990). Telomeres, telomerase and senescence. *BioEssays* **12**, 363–369.

Greider, C. W., and Blackburn, E. H. (1985). Identification of a specific telomere terminal transferase activity in *Tetrahymena* extracts. *Cell (Cambridge, Mass.)* **43**, 405–413.

Greider, C. W., and Blackburn, E. H. (1987). The telomere terminal transferase of *Tetrahymena* is a ribonucleoprotein enzyme with two kinds of primer specificity. *Cell (Cambridge, Mass.)* **51**, 887–898.

Greider, C. W., and Blackburn, E. H. (1989). A telomeric sequence in the RNA of *Tetrahymena* telomerase required for telomere repeat synthesis. *Nature (London)* **337**, 331–337.

Guerrini, A. M., Ascenzioni, F., Pisani, G., Rappazzo, G., Della, V. G., and Donini, P. (1990). Cloning a fragment from the telomere of the long arm of human chromosome 9 in a YAC vector. *Chromosoma* **99**, 138–142.

Haber, J. E., and Thorburn, P. C. (1984). Healing of broken linear dicentric chromosomes in yeast. *Genetics* **106**, 207–226.

Haber, J. E., Thorburn, P. C., and Rogers, D. (1984). Meiotic and mitotic behavior of dicentric chromosomes in *Saccharomyces cerevisiae. Genetics* **106**, 185–205.

Hardin, C. C., Henderson, E., Watson, T., and Prosser, J. K. (1991). Monovalent cation induced structural transitions in telomeric DNAs: G- DNA folding intermediates. *Biochemistry* **30**, 4460–4472.

induced structural transitions in telomeric DNAs: G- DNA folding intermediates. *Biochemistry* **30**, 4460–4472.

Harley, C. B., Futcher, A. B., and Greider, C. W. (1990). Telomeres shorten during ageing of human fibroblasts. *Nature (London)* **345**, 458–460.

Harrington, L. A., and Greider, C. W. (1991). Telomerase primer specificity and chromosome healing. *Nature (London)* **353**, 451–454.

Harris, P. C., Barton, N. J., Higgs, D. R., Reeders, S. T., and Wilkie, A. O. (1990). A long-range restriction map between the alpha-globin complex and a marker closely linked to the polycystic kidney disease 1 (PKD1) locus. *Genomics* **7**, 195–206.

Hastie, N. D., and Allshire, R. C. (1989). Human telomeres: Fusion and interstitial sites. *Trends Genet.* **5**, 326–331.

Hastie, N. D., Dempster, M., Dunlop, M. G., Thompson, A. M., Green, D. K., and Allshire, R. C. (1990). Telomere reduction in human colorectal carcinoma and with ageing. *Nature (London)* **346**, 866–868.

Hazelrigg, T., Levis, R. W., and Rubin, G. M. (1984). Transformation of *white* locus DNA in *Drosophila:* Dosage compensation, *zeste* interaction, and position effects. *Cell (Cambridge, Mass.)* **36**, 469–481.

Henderson, E., Hardin, C. C., Walk, S. K., Tinoco, I. J., and Blackburn, E. H. (1987). Telomeric DNA oligonucleotides form novel intramolecular structures containing guanine–guanine base pairs. *Cell (Cambridge, Mass.)* **51**, 899–908.

Henderson, E. R., Moore, M., and Malcolm, B. A. (1990). Telomere G-strand structure and function analyzed by chemical protection, base analogue substitution, and utilization by telomerase *in vitro. Biochemistry* **29**, 732–737.

Herrick, G., Hunter, D., Williams, K., and Kotter, K. (1987). Alternative processing during development of a macronuclear chromosome family in *Oxytricha fallax. Genes Dev.* **1**, 1047–1058.

Hicke, B. J., Celander, D. W., MacDonald, G. H., Price, C. M., and Cech, T. R. (1990). Two versions of the gene encoding the 41-kilodalton subunit of the telomere binding protein of *Oxytricha nova. Proc. Natl. Acad. Sci. U.S.A.* **87**, 1481–1485.

Hiraoka, Y., Agard, D. A., and Sedat, J. W. (1990). Temporal and spatial coordination of chromosome movement, spindle formation, and nuclear envelope breakdown during prometaphase in *Drosophila melanogaster* embryos. *J. Cell Biol.* **111**, 2815–2828.

Hochstrasser, M., and Sedat, J. W. (1987). Three-dimensional organization of *Drosophila melanogaster* interphase nuclei. II. Chromosome spatial organization and gene regulation. *J. Cell Biol.* **104**, 1471–1483.

Hochstrasser, M. Mathog, D., Gruenbaum, Y., Saumweber, H., and Sedat, J. W. (1986). Spatial organization of chromosomes in the salivary gland nuclei of *Drosophila melanogaster. J. Cell Biol.* **102**, 112–123.

Hofmann, J. F.-X., Laroche, T., Brand, A. H., and Gasser, S. M. (1989). RAP-1 factor is necessary for DNA loop formation *in vitro* at the silent mating type locus HML. *Cell (Cambridge, Mass.)* **57**, 725–737.

Horowitz, H., and Haber, J. E. (1984). Subtelomeric regions of yeast chromosomes contain a 36 base-pair tandemly repeated sequence. *Nucleic Acids Res.* **12**, 7105–7121.

Horowitz, H., Thorburn, P., and Haber, J. E. (1984). Rearrangements of highly polymorphic regions near telomeres of *Saccharomyces cerevisiae. Mol. Cell. Biol.* **4**, 2509–2517.

Jäger, D., and Philippsen, P. (1989a). Many yeast chromosomes lack the telomere-specific Y' sequence. *Mol. Cell. Biol.* **9**, 5754–5757.

Jäger, D., and Philippsen, P. (1989b). Stabilization of dicentric chromosomes in *Saccha-*

romyces cerevisiae by telomere addition to broken ends or by centromere deletion. *EMBO J.* **8**, 247–254.

Johnson, E. M. (1980). A family of inverted repeat sequences and specific single-strand gaps at the termini of the *Physarum* rDNA palindrome. *Cell (Cambridge, Mass.)* **22**, 875–886.

Kemble, R. J., and Thompson, R. D. (1982). S1 and S2, the linear mitochondrial DNAs present in a male sterile line of maize, possess terminally attached proteins. *Nucleic Acids Res.* **10**, 8181–8190.

Kikuchi, Y., Hirai, K., and Hishinuma, F. (1984). The yeast linear DNA killer plasmids pGKL1 and pGKL2, possess terminally attached proteins. *Nucleic Acids Res.* **12**, 5685–5692.

Kimler, B. F., Leeper, D. B., and Schneiderman, M. H. (1981). Radiation-induced division delay in Chinese hamster ovary fibroblast and carcinoma cells: Dose effect and ploidy. *Radiat. Res.* **85**, 270–280.

Kipling, D., and Cooke, H. J. (1990). Hypervariable ultra-long telomeres in mice. *Nature (London)* **347**, 400–402.

Kitani, T., Yoda, K., and Okazaki, T. (1984). Discontinuous DNA replication of *Drosophila melanogaster* is primed by octaribonucleotide primer. *Mol. Cell. Biol.* **4**, 1591–1596.

Klobutcher, L. A., Swanton, M. T., Donini, P., and Prescott, D. M. (1981). All gene-sized DNA molecules in four species of hypotrichs have the same terminal sequence and an unusual 3′ terminus. *Proc. Natl. Acad. Sci. U.S.A.* **78**, 3015–3019.

Knudson, A. G. J. (1986). Genetics of human cancer. *Annu. Rev. Genet.* **20**, 231–251.

Konopa, J. (1988). G2 block induced by DNA crosslinking agents and its possible consequences. *Biochem. Pharmacol.* **37**, 2303–2309.

Koufos, A., Hansen, M. F., Lampkin, B. C., Workman, M. L., Copeland, N. G., Jenkins, N. A., and Cavenee, W. K. (1984). Loss of alleles at loci on human chromosome 11 during genesis of Wilms' tumour. *Nature (London)* **309**, 170–172.

Kovacs, G., Müller-Brechlin, R., and Szucs, S. (1987). Telomeric association in two human renal tumors. *Cancer Genet. Cytogenet.* **28**, 363–366.

Larson, D. D., Spangler, E. A., and Blackburn, E. H. (1987). Dynamics of telomere length variation in *Tetrahymena thermophila*. *Cell (Cambridge, Mass.)* **50**, 477–483.

Le Blancq, S. M., Korman, S. H., and Van der Ploeg, L. H. T. (1991). Frequent rearrangements of rRNA-encoding chromosomes in *Giardia*. *Nucleic Acids Res.* **19**, 4405–4412.

Leigh, B. (1978). The formation and recovery of two-break chromosome rearrangements from irradiated spermatozoa of *Drosophila melanogaster*. *Mutat. Res.* **49**, 45–54.

Levis, R., Hazelrigg, T., and Rubin, G. M. (1985). Effects of genome position on the expression of transduced copies of the *white* gene of *Drosophila*. *Science* **229**, 558–561.

Levis, R. W. (1989). Viable deletions of a telomere from a *Drosophila* chromosome. *Cell (Cambridge, Mass.)* **58**, 791–801.

Levy, M. Z., Allsopp, R. C., Futcher, A. B., Greider, C. W., and Harley, C. B. (1992). Telomere end-replication problem and cell aging. *J. Mol. Biol.* **225**, 951–960.

Lima-de-Faria, A., and Jaworska, H. (1968). Late DNA synthesis in heterochromatin. *Nature (London)* **217**, 138–142.

Lindsley, D. L., and Tokuyasu, K. T. (1980). Spermatogenesis. *In* "The Genetics and Biology of *Drosophila*" (M. Ashburner and T. R. F. Wright, eds.), Vol. 2, pp. 225–294. Academic Press, London.

Liu, Z., and Tye, B.-K. (1991). A yeast protein that binds to vertebrate telomeres and conserved yeast telomeric junction. *Genes Dev.* **5**, 49–59.

Lock, R. B., and Ross, W. E. (1990). Inhibition of p34[cdc2] kinase activity by etoposide or irradiation as a mechanism of G_2 arrest in Chinese hamster ovary cells. *Cancer Res.* **50**, 3761–3766.

Longtine, M. S., Wilson, N. M., Petracek, M. E., and Berman, J. (1989). A yeast telomere binding activity binds to two related telomere sequence motifs and is indistinguishable from RAP1. *Curr. Genet.* **16**, 225–239.

Louis, E. J., and Haber, J. E. (1990a). The subtelomeric Y' repeat family in *Saccharomyces cerevisiae:* An experimental system for repeated sequence evolution. *Genetics* **124**, 533–545.

Louis, E. J., and Haber, J. E. (1990b). Mitotic recombination among subtelomeric Y' repeats in *Saccharomyces cerevisiae. Genetics* **124**, 547–559.

Louis, E. J., and Haber, J. E. (1991). Evolutionarily recent transfer of a group I mitochondrial intron to telomere regions in *Saccharomyces cerevisiae. Curr. Genet.* **20**, 411–415.

Louis, E. J., and Haber, J. E. (1992). The structure and evolution of subtelomeric Y' repeats in *Saccharomyces cerevisiae. Genetics* **131**, 559–574.

Lucchini, R., Pauli, U., Braun, R., Koller, T., and Sogo, J. M. (1987). Structure of the extrachromosomal ribosomal RNA chromatin of *Physarum polycephalum. J. Mol. Biol.* **196**, 829–843.

Lundblad, V., and Blackburn, E. H. (1990). RNA-dependent polymerase motifs in EST1: Tentative identification of a protein component of an essential yeast telomerase. *Cell (Cambridge, Mass.)* **60**, 529–530.

Lundblad, V., and Szostak, J. W. (1989). A mutant with a defect in telomere elongation leads to senescence in yeast. *Cell (Cambridge, Mass.)* **57**, 633–643.

Lustig, A. J., and Petes, T. D. (1986). Identification of yeast mutants with altered telomere structure. *Proc. Natl. Acad. Sci. U.S.A.* **83**, 1398–1402.

Lustig, A. J., Kurtz, S., and Shore, D. (1990). Involvement of the silencer and UAS binding protein RAP1 in regulation of telomere length. *Science* **250**, 549–553.

Maddern, R. H., and Leigh, B. (1976). The timing of the restitution of chromosome breaks induced by x-rays in the mature sperm of *Drosophila melanogaster. Mutat. Res.* **41**, 255–268.

Maguire, M. (1984). The mechanism of meiotic homologue pairing. *J. Theor. Biol.* **106**, 605–615.

Mandahl, N., Heim, S., Kristoffersson, U., Mitelman, F., Rooser, B., Rydholm, A., and Willen, H. (1985). Telomeric association in a malignant fibrous histiocytoma. *Hum. Genet.* **71**, 321–324.

Mandahl, N., Heim, S., Arheden, K., Rydholm, A., Willen, H., and Mitelman, F. (1988). Rings, dicentrics, and telomeric association in histiocytomas. *Cancer Genet. Cytogenet.* **30**, 23–33.

Mason, J. M., Strobel, E., and Green, M. M. (1984). *mu-2:* Mutator gene in *Drosophila* that potentiates the induction of terminal deficiencies. *Proc. Natl. Acad. Sci. U.S.A.* **81**, 6090–6094.

Mathog, D., and Sedat, J. W. (1989). The three-dimensional organization of polytene nuclei in male *Drosophila melanogaster* with compound XY or ring X chromosomes. *Genetics* **121**, 293–311.

Mathog, D., Hochstrasser, M., Gruenbaum, Y., Saumweber, H., and Sedat, J. W. (1984). Characteristic folding pattern of polytene chromosomes in *Drosophila* salivary gland nuclei. *Nature (London)* **308**, 414–421.

Matsumoto, T., Fukui, K., Niwa, O., Sugawara, N., Szostak, J. W., and Yanagida, M. (1987). Identification of healed terminal DNA fragments in linear minichromosomes of *Schizosaccharomyces pombe. Mol. Cell. Biol.* **7**, 4424–4430.

McCarroll, R. M., and Fangman, W. L. (1988). Time of replication of yeast centromeres and telomeres. *Cell (Cambridge, Mass.)* **54,** 505–513.

McClintock, B. (1938). The fusion of broken ends of sister half-chromatids following chromatid breakage at meiotic anaphase. *Res. Bull.—Mo., Agric. Exp. Stn.* **290,** 1–48.

McClintock, B. (1939). The behavior of successive nuclear divisions of a chromosome broken at meiosis. *Proc. Natl. Acad. Sci. U.S.A.* **25,** 405–416.

McClintock, B. (1941). The stability of broken ends of chromosomes in *Zea mays. Genetics* **26,** 234–282.

McClintock, B. (1984). The significance of responses of the genome to challenge. *Science* **226,** 792–801.

Mechler, B. M., McGinnis, W., and Gehring, W. J. (1985). Molecular cloning of *lethal(2)giant larvae*, a recessive oncogene of *Drosophila melanogaster. EMBO J.* **4,** 1551–1557.

Meinhardt, E., Kempken, F., and Esser, K. (1986). Proteins are attached to the ends of a linear plasmid in the filamentous fungus *Ascobolus immersus. Curr. Genet.* **11,** 243–246.

Meyne, J., Ratliff, R. L., and Moyzis, R. K. (1989). Conservation of the human telomere sequence (TTAGGG)$_n$ among vertebrates. *Proc. Natl. Acad. Sci. U.S.A.* **86,** 7049–7053.

Meyne, J., Baker, R. J., Hobart, H. H., Hsu, T. C., Ryder, O. A., Ward, O. G., Wiley, J. E., Wurster-Hill, D. H., Yates, T. L., and Moyzis, R. K. (1990). Distribution of non-telomeric sites of the (TTAGGG)$_n$ telomeric sequence in vertebrates. *Chromosoma* **99,** 3–10.

Mimori, T., and Hardin, J. A. (1986). Mechanism of interaction between Ku protein and DNA. *J. Biol. Chem.* **261,** 10375–10379.

Mimori, T., Hardin, J. A., and Steitz, J. A. (1986). Characterization of the DNA-binding protein antigen Ku recognized by autoantibodies from patients with rheumatic disorders. *J. Biol. Chem.* **261,** 2274–2278.

Morawetz, C. (1987). Effect of irradiation and mutagenic chemicals on the generation of ADH2-constitutive mutants in yeast. Significance for the inducibility of Ty transposition. *Mutat. Res.* **177,** 53–60.

Morawetz, C., and Hagen, U. (1990). Effect of irradiation and mutagenic chemicals on the generation of ADH2- and ADH4-constitutive mutants in yeast: The inducibility of Ty transposition by UV and ethyl methane sulfonate. *Mutat. Res.* **229,** 69–77.

Morgan, R., Jarzabek, V., Jaffe, J. P., Hecht, B. K., Hecht, F., and Sandberg, A. A. (1986). Telomeric fusion in pre-T-cell acute lymphoblastic leukemia. *Hum. Genet.* **73,** 260–263.

Morin, G. B. (1989). The human telomere terminal transferase enzyme is a ribonucleoprotein that synthesizes TTAGGG repeats. *Cell (Cambridge, Mass.)* **59,** 521–529.

Morin, G. B. (1991). Recognition of a chromosome truncation site associated with alpha-thalassaemia by human telomerase. *Nature (London)* **353,** 454–456.

Morin, G. B., and Cech, T. R. (1986). The telomeres of the linear mitochondrial DNA of *Tetrahymena thermophila* consist of 53 bp tandem repeats. *Cell (Cambridge, Mass.)* **46,** 873–883.

Morin, G. B., and Cech, T. R. (1988). Mitochondrial telomeres: Surprising diversity of repeated telomeric DNA sequences among six species of *Tetrahymena. Cell (Cambridge, Mass.)* **52,** 367–374.

Moyer, R. W., and Graves, R. L. (1981). The method of cytoplasmic orthopox replication. *Cell (Cambridge, Mass.)* **27,** 391–401.

Moyzis, R. K. (1991). The human telomere. *Sci. Am.* **265,** 48–55.

Müller, F., Wicky, C., Spicher, A., and Tobler, H. (1991). Telomere formation after developmentally regulated chromosome breakage during the process of chromatin diminution in *Ascaris lumbricoides*. *Cell (Cambridge, Mass.)* **67**, 815–822.

Muller, H. J. (1938). The remaking of chromosomes. *Collect. Net* **8**, 182–195.

Muller, H. J. (1940). An analysis of the process of structural change in chromosomes of *Drosophila*. *J. Genet.* **40**, 1–66.

Muller, H. J., and Herskowitz, I. H. (1954). Concerning the healing of chromosome ends produced by breakage in *Drosophila melanogaster*. *Am. Nat.* **88**, 177–208.

Murray, A. W., and Szostak, J. W. (1986). Construction and behavior of circularly permuted and telocentric chromosomes in *Saccharomyces cerevisiae*. *Mol. Cell. Biol.* **6**, 3166–3172.

Murray, A. W., Claus, T. E., and Szostak, J. W. (1988). Characterization of two telomeric DNA processing reactions in *Saccharomyces cerevisiae*. *Mol. Cell. Biol.* **8**, 4642–4650.

Nielsen, L., Schmidt, E. R., and Edström, J. E. (1990). Subrepeats result from regional DNA sequence conservation in tandem repeats in *Chironomus* telomeres. *J. Mol. Biol.* **216**, 577–584.

Novitski, E. (1952). The genetic consequences of anaphase bridge formation in *Drosophila*. *Genetics* **37**, 270–287.

Novitski, E. (1955). Genetic measures of centromere activity in *Drosophila melanogaster*. *J. Cell. Comp. Physiol.* **45**, Suppl. 2, 151–169.

Olovnikov, A. M. (1973). A theory of marginotomy. *J. Theor. Biol.* **41**, 181–190.

Oquendo, P., Goman, M., Mackay, M., Langsley, G., Walliker, D., and Scaife, J. (1986). Characterisation of a repetitive DNA sequence from the malaria parasite, *Plasmodium falciparum*. *Mol. Biochem. Parasitol.* **18**, 89–101.

Orkin, S. H., Goldman, D. S., and Sallan, S. E. (1984). Development of homozygosity for chromosome 11p markers in Wilms' tumour. *Nature (London)* **309**, 172–174.

Orr-Weaver, T. L., Szostak, J. W., and Rothstein, R. J. (1981). Yeast transformation: A model system for the study of recombination. *Proc. Natl. Acad. Sci. U.S.A.* **78**, 6354–6358.

Pace, T., Ponzi, M., Dore, F., and Fontali, C. (1987). Telomeric motifs are present in a highly repetitive element in the *Plasmodium berghei* genome. *Mol. Biochem. Parasitol.* **24**, 193–202.

Pace, T., Ponzi, M., Dore, E., Janse, C., Mons, D., and Fontali, C. (1990). Long insertions within telomeres contribute to chromosome size polymorphism in *Plasmodium berghei*. *Mol. Cell. Biol.* **10**, 6759–6764.

Paddy, M. R., Belmont, A. S., Saumweber, H., Agard, D. A., and Sedat, J. W. (1990). Interphase nuclear envelope lamins form a discontinuous network that interacts with only a fraction of the chromatin in the nuclear periphery. *Cell (Cambridge, Mass.)* **62**, 89–106.

Parker, D. R., and Hammond, A. E. (1958). The production of translocations in *Drosophila* oocytes. *Genetics* **43**, 92–100.

Pathak, S., Wang, Z., Dhaliwal, M. K., and Sacks, P. C. (1988). Telomeric association: Another characteristic of cancer chromosomes? *Cytogenet. Cell Genet.* **47**, 227–229.

Pays, E., and Steinert, M. (1988). Control of antigen expression in African trypanosomes. *Annu. Rev. Genet.* **22**, 107–126.

Petit, C., Levilliers, J., and Weissenbach, J. (1988). Physical mapping of the human pseudo-autosomal region; comparison with genetic linkage map. *EMBO J.* **7**, 2369–2376.

Petit, C., Levilliers, J., Rouyer, F., Simmler, M. C., Herouin, E., and Weissenbach, J.

(1990). Isolation of sequences from Xp22.3 and deletion mapping using sex chromosome rearrangements from human X–Y interchange sex reversals. *Genomics* **6**, 651–658.

Pluta, A. F., and Zakian, V. A. (1989). Recombination occurs during telomere formation in yeast. *Nature (London)* **337**, 429–433.

Pluta, A. F., Kaine, B. P., and Spear, B. B. (1982). The terminal organization of macronuclear DNA in *Oxytricha fallax*. *Nucleic Acids Res.* **10**, 8145–8154.

Pluta, A. F., Dani, G. M., Spear, B. B., and Zakian, V. A. (1984). Elaboration of telomeres in yeast: Recognition and modification of termini from *Oxytricha* macronuclear DNA. *Proc. Natl. Acad. Sci. U.S.A.* **81**, 1475–1479.

Pologe, L. G., and Ravetch, J. V. (1986). A chromosomal rearrangement in a *P. falciparum* histidine-rich protein gene is associated with the knobless phenotype. *Nature (London)* **322**, 474–477.

Pologe, L. G., and Ravetch, J. V. (1988). Large deletions result from breakage and healing of *P. falciparum* chromosomes. *Cell (Cambridge, Mass.)* **55**, 869–874.

Ponzi, M., Janse, C. J., Dore, E., Scotti, R., Pace, T., Reterink, T. J. F., van der Berg, F. M., and Mons, B. (1990). Generation of chromosome size polymorphism during *in vivo* mitotic multiplication of *Plasmodium berghei* involves both loss and addition of subtelomeric repeat sequences. *Mol. Biochem. Parasitol.* **41**, 73–82.

Price, C. M. (1990). Telomere structure in *Euplotes crassus:* Characterization of DNA–protein interactions and isolation of a telomere-binding protein. *Mol. Cell. Biol.* **10**, 3421–3431.

Price, C. M., and Cech, T. R. (1987). Telomeric DNA-protein interactions of *Oxytricha* macronuclear DNA. *Genes Dev.* **1**, 783–793.

Price, C. M., and Cech, T. R. (1989). Properties of the telomeric DNA-binding protein from *Oxytricha nova*. *Biochemistry* **28**, 769–774.

Rabl, C. (1885). Über Zellteilung. *Morphol. Jahrb.* **10**, 214–330.

Raghuraman, M. K., and Cech, T. R. (1989). Assembly and self-assembly of *Oxytricha* telomeric nucleoprotein complexes. *Cell (Cambridge, Mass.)* **59**, 719–728.

Raghuraman, M. K., and Cech, T. R. (1990). Effect of monovalent cation-induced telomeric DNA structure on the binding of *Oxytricha* telomeric protein. *Nucleic Acids Res.* **18**, 4543–4552.

Raghuraman, M. K., Dunn, C. J., Hicke, B. J., and Cech, T. R. (1989). *Oxytricha* telomeric nucleoprotein complexes reconstituted with synthetic DNA. *Nucleic Acids Res.* **17**, 4235–4253.

Rawlins, D. J., Highett, M. I., and Shaw, P. J. (1991). Localization of telomeres in plant interphase nuclei by *in situ* hybridization and 3D confocal microscopy. *Chromosoma* **100**, 424–431.

Rekosh, D. M., Russell, W. C., Bellet, A. J., and Robinson, A. J. (1977). Identification of a protein linked to the ends of adenovirus DNA. *Cell (Cambridge, Mass.)* **11**, 283–295.

Renkawitz-Pohl, R., and Bialojan, S. (1984). A DNA sequence of *Drosophila melanogaster* with a differential telomeric distribution. *Chromosoma* **89**, 206–211.

Reynolds, A. E., McCarroll, R. M., Newlon, C. S., and Fangman, W. L. (1989). Time of replication of ARS elements along yeast chromosome III. *Mol. Cell. Biol.* **9**, 4488–4494.

Richards, E. J., and Ausubel, F. M. (1988). Isolation of a higher eukaryotic telomere from *Arabidopsis thaliana*. *Cell (Cambridge, Mass.)* **53**, 127–136.

Richards, E. J., Goodman, H. M., and Ausubel, F. M. (1991). The centromere region of *Arabidopsis thaliana* chromosome 1 contains telomere-similar sequences. *Nucleic Acids Res.* **19**, 3351–3357.

Riethman, H. C., Moyzis, R. K., Meyne, J., Burke, D. T., and Olson, M. V. (1989). Cloning human telomeric DNA fragments into *Saccharomyces cerevisiae* using a yeast-artificial-chromosome vector. *Proc. Natl. Acad. Sci. U.S.A.* **86**, 6240–6244.

Ripley, L. S. (1990). Frameshift mutation: Determinants of specificity. *Annu. Rev. Genet.* **24**, 189–213.

Roberts, P. A. (1975). In support of the telomere concept. *Genetics* **80**, 135–142.

Romero, D. P., and Blackburn, E. H. (1991). A conserved secondary structure for telomerase RNA. *Cell (Cambridge, Mass.)* **67**, 343–353.

Rouyer, F., Simmler, M. C., Johnsson, C., Vergnaud, G., Cooke, H. J., and Weissenbach, J. (1986). A gradient of sex linkage in the pseudoautosomal region of the human sex chromosomes. *Nature (London)* **319**, 291–295.

Rouyer, F., de la Chapelle, A., Andersson, M., and Weissenbach, J. (1990). An interspersed repeated sequence specific for human subtelomeric regions. *EMBO J.* **9**, 505–514.

Rubin, G. M. (1978). Isolation of a telomeric DNA sequence from *Drosophila melanogaster*. *Cold Spring Harbor Symp. Quant. Biol.* **42**, 1041–1046.

Runge, K. W., and Zakian, V. A. (1989). Introduction of extra telomeric DNA sequences into *Saccharomyces cerevisiae* results in telomere elongation. *Mol. Cell. Biol.* **9**, 1488–1497.

Saiga, H., and Edström, J. E. (1985). Long tandem arrays of complex repeat units in *Chironomus* telomeres. *EMBO J.* **4**, 799–804.

Sen, D., and Gilbert, W. (1988). Formation of parallel four-stranded complexes by guanine-rich motifs in DNA and its implications for meiosis. *Nature (London)* **334**, 364–366.

Sen, D., and Gilbert, W. (1990). A sodium-potassium switch in the formation of four-stranded G4-DNA. *Nature (London)* **344**, 410–414.

Shampay, J., and Blackburn, E. H. (1988). Generation of telomere-length heterogeneity in *Saccharomyces cerevisiae*. *Proc. Natl. Acad. Sci. U.S.A.* **85**, 534–538.

Shampay, J., Szostak, J. W., and Blackburn, E. H. (1984). DNA sequences of telomeres maintained in yeast. *Nature (London)* **310**, 154–157.

Sherwood, S. W., Rush, D., Ellsworth, J. L., and Schimke, R. T. (1988). Defining cellular senescence in IMR-90 cells: A flow cytometric analysis. *Proc. Natl. Acad. Sci. U.S.A.* **85**, 9086–9090.

Shippen-Lentz, D., and Blackburn, E. H. (1989). Telomere terminal transferase activity from *Euplotes crassus* adds large numbers of TTTTGGGG repeats onto telomeric primers. *Mol. Cell. Biol.* **9**, 2761–2764.

Shippen-Lentz, D., and Blackburn, E. H. (1990). Functional evidence for an RNA template in telomerase. *Science* **247**, 546–552.

Shoeman, R. L., and Traub, P. (1990). The *in vitro* DNA-binding properties of purified nuclear lamin proteins and vimentin. *J. Biol. Chem.* **265**, 9055–9061.

Shoeman, R. L., Wadle, S., Scherbarth, A., and Traub, P. (1988). The binding *in vitro* of the intermediate filament protein vimentin to synthetic oligonucleotides containing telomere sequences. *J. Biol. Chem.* **263**, 18744–18749.

Simmler, M. C., Rouyer, F., Vergnaud, G., Nystrom, L. M., Ngo, K. Y., de la Chapelle, A., and Weissenbach, J. (1985). Pseudoautosomal DNA sequences in the pairing region of the human sex chromosomes. *Nature (London)* **317**, 692–697.

Simmler, M. C., Johnsson, C., Petit, C., Rouyer, F., Vergnaud, G., and Weissenbach, J. (1987). Two highly polymorphic minisatellites from the pseudoautosomal region of the human sex chromosomes. *EMBO J.* **6**, 963–969.

Singh, H., and Dumas, L. B. (1984). A DNA primase that copurifies with the major DNA polymerase from the yeast *Saccharomyces cerevisiae*. *J. Biol. Chem.* **259**, 7936–7940.

Smith, D. A., Baker, B. S., and Gatti, M. (1985). Mutations in genes encoding essential mitotic functions in *Drosophila melanogaster*. *Genetics* **110**, 647–670.

Sobels, F. H. (1974). The persistence of chromosome breaks in different stages of spermatogenesis of *D. melanogaster*. *Mutat. Res.* **23**, 361–368.

Spofford, J. B. (1976). Position-effect variegation in *Drosophila*. *In* "The Genetics and Biology of *Drosophila*" (M. Ashburner and E. Novitski, eds.), Vol. 1c, pp. 469–481. Academic Press, London.

Spradling, A. C., and Karpen, G. H. (1990). 60 years of mystery. *Genetics* **126**, 779–784.

Stanbridge, E. (1990). Human tumor suppressive genes. *Annu. Rev. Genet.* **24**, 615–657.

Starling, J. A., Maule, J., Hastie, N. D., and Allshire, R. C. (1990). Extensive telomere repeat arrays in mouse are hypervariable. *Nucleic Acids Res.* **18**, 6881–6888.

Steinemann, M. (1984). Telomere repeats within the neo-Y-chromosome of *Drosophila miranda*. *Chromosoma* **90**, 1–5.

Steinemann, M., and Nauber, U. (1986). Frequency of telomere repeat units in the *Drosophila miranda* genome. *Genetica* **69**, 47–57.

Streisinger, G., Okada, Y., Emrich, J., Newton, J., Tsugita, A., Terzaghi, E., and Inouye, M. (1966). Frameshift mutations and the genetic code. *Cold Spring Harbor Symp. Quant. Biol.* **31**, 77–84.

Stuiver, M. H., Coenjaerts, F. E., and van der Vliet, P. C. (1990). The autoantigen Ku is indistinguishable from NF IV, a protein forming multimeric protein-DNA complexes. *J. Exp. Med.* **172**, 1049–1054.

Stuiver, M. H., Celis, J. E., and van der Vliet, P. C. (1991). Identification of nuclear factor IV/Ku autoantigen in a human 2D-gel protein database. Modification of the large subunit depends on cellular proliferation. *FEBS Lett.* **282**, 189–192.

Sturtevant, A. A., and Beadle, G. W. (1936). The relations of invertions in the X chromosome of *Drosophila melanogaster* to crossing over and disjunction. *Genetics* **21**, 554–604.

Sundquist, W. I., and Klug, A. (1989). Telomeric DNA dimerizes by formation of guanine tetrads between hairpin loops. *Nature (London)* **342**, 825–829.

Szostak, J. W., and Blackburn, E. H. (1982). Cloning yeast telomeres on linear plasmid vectors. *Cell (Cambridge, Mass.)* **29**, 245–255.

Tartof, K. D., Bishop, C., Jones, M., Hobbs, C. A., and Locke, J. (1989). Towards an understanding of position effect variegation. *Dev. Genet.* **10**, 162–176.

Ten-Hagen, K. G., Gilbert, D. M., Willard, H. F., and Cohen, S. N. (1990). Replication timing of DNA sequences associated with human centromeres and telomeres. *Mol. Cell. Biol.* **10**, 6348–6355.

Traverse, K. L., and Pardue, M. L. (1988). A spontaneously opened ring chromosome of *Drosophila melanogaster* has acquired He-T DNA sequences at both new telomeres. *Proc. Natl. Acad. Sci. U.S.A.* **85**, 8116–8120.

Traverse, K. L., and Pardue, M. L. (1989). Studies of He-T DNA sequences in the pericentric regions of *Drosophila* chromosomes. *Chromosoma* **97**, 261–271.

Valgeirsdottir, K., Traverse, K. L., and Pardue, M. L. (1990). HeT DNA: A family of mosaic repeated sequences specific for heterochromatin in *Drosophila melanogaster*. *Proc. Natl. Acad. Sci. U.S.A.* **87**, 7998–8002.

Van der Ploeg, L. H., Liu, A. Y., and Borst, P. (1984). Structure of the growing telomeres of Trypanosomes. *Cell (Cambridge, Mass.)* **36**, 459–468.

Van der Werf, A., Van Assel, S., Aerts, D., Steinert, M., and Pays, E. (1990). Telomere interactions may condition the programming of antigen expression in *Trypanosoma brucei*. *EMBO J.* **9**, 1035–1040.

Venter, U., and Hörz, W. (1989). The acid phosphatase genes PHO10 and PHO11 in *S. cerevisiae* are located at the telomeres of chromosomes VIII and I. *Nucleic Acids Res.* **17**, 1353–1369.

Vogelstein, B., Fearon, E. R., Kern, S. E., Hamilton, S. R., Preisinger, A. C., Nakamura, Y., and White, R. (1989). Allelotype of colorectal carcinomas. *Science* **244**, 207–211.

von Wettstein, D., Rasmussen, S. W., and Holm, P. B. (1984). The synaptonemal complex in genetic segregation. *Annu. Rev. Genet.* **18**, 331–413.

Walker, C. L., Cargile, C. B., Floy, K. M., Delannoy, M., and Migeon, B. R. (1991). The Barr body is a looped X chromosome formed by telomere association. *Proc. Natl. Acad. Sci. U.S.A.* **88**, 6191–6195.

Walmsley, R. M., and Petes, T. D. (1985). Genetic control of chromosome length in yeast. *Proc. Natl. Acad. Sci. U.S.A.* **82**, 506–510.

Walmsley, R. M., Chan, C. S., Tye, B. K., and Petes, T. D. (1984). Unusual DNA sequences associated with the ends of yeast chromosomes. *Nature (London)* **310**, 157–160.

Wang, S. S., and Zakian, V. A. (1990a). Sequencing of *Saccharomyces* telomeres cloned using T4 DNA polymerase reveals two domains. *Mol. Cell. Biol.* **10**, 4415–4419.

Wang, S. S., and Zakian, V. A. (1990b). Telomere–telomere recombination provides an express pathway for telomere acquisition. *Nature (London)* **345**, 456–458.

Watson, J. D. (1972). Origin of concatameric T7 DNA. *Nature (London), New Biol.* **239**, 197–201.

Weiner, A. M. (1988). Eukaryotic nuclear telomeres: Molecular fossils of the RNP world? *Cell (Cambridge, Mass.)* **52**, 155–158.

Weinert, T. A., and Hartwell, L. H. (1988). The *RAD9* gene controls the cell cycle response to DNA damage in *Saccharomyces cerevisiae*. *Science* **241**, 317–322.

Westergaard, M., and von Wettstein, D. (1972). The synaptonemal complex. *Annu. Rev. Genet.* **6**, 71–110.

Wilkie, A. O., Buckle, V. J., Harris, P. C., Lamb, J., Barton, N. J., Reeders, S. T., Lindenbaum, R. H., Nicholls, R. D., Barrow, M., Bethlenfalvay, N. C., Hutz, M. H., Tolmie, J. L., Weatherall, D. J., and Higgs, D. R. (1990a). Clinical features and molecular analysis of the alpha thalassemia/mental retardation syndromes. I. Cases due to deletions involving chromosome band 16p13.3. *Am. J. Hum. Genet.* **46**, 1112–1126.

Wilkie, A. O., Lamb, J., Harris, P. C., Finney, R. D., and Higgs, D. R. (1990b). A truncated human chromosome 16 associated with alpha thalassaemia is stabilized by addition of telomeric repeat (TTAGGG)$_n$. *Nature (London)* **346**, 868–871.

Wilkie, A. O., Higgs, D. R., Rack, K. A., Buckle, V. J., Spurr, N. K., Fischel-Ghodsian, N., Ceccherini, I., Brown, W. R., and Harris, P. C. (1991). Stable length polymorphism of up to 260 kb at the tip of the short arm of human chromosome 16. *Cell (Cambridge, Mass.)* **64**, 595–606.

Williamson, J. R., Raghuraman, M. K., and Cech, T. R. (1989). Monovalent cation-induced structure of telomeric DNA: The G-quartet model. *Cell (Cambridge, Mass.)* **59**, 871–880.

Würgler, F. E., and Matter, B. E. (1968). Split-dose experiments with stage-14 oocytes of *Drosophila melanogaster*. *Mutat. Res.* **6**, 484–486.

Yao, M.-C. (1989). Site-specific chromosome breakage and DNA deletion in ciliates. *In* "Mobile DNA" (D. E. Berg and M. M. Howe, eds.), pp. 715–733. Am. Soc. Microbiol., Washington, DC.

Yokoyama, R., and Yao, M.-C. (1986). Sequence characterization of *Tetrahymena* macronuclear DNA ends. *Nucleic Acids Res.* **14**, 2109–2122.

Young, B. S., Pession, A., Traverse, K. L., French, C., and Pardue, M. L. (1983). Telomere regions in *Drosophila* share complex DNA sequences with pericentric heterochromatin. *Cell (Cambridge, Mass.)* **34**, 85–94.

Yu, G. L., and Blackburn, E. H. (1991). Developmentally programmed healing of chromosomes by telomerase in *Tetrahymena. Cell (Cambridge, Mass.)* **67**, 823–832.

Yu, G. L., Bradley, J. D., Attardi, L. D., and Blackburn, E. H. (1990). *In vivo* alteration of telomere sequences and senescence caused by mutated *Tetrahymena* telomerase RNAs. *Nature (London)* **344**, 126–132.

Zahler, A. M., Williamson, J. R., Cech, T. R., and Prescott, D. M. (1991). Inhibition of telomerase by G-quartet DNA structures. *Nature (London)* **350**, 718–720.

Zakian, V. A. (1989). Structure and function of telomeres. *Annu. Rev. Genet.* **23**, 579–604.

Zakian, V. A., and Blanton, H. M. (1988). Distribution of telomere-associated sequences on natural chromosomes in *Saccharomyces cerevisiae. Mol. Cell. Biol.* **8**, 2257–2260.

Zakian, V. A., Blanton, H. M., and Dani, G. M. (1985). Formation and stability of linear plasmids in recombination deficient strain of yeast. *Curr. Genet.* **9**, 441–445.

Zakian, V. A., Runge, K., and Wang, S.-S. (1990). How does the end begin? Formation and maintenance of telomeres in ciliates and yeast. *Trends Genet.* **6**, 12–16.

MOLECULAR GENETICS OF SUPEROXIDE DISMUTASES IN YEASTS AND RELATED FUNGI

Edith Butler Gralla* and Daniel J. Kosman†

*Department of Chemistry and Biochemistry, University of California, Los Angeles, Los Angeles, California 90024
†Department of Biochemistry, The State University of New York at Buffalo, Buffalo, New York 14214

If the atmospheric air were perfectly pure, the life of animals breathing it would be much more energetic, better and more pleasant in many ways; but at the same time it might be proportionately shortened, and being rapidly consumed by such active air, they might live only one quarter of the time that they live in the ordinary air of our atmosphere, impure though it may be.

Macquer (1777)

251

Copyright © 1992 by Academic Press, Inc.
All rights of reproduction in any form reserved.

I. Introduction

The accumulation of H_2O in the troposphere in sufficient quantities to condense into liquid water on the Earth's surface must be viewed as the most significant event in the biologic history of this planet. Arguably, the appearance of dioxygen, O_2, in the Earth's atmosphere is equally significant. Although liquid water provided the solvent for life's genesis, it was O_2, and its allotrope, ozone, O_3, that fueled the explosion of new life forms at the beginning of the Cambrian era, about 550 million years ago. At that time, the Earth's atmosphere contained about 2% (v/v) O_2, or 10% of its current value of 20.95% (Cloud, 1983). This level of O_2 could sustain a level of O_3 in the upper atmosphere sufficient to shield the surface from *lethal* levels of ultraviolet (UV) radiation, 200–300 nm (Berkner and Marshall, 1965; Levine, 1982, 1988). As a result, life was sustainable on land. As importantly, cellular respiration based on O_2 as the terminal electron acceptor increased the efficiency of energy utilization in the new aerobes (Koppenol and Butler, 1985).

These new life forms were exposed to two potentially cytotoxic elements: the residual UV radiation and O_2. Although not sufficient to cause significant cell killing, the far-UV flux at the surface was still mutagenic relative to present-day levels. This atmospheric condition may have stimulated (or supported) the high rate of evolutionary change between 500 and 300 million years ago (Boveris, 1978). However, for both anaerobes and the new facultative aerobes, O_2 was a more serious environmental threat. These organisms had evolved in the presence of the damaging effects of UV radiation and thus had developed mechanisms of UV protection and UV-damage repair. To become aerotolerant, organisms had to devise entirely new strategies in order to thrive. This implies a paradox: atmospheric O_2 provided an environment that supported a completely new type and quantity of life, but was itself cytotoxic (Fridovich, 1983, 1989; Bilinski, 1991). The purpose of this article is twofold: to show how study of the genetics and physiology of oxidant stress in yeasts and fungi has provided significant insight into this intriguing and biologically important paradox, and to identify important areas of research that are ripe for further investigation.

II. Cellular Physiology of Superoxide and Superoxide Dismutases

A. The Chemistry of Dioxygen, O_2

Unlike sulfate and nitrate, which served as terminal electron acceptors in the early and middle pre-Cambrian anaerobic milieu (and still do for a variety of prokaryotes), O_2 reduction can proceed in single, one-electron steps, generating the reactive radicals, $O_2^{\cdot-}$ (superoxide) and $OH\cdot$ (hydroxyl) (Koppenol and Butler, 1985). In contrast, sulfate and nitrate ordinarily undergo only two-electron reductions. Indeed, even if reactive free radicals were generated in sulfate- and nitrate-reducing bacteria, they would be scavenged by the products of these respiratory processes, e.g., sulfide and nitrite, which are efficient radical scavengers. Thus use of O_2 as a terminal oxidant presented a new type of biologic hazard. The one-electron (termed *univalent*) reduction steps for O_2, and the two- and four-electron pathways associated with specific and well-characterized enzyme-catalyzed reactions, are given in Scheme I. Shown also are the standard reduction potentials for these several reactions (Koppenol and Butler, 1985).

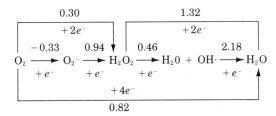

SCHEME 1. Dioxygen uni-, di-, and tetravalent reduction steps and standard reduction potentials. Values in volts (per electron) relative to the hydrogen electrode.

These values show that H_2O_2 and $OH\cdot$ are thermodynamically powerful oxidants, as is O_2 in two- and four-electron reduction reactions. However, the reactions of O_2 are kinetically slow, relative to those involving H_2O_2 and $OH\cdot$, because of the nature of the electronic structure of O_2 (Taube, 1965). Ground-state dioxygen is a triplet-state molecule with two unpaired electrons in orthogonal p orbitals and thus in parallel spin states. Organic reductants (—SH, NADH, $FADH_2$, etc.) are singlet ground-state molecules, as are the two- and four-electron reduction products of O_2, H_2O_2, and H_2O, respectively. For O_2 to undergo reduction by two (or four) electrons—that is, to go from a triplet state to a singlet state—the spin of the system oxidant + reductant must go from triplet/singlet to singlet/singlet. This requires an inversion of the spin of one electron (formally), a *kinetically* slow process because it is quantum mechanically of low probability (Taube, 1965). Formally, then, superoxide, $O_2^{\cdot-}$, is an intermediate in the reduction of O_2 (Brunori and Rotilio, 1984; Fridovich, 1984; Sies and Cadenas, 1983; Halliwell, 1984; Cadenas, 1989), whereas H_2O_2 is both the product of a number of enzyme reactions and a formal intermediate in a number of others that involve the equivalent of four electron transfers (Brunori and Rotilio, 1984; Sies and Cadenas, 1983).

The chemical data indicate that O_2^- is a more efficient reductant than oxidant, probably due to the fact that a proton is required in the transition state for O_2^- reduction ($O_2^- + e^- + H^+ \rightarrow HO_2^-$), whereas H_2O_2 is thermodynamically a strong oxidant as well as a kinetically efficient one. Consequently, the potential for cytotoxic *abiologic* oxidation within an aerobic cell would appear to be associated with the steady-state level of peroxide, and not with O_2 or O_2^-, which are slow oxidants at best. However, there is substantial evidence (*vide infra*) that the steady-state levels of O_2^- *do* correlate with the manifestations of cellular oxidant stress. This apparent contradiction has been most often explained by reference to the reaction shown in Eq. (1).

$$H^+ + O_2^- + H_2O_2 \rightarrow OH\cdot + H_2O + O_2 \tag{1}$$

$$O_2^- + M^{(n+1)+} \rightarrow M^{n+} + O_2 \tag{2}$$

$$M^{n+} + H_2O_2 \rightarrow M^{(n+1)+} + OH\cdot + H_2O \tag{3}$$

Equation (1) describes the reduction of H_2O_2 (a good oxidant) by $(H)O_2^-$ (relatively, a reductant), generating the extremely reactive hydroxyl radical, $OH\cdot$. There is a problem with this explanation of O_2^- toxicity, however (Fee, 1982). Although extremely exergonic (-18.2 kcal/mol) (Koppenol and Butler, 1985), the reaction described by Eq. (1) is slow (≤ 0.13 M^{-1} sec^{-1}) (Koppenol *et al.*, 1978) if it occurs at all. However, it is strongly catalyzed by transition metal ions, e.g., $Fe^{2+/3+}$ and $Cu^{1+/2+}$ (Haber and Weiss, 1934). This catalysis is illustrated by Eqs. (2) and (3); both reactions are exergonic (~ -10 kcal/mol, depending on M^{n+}) and both have significant rate constants ($\geq 10^2$ M^{-1} sec^{-1}) (Allen and Bielski, 1982; Koppenol and Butler, 1985). Equation (3) represents the Fenton reaction (Fenton, 1894), whereas the generation of $OH\cdot$ from H_2O_2 and O_2^- is associated with the Haber–Weiss (and Willstätter) cycle (Haber and Willstätter, 1931; Haber and Weiss, 1934). Thus, Eqs. (2) and (3) have been referred to as the Fenton-driven Haber–Weiss reaction, or, alternatively, O_2^--dependent Fenton chemistry.

The hydroxyl radical may be the primary causative agent of cellular oxidant damage. Indeed, it was this species whose cellular reactivity was proposed to link "oxygen poisoning" and radiation injury to a common free-radical mechanism (Gerschman *et al.*, 1954). With an $E^0 = 2.18$ V, it is not surprising that the reduction rate constants for $OH\cdot$ are near diffusion limits. For example, $OH\cdot + H_2O_2 \rightarrow HO_2^- + H_2O$ (part of the Haber–Weiss cycle) has $k = 4.5 \times 10^7$ M^{-1} sec^{-1} (Anbar and Neta, 1967) and the reaction of $OH\cdot$ with EDTA (as a characteristic scavenger of OH·) has $k = 2.8 \times 10^9\ M^{-1}\ \text{sec}^{-1}$ (Buxton et al., 1988). It is assumed that OH· reacts in a single encounter and thus does not diffuse away from its site of generation (Pryor, 1986). Given its reduction potential, it can strip an electron or a hydrogen atom from any biologic compound.

B. Oxygen Defense Mechanisms—General Considerations

This brief survey of the redox reactions of O_2 and its reduction products provides the context for understanding the cellular mechanisms that deal with these species. The basic principles of mass action and kinetics suggest that maintaining a diminishingly small concentration of free oxygen in the cell by itself is protective. The evolution of O_2 carriers achieves this end. (Probably the best example of this strategy is leghemoglobin, which, while delivering O_2 to sustain the high rate of respiration necessary for the reduction of N_2 to NH_3 in N_2-fixing bacteria, protects the O_2-sensitive nitrogenase by maintaining an effective anaerobiosis in the peribacteroid.) Because the most reactive intermediate in the univalent reduction of O_2 is OH·, reducing its steady-state level to zero is clearly desirable. However, because OH· is so reactive that it does not diffuse from its site of generation, an "OH· reductase" would have to be at sufficient concentration in the cell so that it was the first cellular component encountered by an OH·. Clearly, this impractical solution would have little selective advantage. In fact, what appears to have allowed aerobes to survive under O_2 is a system that (1) scavenges $O_2^{\cdot-}$, (2) scavenges H_2O_2, and (3) binds free metal ions. Obviously, this scheme works because it eliminates the reactants and catalysts for the superoxide-driven Fenton chemistry that produces OH· [Eqs. (1)–(3)]. The enzyme and protein components of this scheme are given in Table 1.

This article focuses on the enzymes responsible for the first component of this defense mechanism, the superoxide dismutases. These enzymes catalyze the disproportionation of $O_2^{\cdot-}$ to H_2O_2 and O_2. As discussed in Section III, eukaryotes contain at least two superoxide dismutases that are catalytically equivalent but evolutionarily, genetically, and structurally distinct (Steinman, 1982; Fridovich, 1986; Bannister et al., 1987). They are also functionally distinct because they are found in different cell compartments, as will be discussed in Section III. The regulation of expression of these enzymes is detailed in Section IV; in Section V their physiologic functions are discussed in

TABLE 1
Protein Components of Cellular Antioxidant Defense

Component and function	Cellular locale, genetic designation[a]

Superoxide dismutases

EC 1.15.1.1: Superoxide ($O_2^{\cdot -}$) dismutation

Cu,Zn containing	Eukaryotes
	Cytosol
	SOD1—*S. cerevisiae*, humans
	sod-1—*N. crassa*
	sod2, 4, 4A, 5—maize
	Microbodies (peroxisomes, glyoxysomes)
	SOD1—humans, *S. cerevisiae*(?)
	Chloroplasts
	Sod1—maize
	Mitochondria (rarely)
	Cu,Zn-SOD I—*Citrullus vulgaria*
	Extracellular
	EC-SOD-1 (human gene cloned)
	Prokaryotes (rarely)
	periplasmic(?)
	sodC—*C. crescentus*, *H. parainfluenzae*, others
Mn containing	Eukaryotes
	Mitochondrial matrix
	SOD2—*S. cerevisiae*, *N. crassa*, humans
	Sod3—maize
	Microbodies—plants
	Prokaryotes—not localized
	sodA—*E. coli*
Fe containing	Eukaryotes
	Chloroplasts (occasionally)
	pSOD2 (FeSOD$_{chl}$)—*N. plumbaginifolia*
	Prokaryotes—not localized
	sodB—*E. coli*

Catalases

EC 1.11.1.6: peroxide dismutation

Heme containing	Eukaryotes
	Cytosol
	CTT1—*S. cerevisiae*
	Cat1, Cat2—maize
	Mitochondria
	Cat3—maize
	Microbodies (peroxisomes, glyoxysomes)
	CTA1—*S. cerevisiae*
	Cat1, Cat2—maize
	Prokaryotes—cytosol, periplasmic space (*kat*)

TABLE 1 *(continued)*

Component and function	Cellular locale, genetic designation[a]
Peroxidases	
EC 1.11.1.7: (alkyl)peroxide reduction	
Heme containing	Eukaryotes
	Cytosol, mitochondrial intermembrane space
	CCP1—S. cerevisiae
Selenocysteine containing	Eukaryotes—cytosol, mitochondrial matrix
Metal storage, detoxification	Eukaryotes, prokaryotes
Fe	Ferritin, transferrin
Cu	Metallothionein, copper thionein
	CUP1—S. cerevisiae

[a]Genetic designations are given using notation common to the respective genus. The designation of the Cu,ZnSOD gene in *Saccharomyces cerevisiae* as *SOD1* is suggested so as to be consistent with all other eukaryotic organisms. The MnSOD gene should be designated *SOD2*.

light of the phenotypes of strains of yeast and fungi that lack either or both activities.

C. CELLULAR PRODUCTION AND CHEMISTRY OF SUPEROXIDE, $O_2^{\cdot-}$

Superoxide can be produced in a wide variety of cellular redox processes (Sies and Cadenas, 1983; Halliwell, 1984; Fridovich, 1983). The simplest type of reaction that can generate $O_2^{\cdot-}$ is autooxidation. Thus, essentially all of the normal cellular reductants [RSH, NADH, $FADH_2$, and (hydro)quinones] can react with O_2, albeit slowly, in the obligate one-electron reduction to $O_2^{\cdot-}$. In *Saccharomyces cerevisiae*, the dominant source of $O_2^{\cdot-}$ appears to be leakage from the mitochrondrial electron transport chain (Boveris, 1978; Boveris and Cadenas, 1982). Low but physiologically important amounts can be generated by other metabolic pathways as well (*vide infra*). This pattern is true also in aerobically growing *Escherichia coli* in that >75% of the $O_2^{\cdot-}$-detected comes from the respiratory chain (Imlay and Fridovich, 1991a). In mitochondria, superoxide can be generated at either the NADH–Q segment (complex I) or the QH_2:cytochrome c segment (complex III) (Fig. 1). The latter is thought to contribute >80% of the total superoxide in yeast, ~4 nmol $O_2^{\cdot-}$/min/mg mitochrondrial protein, which represents ~2% of total O_2 consumption (Boveris, 1978; Boveris and Cadenas, 1982).

Metal centers in low-valent states can also react with O_2. The for-

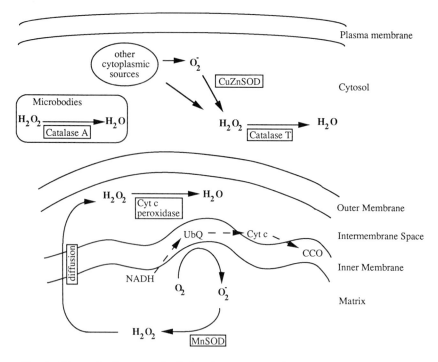

FIG. 1. Schematic diagram of cellular O_2^- sources and oxyradical defenses. The site of major cellular production of O_2^- and H_2O_2 in respiring yeast cells is the mitochondria (Boveris, 1978). The broken arrows represent the flow of electrons during respiration; CCO is cytochrome oxidase, Cyt c is cytochrome c. The O_2^- is generated by leakage from ubisemiquinone (UbQ) in the inner mitochondrial membrane and is disproportionated to H_2O_2 and O_2 by MnSOD in the matrix space. The H_2O_2 diffuses into the intermembrane space where it is reduced to H_2O by cytochrome c peroxidase. Superoxide or peroxide which escapes the mitochondrion can be handled in a similar manner by the Cu,ZnSOD and catalase T activities in the cytoplasm. In addition, in many yeasts oxygen radicals are generated in other locations, such as microbodies (or peroxisomes), and by adventitious redox reactions between molecular oxygen and cellular reductants. These radicals are handled by the cytoplasmic enzymes and the microbody specific Catalase A.

mation of methemoglobin from oxyhemoglobin is prototypical of this type of O_2^- generator. In fact, many low-valent metal ions are kinetically excellent reductants of O_2. Ferrous and cuprous ions, free or in a complex with an available coordination site, are exceptionally efficient (Halliwell and Gutteridge, 1989). Consequently, the combination of cellular reductant and "free" iron or copper is an efficient O_2^- generator. Enzymes that catalyze nominal two- or four-electron reduction reactions can "leak" the intermediate O_2^- also (Seis and Cadenas, 1983). This is suggested by several observations that catalytic substrate turnover in such enzyme reactions can be suppressed by addition of O_2^- scavengers.

Indeed, there is an evolutionary development that, in effect, confirms the hypothesis that oxygen free radicals and O_2^- are cytotoxic, and that O_2^- can be generated by an NAD(P)H-driven, metal-catalyzed reaction. This is the respiratory burst of stimulated macrophages in which O_2^- is produced by a membrane-bound b-type cytochrome NADPH:O_2 oxidoreductase (Johnson and Ward, 1985). Superoxide radical has been demonstrated to be part of the cell-killing process of these and other types of phagocytic cells. Whether it acts directly or via the H_2O_2 or OH· that can be subsequently generated is not yet clear.

A number of mediators of electron transfer reactions, the viologens chief among them, are both substrate for cellular reductases and, in the reduced form, excellent one-electron reductants of O_2. Thus, methyl viologen, more commonly known as paraquat, is a strong O_2^- generator when added to cells or cell extracts supplemented with reducing equivalents. This redox cycling with O_2, and the production of O_2^-, is the mechanism by which paraquat, streptonigrin, etc. kill cells (Halliwell, 1984; Halliwell and Gutteridge, 1989) and underlies the use of paraquat to screen for $sod1$ mutants in $S.$ $cerevisiae$ (Bilinski et al., 1985). Intra- and extracellular nonspecific reductases that can act on paraquat have been identified in both $S.$ $cerevisiae$ (Lesuisse et al., 1990) and $Debaromyces$ $hansonii$ (Yamashoji et al., 1991). A plasma membrane redutase activity in the former appears to contribute to the sensitivity of $S.$ $cerevisiae$ to paraquat (B. Crawford and D. Kosman, unpublished).

In yeast growing on normal media (glucose- and ammonia-containing media that do not induce peroxisome proliferation), mitochondrial O_2^- is probably the major intermediate in the cellular generation of H_2O_2 as well, because the rate of formation of H_2O_2 in well-washed submitochondrial particles is roughly half the rate of formation of O_2^- (Boveris, 1978). This inference follows from the superoxide dismutation reactions given in Eqs. (4a)–(4c) below. The uncatalyzed rate of this reaction depends on the protic equilibrium between the hydroperoxy radical, HO_2^-, and superoxide anion radical, O_2^-, Eq. (4d). Thus, the uncatalyzed dismutation reaction at neutral (physiologic) and alkaline pH is relatively slow due to the predominance of the ionized form of superoxide (O_2^- rather than HO_2^-) (Allen and Bielski, 1982).

$$O_2^{\cdot-} + O_2^{\cdot-} + H_2O \xrightarrow{<0.3 \ M^{-1} \ \mathrm{sec}^{-1}} HO_2^- + O_2 + OH^- \qquad (4a)$$

$$HO_2^{\cdot} + O_2^{\cdot-} \xrightarrow{10^8 \ M^{-1} \ \mathrm{sec}^{-1}} HO_2^- + O_2 \qquad (4b)$$

$$HO_2^{\cdot} + HO_2^{\cdot} \xrightarrow{10^6 \ M^{-1} \ \mathrm{sec}^{-1}} H_2O_2 + O_2 \qquad (4c)$$

$$HO_2^{\cdot} \overset{pK_a = 4.7}{\rightleftharpoons} O_2^{\cdot-} + H^+ \qquad (4d)$$

In comparison to the rate constants for the dismutation reactions indicated in these equations, the superoxide dismutases catalyze this reaction with rate constants of $10^8 – 10^9 \ M^{-1} \ \mathrm{sec}^{-1}$ (Steinman, 1982).

MnSOD is poised in the mitochondrial matrix to scavenge the $O_2^{\cdot-}$ leaked from complex III (Fig. 1). In addition, the intermembrane space contains cytochrome c peroxidase (Kaput $et \ al.$, 1982, 1989), a member of the second-line defense, which scavenges the H_2O_2 produced in the $O_2^{\cdot-}$ dismutation reaction (Boveris, 1978; Boveris and Cadenas, 1982). A neutral molecule, H_2O_2 readily diffuses across lipid bilayers and thus can be expected to cross the inner mitochondrial membrane, which is the outer membrane in submitochondrial particles. Confirming this idea, $intact$ mitochondria generate less than 50% of the H_2O_2 seen with submitochondrial particles (Boveris, 1978). And, significantly more H_2O_2 is leaked from intact mitochondria that lack a peroxidase activity (Kusel $et \ al.$, 1973). In this context, the contribution of microbodies (peroxisomes/glyoxysomes) to cellular H_2O_2 generation in yeasts and fungi has not been quantitated. These organelles can be strong producers of peroxide as a result of their oxidative assimilation of several carbon and nitrogen sources (Veenhuis and Harder, 1991). Although they contain large amounts of catalase, it is not known whether significant peroxide leaks occur. Peroxisomal metabolism may also produce $O_2^{\cdot-}$; this radical is leaked from plant glyoxysomes (Sandalio $et \ al.$, 1988; Palma $et \ al.$, 1991). Such studies in yeasts are sorely needed.

How much cellular $O_2^{\cdot-}$ and H_2O_2 matters? It is estimated that a cell (the hepatocyte) has $10^{-11} \ M \ O_2^{\cdot-}$ and $10^{-8} \ M \ H_2O_2$ in the steady state (Boveris, 1978). These concentrations are too low to sustain appreciable rates of OH· generation by superoxide-driven Fenton chemistry, for example. However, a 10-fold increase in H_2O_2, in particular, can be significant. Such concentrations could generate as many as 40–50 OH· per cell per second. It is not surprising, therefore, that (1) aerotolerant organisms have dismutation activities for both $O_2^{\cdot-}$ and H_2O_2 in any cellular compartment in which significant O_2 consumption occurs, and (2) the level of these activities correlates positively with the level of oxygen utilization. Additionally, the role of metal ions in oxygen cytotoxicity should not be overlooked. No truly free metal ions are present in cells; metal ions are usually complexed in a sequestered, relatively inert form. However, they can also be bound to macromolecules in a coordination complex which makes them accessible to oxygen-based redox chemistry. These latter situations are thought to be extremely dangerous due to their potential for generation of OH· in proximity to a biomolecule (Chevion, 1988; Saltman, 1989). This site-specific model of oxygen-induced pathology can help to explain why seemingly low levels of the reactants can be cytotoxic.

The steady state $[O_2^-]$ in wild-type $E.$ $coli$ has been calculated to be 2×10^{-10} M (Imlay and Fridovich, 1991a). In the absence of $catalyzed$ superoxide dismutation, that is, in a strain that is sod^- this value was calculated to be 6.7×10^{-6} M, or comparable to levels of O_2^- effective in the in $vitro$ inactivation reactions of α,β-dihydroxyisovalerate dehydratase, aconitase, and 6-phosphogluconate dehydratase (Kuo et $al.,$ 1987; Gardner and Fridovich, 1991a,b; Imlay and Fridovich, 1991a). Of interest is the fact that these three enzymes share a structural and catalytic element in common that is likely to be the target of the oxidative inactivation: each is an Fe–S protein with a [4Fe–4S] cluster. Aerobic auxotrophies in $Escherichia$ $coli$ have been correlated to the inactivation by O_2^- of these enzymes. For example, branched-chain amino acid biosynthesis is inhibited because of the oxygen-dependent inactivation of α,β-dihydroxyisovalerate (dihydroxyacid) dehydratase (Brown and Yein, 1978; Kuo et $al.,$ 1987). With respect to inactivation of 6-phosphogluconate dehydratase, sod^- $E.$ $coli$ grow poorly on gluconate under air (Gardner and Fridovich, 1991a).

Yeasts, in particular, are unique among eukaryotic systems to characterize the antioxidant defense system in aerotolerant organisms. Many yeasts are facultative anaerobes, and many can grow on various carbon and nitrogen sources that require either no oxygen consumption, mitochondrial respiration, or nonmitochondrial oxygen consumption (microbody metabolism) for their utilization. Therefore, the degree and locale of cellular oxygen consumption and electron transfer activity in some yeasts can be manipulated with prior knowledge of the physiologic and biochemical adaptations occurring in the cell. As a result, the potential for any one metabolic reaction to generate an oxy radical or to be sensitive to it, or for a specific cellular component to protect against it, can be most systematically evaluated in yeasts and fungi. In addition, those yeasts and fungi having appreciable or developing genetics, e.g., $S.$ $cerevisiae,$ $S.$ $pombe,$ $Candida$ $glabrata,$ and $Neurospora$ $crassa,$ provide the essentially unlimited potential of recombinant DNA technology for the elucidation of the mechanisms underlying

cellular oxidant stress and their regulation. This review will attempt
to demonstrate this truth.

D. SUPEROXIDE DISMUTASES AS AN ANTIOXIDANT DEFENSE

The superoxide dismutation activity of Cu,ZnSOD, a protein isolated originally from red blood cells as hemocuprein or erythrocuprein on the basis of its copper content, was what first suggested its
physiologic role (McCord and Fridovich, 1969). As noted above, a variety of reactions produce $O_2^{\cdot-}$, which can be detected by coupling it to
the formation of a chromogenic product. Equation (5) illustrates one
of the most widely used of these reactions (McCord and Fridovich,
1969).

$$HO_2^{\cdot} + Fe^{3+}\text{-cytochrome } c \rightarrow O_2 + H^+ + Fe^{2+}\text{-cytochrome } c \qquad (5)$$

Cu,ZnSOD was first identified because of its inhibition of this reaction.

The human Cu,ZnSOD gene was the first of all eukaryotic SOD
genes cloned (Sherman et al., 1983), and has been historically referred
to genetically as *SOD1* (Tan et al., 1973; Sherman et al., 1983). [The
genetic notation for *S. cerevisiae* is used herein, in which dominant
alleles are uppercase, recessive alleles are lowercase (see Rose et al.,
1990).] The MnSOD, as the second of these proteins to be isolated
(Keele et al., 1970; Ravindranath and Fridovich, 1975) and cloned
(Marres et al., 1985), should be designated genetically *SOD2*. Unfortunately, the MnSOD locus in *S. cerevisiae* has been referred to as
SOD1 (van Loon et al., 1986), a nomenclature that has been recently
perpetuated (Turi et al., 1991). For consistancy among organisms,
this designation should be discontinued. The genetic designations of
various pro- and eukaryotic SODs are shown in Table 1.

In addition to these enzymes, an extracellular Cu,ZnSOD has been
characterized in eukaryotes. The human enzyme, designated EC-
SOD1, is a secreted glycoprotein containing Cu and Zn; the gene for
this SOD has been cloned (Hjalmarsson et al., 1987). Although EC-
SOD and the intracellular SOD1 differ in primary sequence, they do
share a number of structural homologies with respect to the metal centers (Hjalmarsson et al., 1987). Some evidence for this type of dismutase in *S. cerevisiae* and *N. crassa* has been presented also (Munkres,
1990a). Although apparent mutants in this activity have been isolated
(Munkres, 1985, and unpublished work), the gene for the EC-SOD1
in neither microbe (designated SOD-4; K. Munkres, unpublished) has
been cloned.

One of the first *in vitro* tests of the hypotheses that (1) $O_2^{\cdot-}$ (or, perhaps, OH· derived from $O_2^{\cdot-}$) could damage a biologic component and (2) that SOD could protect against this damage was the demonstration that $O_2^{\cdot-}$ caused the depolymerization of hyaluronic acid, the "lubricant" in synovial fluid, and that SOD1 inhibited this depolymerization (McCord, 1974). Similar *in vitro* chemical modification of proteins, nucleic acids, and lipids (fatty acids) as a result of $O_2^{\cdot-}$ generation and its inhibition by a superoxide dismutation activity have been extensively described (Halliwell and Gutteridge, 1989).

In vivo evidence for the role of $O_2^{\cdot-}$ in the chemical modification of biologic substrates has, in general, been more circumstantial. Recent genetic manipulation in a variety of organisms has led to strong support of the *in vitro* results, however. Generally, mutants in both prokaryotic and eukaryotic organisms that lack SOD activity are less vigorous, are more sensitive to O_2, and often exhibit specific oxygen-dependent phenotype(s). Systematic study of these several mutants in yeast and fungi has led to an understanding of many of these phenotypes, which are discussed fully in Sections IV and V.

III. Structural, Functional, and Evolutionary Relations in Cu,Zn and Mn Superoxide Dismutases in Yeast and Fungi

A. STRUCTURAL CONSIDERATIONS

Table 2 summarizes the basic structural features of the Cu,Zn and Mn superoxide dismutases from *S. cerevisiae* (Goscin and Fridovich, 1972; Ravindranath and Fridovich, 1975; Bjerrum, 1987); the proteins from *N. crassa* are quite similar (Misra and Fridovich, 1972; Henry *et al.*, 1980; Chary *et al.*, 1990). Several excellent reviews have detailed the properties of these two types of SODs (Steinman, 1982; Fridovich, 1986; Bannister *et al.*, 1987; Beyer *et al.*, 1991). This section will discuss them only briefly while focusing on a few specific aspects that merit further comment.

1. Cu,ZnSOD

The SOD1 proteins from *S. cerevisiae* (Steinman, 1980) and *N. crassa* (Lerch and Schenk, 1985) have been sequenced. The corresponding genes have been cloned (Bermingham-McDonogh *et al.*, 1988; Chary *et al.*, 1990). A distinguishing feature of these proteins is their lack of Trp and limited number of Tyr (two). Thus, unlike most proteins (including SOD2) they absorb relatively weakly in the near-UV; they also exhibit visible absorbance due to the Cu(II) prosthetic group

TABLE 2

Properties of *Saccharomyces cerevisiae* Superoxide Dismutases

Cu,ZnSOD, *SOD1*	MnSOD, *SOD2*
Properties of monomer	
Single chain, one disulfide	Single chain, no disulfide
Eight-strand β barrel	α helix, β sheet
153 amino acids	214 amino acids $(241)^a$
M_r 15,700	M_r 23,059 $(26,123)^a$
pI = 4.6	p$I \approx 6$
1 Cu, 1 Zn	1 Mn
$\lambda_{max}(\varepsilon)$	$\lambda_{max}(\varepsilon)$
258 nm (5650 M^{-1} cm^{-1})	282 nm (48,000 M^{-1} cm^{-1})
670 nm (230 M^{-1} cm^{-1})	480 nm (280 M^{-1} cm^{-1})
Functional unit (quaternary structure)	
Dimer	Tetramer

^{a} Values in parentheses for precurser polypeptide prior to cleavage of mitochondrial leader sequence. M_r based on predicted sequence. See text for details and references.

(Table 2). Both microbial Cu,ZnSODs contain 153 amino acid residues. (These sequences are used as the basis for numbering residues below.) The SOD1 monomer is an eight-stranded antiparallel β-barrel with two "Greek key" loops (residues 36–40 and 100–112). In addition, the structure contains two larger loops, residues 47–84 and 120–140, respectively; neither loop exhibits ordered secondary structure. The first is called the *Zn loop* and the second is called the *electrostatic channel loop* for reasons made clear below (Tainer *et al.*, 1982).

The Cu,ZnSODs (SOD1) from yeast and fungi are 68.6% homologous in primary sequence and exhibit 50–56% homology with Cu,ZnSODs from higher eukaryotes, which themselves are 60–65% homologous (Steinman, 1980; Lerch and Schenk, 1985; Bannister *et al.*, 1987; Chary *et al.*, 1990). There are numerous regions in the sequence that are absolutely conserved among all SOD1. Primarily these involve three elements: (1) the metal-binding ligands, (2) the interface between the subunits, and (3) the residues that provide the electric field gradient that channels the substrate O_2^- and other small anions to the active site Cu(II).

The Cu(II) is liganded to four imidazoles (His-46, -48, -63, and -120) in a tetrahedrally distorted square-planar conformation in which a solvent water is a fifth ligand (Tainer *et al.*, 1982, 1983). The imidazole of His-63 is also a bridging ligand to the adjacent Zn(II). The Cu(II) and Zn(II) are ~6 Å apart. The other three ligands to the Zn(II) come from His-71 and -80, and Asp-83. The Zn(II) is in a tetrahedral ligand field, distorted toward trigonal-pyramidal geometry with Asp-83 at the apex. The region between residues 44–71 (28 residues) contains 17 identities and 7 strongly conservative substitutions and forms the core of the zinc-binding loop (Chary et al., 1990).

The two active sites in the dimeric protein are oriented away from each other with the two Cu atoms >30 Å apart. There is no evidence for any interaction between the two active sites catalytically (kinetically). The subunit interface exhibits tight packing; it encompasses about 9% of the monomer surface area. It involves residues 1–20, 50–70, and 100–110. These regions are strongly conserved. Note that the second region is part of the Zn loop. The extensive and tight interface packing contributes to the remarkable stability of Cu,ZnSODs, as do a single intrachain disulfide bond between conserved Cys residues 57 and 146, and the extensive hydrogen-bonding network. The protein resists heat, organic solvents, and even 8 M urea, although the yeast enzyme is somewhat less stable than other Cu,ZnSODs (Barra et al., 1979).

Cu,ZnSOD reacts with O_2^- at 10% of the diffusion limit of two molecules colliding in solution, despite the fact that the Cu(II) at the active site represents less than 0.1% of the protein surface. This paradox has led to the suggestion (Koppenol, 1981), since essentially demonstrated (Getzoff et al., 1983), that the O_2^- is guided to the active site by an electric field gradient established by the distribution of charges on the protein surface, including residues in the active site. These residues are found in what has been termed the electrostatic channel, residues 121–143. The last of these is Arg-143, which provides an essential catalytic element, because its chemical modification leads to loss of enzymatic activity (Borders et al., 1985). Substitution by Lys is catalytically conservative, however (Bertini et al., 1988). In addition, 12 or 13 (the yeast and Neurospora proteins differ) of the 23 residues in this channel are polar; of the 23 residues, 20 are conserved. The spatial organization of these residues results in a dipole gradient oriented to draw a negative point charge (O_2^-) from the bulk medium toward the copper (Koppenol, 1981; Getzoff et al., 1983). This effectively increases the protein cross-section which results in a productive collision between the enzyme and substrate and explains the magnitude of the catalytic rate constant for these enzymes, $\sim 10^9\ M^{-1}\ sec^{-1}$. Because the p$I$ of SOD1 is 4.6 (Table 2), this positive field gradient exists despite the overall net negative charge on the protein at neutral pH.

2. MnSOD

The MnSODs share no structural motifs in common with the Cu,Zn proteins, but are themselves highly homologous (Beyer et al., 1991). The MnSOD of S. cerevisiae has been sequenced (Ditlow et al., 1984) and its gene cloned (Marres *et al.*, 1985); a complete *N. crassa* protein sequence has not been reported, although its gene is being cloned (D. Natvig, personal communication). MnSODs exhibit visible absorbance due to the associated Mn(III) (Table 2). In eukaryotes, MnSODs are tetrameric. In general, the MnSODs show more divergence in primary sequence than do the Cu,Zn proteins. Identities range from 94% (human and mouse) to 39% (*E. coli* and yeast), with a 27% homology over all species (Beyer *et al.*, 1991; Bjerrum, 1987). Nonetheless, spectroscopic and energy minimization analyses indicate strong conservation of secondary structural elements in all MnSODs sequenced to date, even for proteins differing most in primary sequence (Bjerrum, 1987). The greater divergence of sequence in MnSODs may reflect their evolutionarily earlier origin (see below).

The MnSOD subunit is folded into two domains; the N-terminal domain is made up of five α helices and the C-terminal one is composed of a three-stranded β sheet and two relatively short helical segments. In *S. cerevisiae*, the Mn ligands are His-26, His-81, Asp-168 and His-172. Based on the X-ray data collected on the MnSODs from *Bacillus stearothermophilus* (Parker and Blake, 1988) and *Thermus thermophilus* (Stallings *et al.*, 1985), His-26 and His-81 are contributed by the first and third helices in the N-terminal domain, respectively, and Asp-168 and His-172 are part of the C terminus of the third strand of the β sheet (numbering based on the yeast sequence). Thus, the Mn is bound in the hinge region between the two domains. These residues are absolutely conserved in all MnSODs (Beyer *et al.*, 1991). The Mn site closely resembles the Zn site in SOD1 because it is composed of three His and one Asp and has a trigonal-pyramidal distortion from tetrahedral geometry; the Zn(II) in Cu,ZnSOD does not participate directly in catalysis, however.

As in Cu,ZnSOD, the catalytic metal ion is found at the bottom of a cavity. The cavity of MnSOD is associated with several basic residues but with only three acidic ones (Parker and Blake, 1988; Stallings *et al.*, 1985). This distribution of charged residues suggests that protein electrostatics serve to promote productive collisions between O_2^- and the MnSODs as they do with the Cu,Zn proteins (Borders *et al.*, 1989). The MnSOD substrate-binding cavity is lined with nonpolar aromatic residues, however. There may (Stallings *et al.*, 1985) or may not (Parker and Blake, 1988) be a coordinated H_2O as there is at the Cu(II) in SOD1. A third distinguishing feature is the presence of the phenolic —OH of a tyrosine residue (Tyr-34, which is absolutely conserved) 5 Å from the Mn (Parker and Blake, 1988; Stallings *et al.*, 1985). These differences suggest that the details of the electron transfer mechanism in the two dismutases may well be different, as is discussed further below (Bannister *et al.*, 1987).

The subunit contacts in the MnSOD from eukaryotic mitochondria can be inferred from the structure of the MnSOD from *T. thermophilus*, which is also tetrameric (Stallings *et al.*, 1985). These data show that the subunit interfaces in MnSODs are much less extensive than in Cu,ZnSODs. The subunits in the tetramer are arrayed in a flattened arrangement around a central hollow core. The tetramer is composed of two dimeric units. The dimer interface involves a short sequence of four residues in the region around one of the Mn ligands, His-172; there is no penetration of the main chain of one subunit into the other. Although the Mn ligand of each dimer pair is at the interface, the two Mn atoms are still separated by 18 Å. The tetramer interface involves a short α-helical segment in the N-terminal domain and two short loops that connect the three strands of the β-sheet in the C-terminal one. These interactions do not lead to the tight packing observed in the structure of Cu,ZnSODs. In addition, MnSODs do not appear to have any intrachain disulfide bonds. These several structural features account for the relative instability of MnSOD (Steinman, 1982).

MnSOD is found in the mitochondrial matrix but is encoded by a nuclear gene. Therefore, it is synthesized as a preprotein containing an N-terminal mitochondrial targeting sequence. The leader sequence for the SOD2 protein from *S. cerevisiae* has been deduced from the DNA sequence, and consists of 27 amino acids, rich in basic amino acid residues, as is characteristic of mitochondrial targeting signals (Marres *et al.*, 1985). It lacks a hydrophobic core, which would target the precursor protein to the intermembrane space, as, for example, cytochrome *c* peroxidase is targeted (Kaput *et al.*, 1982).

B. Catalytic Considerations

Cu,ZnSODs and MnSODs catalyze $O_2^{\cdot-}$ dismutation by essentially the same overall mechanism as illustrated by Eqs. (6a) and (6b), where M^{n+} represents Cu(II) or Mn(III), respectively (Steinman, 1982; Bannister *et al.*, 1987).

$$\text{E–M}^{n+} + O_2^{\cdot-} \rightarrow \text{E–M}^{(n-1)+} + O_2 \tag{6a}$$

$$\text{E–M}^{(n-1)+} + O_2^{\cdot-} + 2H^+ \rightarrow \text{E–M}^{n+} + H_2O_2 \tag{6b}$$

As noted above, there may be differences in the details of the electron transfer steps. The major difference may be that the electron transfer to Cu(II) from O_2^- is an inner-sphere process involving the direct ligation of O_2^- to the metal (Gabor *et al.*, 1972). In the MnSOD reaction, this electron transfer may be outer sphere (McAdam *et al.*, 1977). Such a mechanism would be facilitated in a channel lined with aromatic residues, as is found in MnSODs. Another difference is the relative pH independence of the Cu,ZnSOD reaction apparently due to the "buffering" effect of the Zn(II) ligated to the bridging His-63 (Steinman, 1982). This allows for a pH-independent ligand exchange involving O_2^- at the Cu(II). In contrast, MnSODs (and FeSODs) are inhibited at alkaline pH. This appears to be due to the ionization of the active site Tyr-34 (see above), which may serve as an acid catalyst in the outer-sphere electron transfer reaction at the Mn(III/II) (Parker and Blake, 1988). All of these suggestions await experimental verification, best provided, perhaps, by structure–function analysis based on site-directed mutagenesis.

C. Functional Considerations: Cellular Locale

As noted in Section II and as will be discussed again (Section V), the cellular roles played by a superoxide dismutation activity must be related to its cellular locale. Some controversy about locale has existed for both SOD1 and SOD2. Aside from the Cu,Zn protein that is found in the chloroplast (gene *Sod1* in maize) (Scandalios, 1990) or plant mitochondria (Cu,ZnSOD I in *Citrullus vulgaria*) (Sandalio and Del Río, 1987), most evidence has indicated that the SOD1s are (primarily) cytoplasmic. Thus, a suggestion (Henry *et al.*, 1980) that a Cu,ZnSOD was mitochondrial in *Neurospora* is inconsistent with the strong likelihood that there is only one *SOD1* gene in this organism, as in yeast, and that it codes for the polypeptide found in the cytosol. The results of L.-Y. Chang *et al.* (1988) support this view. Using quantitative immunocytochemistry, they demonstrated that in rat hepatocytes, 73% of the enzyme was in the cytoplasm and, of some interest, 12% was in the nucleus. In contrast to plants, there is no evidence for another *SOD1* gene in *Neurospora* or *S. cerevisiae* that includes a sequence for a cleavable mitochondrial leader polypeptide (Chary *et al.*, 1990; E. Gralla, unpublished observations).

Recently, however, data have been presented from human (Keller *et al.*, 1991) and plant (Sandalio and Del Río, 1988) studies that suggest that Cu,ZnSOD may also be localized to microbodies in some cases. This localization was demonstrated by immunologic staining of human SOD1 in peroxisomes of cultured human cells, and in yeast expressing the human *SOD1* gene (Keller *et al.*, 1991). This observation does not mean that SOD1 is not cytoplasmic. In yeast, microbodies are essentially absent in cells grown in normal media (glucose plus ammonia), yet the cells have significant levels of Cu,ZnSOD. Growth of *S. cerevisiae* on oleic acid induces proliferation of peroxisomes (Veenhuis *et al.*, 1987) and induces the expression of peroxisomal catalase (*CTA1*) 20-fold. In contrast, such growth induces SOD1 only 3-fold (see Table 3, Section IV). Nonetheless, the role of SOD1 (and SOD2; see below) in dealing with the oxidative metabolism that occurs in microbodies (peroxisomes and glyoxysomes) is an important question deserving close examination.

Although there is no direct evidence that MnSOD is found in the cytosol of yeasts and fungi, it is certainly cytosolic in bacteria, and is suggested to be so in numerous other organisms (Bannister *et al.*, 1987). Evidence against a significant cytosolic locale for endogenous SOD2 in *S. cerevisiae* is that overexpression of the cloned *SOD2* gene does not complement a *sod1* deletion (J. Lee and D. Kosman, unpublished). Although *sod1* mutations can be complemented by heterologous MnSODs lacking a mitochondrial targeting sequence (either natural or engineered), it has not been possible to complement *sod1* mutants with the homologous yeast *SOD2* coding sequence in such constructions (Bowler *et al.*, 1990; C. Bowler, personal communication). Furthermore, the phenotype of a *sod1 sod2* double mutant is not markedly different from that of an *sod1-* single mutant (see Table 4, Section IV) (E. Gralla and J. Valentine, unpublished observations). This also suggests that MnSOD does not compensate for the lack of Cu,ZnSOD activity.

On the other hand, the increased MnSOD observed in Cu-deficient *Dactylium dendroides,* a basidiomycete, was in the nonmitochondrial fraction, apparently compensating for a decrease in Cu,ZnSOD caused by Cu deprivation (Shatzman and Kosman, 1979). Thus it may be that species differ considerably with respect to the fidelity of mitochondrial targeting. For example, there is some evidence from *in vitro* mitochondrial import analyses that some fraction of the processed, mature yeast cytochrome *c* peroxidase polypeptide is released from intact mitochondria (J. Kaput, personal communication). Another aspect of this question is what is meant by "cytoplasmic." In some plants, there is suggestive evidence that MnSOD may be bound to the outer surface of some cytoplasmic microbodies, e.g., glyoxysomes (Sandalio and Del Río, 1988). Whether this is true of SOD2 in yeast and fungi has not been carefully investigated. Thus, for many reasons, the potential for SOD2 to fulfill a nonmitochondrial role in such organisms remains an open question.

D. Evolutionary Considerations

The brief discussion above makes clear that the Cu,ZnSODs and MnSODs are quite distinct proteins and must be evolutionarily unrelated. On the other hand, MnSODs and FeSODs are as homologous to one another as are the MnSODs among themselves, across species (Bannister et al., 1987; Beyer et al., 1991). Consequently, these latter two SODs are thought to have evolved from a common ancestor, probably FeSOD (Asada et al., 1980). This conclusion is based primarily on the distribution patterns of these two proteins. First, obligate anaerobes contain only FeSOD, suggesting that this protein evolved prior to significant aerobiosis, which event occurred 1.5–2 billion years ago in the middle Precambrian era (Section I). MnSOD appeared at that time along with aerotolerant organisms. Significantly, though FeSOD is constitutive in most organisms (but see below), MnSOD is inducible, indicating that it did represent an adaptation to potentially cytotoxic levels of dioxygen. The first cellular locale in which toxic levels of superoxide might have been reached was the thylakoid membranes of blue-green algae, organisms first seen in the fossil record about 2 billion years ago; indeed, MnSOD is found in those membranes, whereas FeSOD is cytosolic in such prokaryotes.

Why the putative gene duplication that resulted in MnSOD also led to use of a different metal is an unanswered question. One consideration is the state and availablity of environmental Fe when FeSOD evolved, in comparison to when MnSOD evolved. Prokaryotic development occurred in a reducing atmosphere in which Fe was present and readily available as the relatively soluble Fe^{2+}. Formation of atmospheric dioxygen resulted in the slow oxidation of this Fe^{2+} to the insoluble Fe^{3+} and maintenance of the partial pressure of O_2 to near zero (Levine, 1988). Thus, at the point when an antioxidant activity became necessary, when MnSOD appeared, Fe (as ferric oxides) had become far more difficult to accumulate and metabolize. Mn may have been a more readily accessible substitute. FeSODs and MnSODs have nearly equal dismutase activity (Beyer et al., 1991). On the other hand, MnSOD (but not FeSOD) is resistant to H_2O_2 (Beyer et al., 1991), giving it a possible selective advantage. The possibility that in the early reducing environment FeSOD served some capacity other than as superoxide dismutase must be considered also; after all, in the absence of O_2 there was no $O_2^{\cdot-}$. The fact that FeSOD is synthesized under strict anaerobic conditions is consistent with this suggestion (Asada *et al.*, 1980).

The origin of the Cu,ZnSOD involves two questions. First is the biologic one of why a third SOD provided selective advantage in eukaryotic development. The SOD1 protein appeared $\sim 10^9$ years ago, in the very late Precambrian era, at a time when atmospheric O_2 reached about 1% of its current level (Section I). The earliest eukaryotes also appeared then. The fossil record indicates that the SOD1 protein and eukaryotes were evolutionarily synchronous.[1] The key to the biologic question is why the FeSOD found in the cytosol of photosynthetic blue-green algae was replaced by the Cu,ZnSOD found in the chloroplasts of most photosynthetic eukaryotes (Asada *et al.*, 1980). Regulation of gene expression may partially explain this substitution: with respect to oxidant stress, FeSODs in prokaryotes seem to be constitutive whereas Cu,ZnSOD expression in eukaryotes is somewhat responsive. In this context, the presence of an FeSOD and not a Cu,ZnSOD in the plastids of *Nicotiana plumbaginifolia* is explicable as well as instructive. This FeSOD is under strong oxidant-dependent regulation (Tsang *et al.*, 1991), suggesting a type of functionally convergent evolution in which the prokaryotic FeSOD gene acquired species-specific cis-acting regulatory elements analogous to those found in Cu,ZnSODs. Sorting how, where, and when these elements evolved should provide a fascinating picture of evolutionary development.

The second question concerning the origin of Cu,ZnSOD focuses on the protein itself; it may have evolved from an ancestral Cu-binding protein that had no catalytic function. This possibility is suggested by a proposed evolution of the SOD1 protein monomer (Getzoff *et al.*, 1989; Chary *et al.*, 1990). First, the tertiary structure of the monomer shows that it contains a twofold axis that relates the α-carbon atoms across an axis of symmetry that has the Cu at the center. This structure could arise from the gene duplication of a smaller unit that, when in a noncovalently linked dimeric form, could have bound Cu in an equivalent position. This Cu-binding precursor would lack the electro-

---

[1] As noted (Table 1), Cu,ZnSODs are found only in eukaryotes except in a few intriguing cases. Apparent gene transfer from eukaryotic host to prokaryotic symbiont may explain some of these (Steinman, 1982). Others involve mammalian pathogens (*Haemophilus, Pseudomonas, Actinobacillus,* etc.) and appear to involve a secreted form of SOD1 (Kroll *et al.*, 1991). This SOD1 could play a role in protecting the bacteria from the O_2^- generated by phagocytic cells. The Cu,ZnSOD targeted to the periplasmic space in *Caulobacter crescentus* (Table 1) may play a similar role in protecting the bacterium from the O_2^- generated by the blue-green algae on which it lives (Steinman and Ely, 1990). Although untested, the hypothesis that these several bacteria acquired an "EC-SOD1" from their hosts cannot be summarily disregarded.

static channel and the Zn-binding loop and thus would be inactive as a productive superoxide dismutase. These two essential elements represented a subsequent evolution: divergence of sequence in one of the original dimer halves leading to the catalytic, electrostatic gradient and, in the other half, addition of an "exon" unit encoding the Zn-binding region (Chary *et al.*, 1990). One might expect that if Cu,ZnSOD had evolved from a Cu-binding protein this precursor would still be extant. No such protein has been reported, although, in general, with the exception of copper thionein (which is completely unrelated to Cu,ZnSOD), Cu-binding proteins have been only poorly characterized in lower eukaryotes.

One intriguing idea is that what gave this ancestral Cu(Zn) "SOD" protein a selective advantage was the fact that it already had some antioxidant "activity" due to its ability to sequester copper, which is an integral part of an antioxidant defense (Table 1). A similar hypothesis can be proposed for the FeSOD (*sodB*) in *E. coli* because *sodB* is part of the *fur* regulon. Fur is an Fe-binding and activated trans-acting factor that positively regulates *sodB* transcription (Fee, 1991). The ancestral SOD gene may even have acquired some regulatory elements associated with oxidative stress proteins. For example, data show that Cu(II) induces transcription from *SOD1* in yeast (Gralla *et al.*, 1991; see Section IV). Clearly, establishing a more complete picture of the evolution of Cu,ZnSODs will require a careful analysis of such potential ancestral Cu-binding proteins and their regulation.

IV. Molecular Genetics of Cu,Zn and Mn Superoxide Dismutases

A. Cu,ZnSOD—*SOD1* in *Saccharomyces cerevisiae*

1. Gene Structure and Mapping of SOD1

The Cu,ZnSOD gene of *S. cerevisiae*, *SOD1*, was cloned (Bermingham-McDonogh *et al.*, 1988) using a unique oligonucleotide whose sequence was deduced from the known amino acid sequence (Steinman, 1980). *SOD1* was recovered from a λgt11 expression library, although it was probably unexpressed, using the oligonucleotide probe. The gene could not be isolated from plasmid libraries, as it proved to be quite unstable in such vectors. However, a plasmid containing the *SOD1* locus can be propagated in *E. coli* if it is freshly transformed in *recA* cells and harvested after the first growth cycle. The biological activity of the clone was demonstrated by transforming it into an *sod1* yeast strain, Dscd1-4A (Bilinski *et al.*, 1985). The presence of the gene on an episome eliminated the oxygen-sensitive phenotypes of the mutant (Bermingham-McDonogh et al., 1988).

SOD1 maps to the right arm of chromosome X in S. cerevisiae (Fig. 2). Genetically, SOD1 maps equidistant between the cyc1-rad6-SUP4-cdc8 cluster and cdc11 (Chang et al., 1991). The gene has been mapped physically by Southern blotting of NotI and SfiI fragments of genomic DNA to the same region (J. Kaput, D. Kosman, and M. Olson, unpublished results). The DNA sequence of the SOD1 locus revealed no surprises and matched the previously reported protein sequence exactly (Johansen et al., 1979; Steinman, 1980). Examination of the 5′ noncoding region revealed possible TATA box and transcription initiation sequences (Bermingham-McDonogh et al., 1988). Transcription initiation occurs at a C residue 44 bp upstream of the translation start site with a minor start at an adjacent A residue (Gralla et al., 1991). Of considerable interest is a 13-base sequence that is essentially identical to a region upstream from the MnSOD TATA box (Fig. 2). The size and strong homology of these two regions suggest that they may be important in regulatory mechanisms specific to SODs. It should be appreciated that these regions stand out because, except for them, the 5′ sequences of the two SOD genes share little similarity (E. Gralla, unpublished results).

Gralla and Valentine (1991) constructed a deletion-replacement mutant at the SOD1 locus by inserting the URA3 gene into this locus, from which 75% of the C-terminal coding region had been removed. This construct was then used to generate a sod1::URA3 integrant by homologous recombination. The resulting deletion mutant has been designated EG1. E. B. Gralla and J. S. Valentine (unpublished results) have also constructed a double SOD deletion mutant with genotype sod1::URA3 sod2::TRP1. The deletion at SOD1 was constructed in a haploid strain since previous study of sod1⁻ strains showed that under nonselective conditions, Cu,ZnSOD was not essential (Bilinski et al., 1985).

UV-induced sod1⁻ strains were selected on the basis of paraquat sensitivity; they all carried recessive mutations and were in a single complementation group. These mutants exhibit no Cu,ZnSOD activity (Bilinski et al., 1985) but do contain normal amounts of SOD1-specific mRNA and immunoreactive Cu,ZnSOD polypeptide. However, this protein fails to bind Cu or to self-associate completely into functional dimers (Chang et al., 1991). Although the molecular basis for this protein phenotype is not known, the mutant gene has been cloned by polymerase chain reaction (PCR) and sequenced. The only changes are upstream from the normal transcription initiation site (-44) at -48,

A

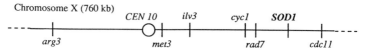

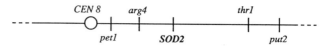

B

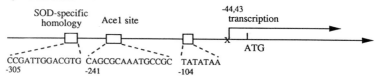

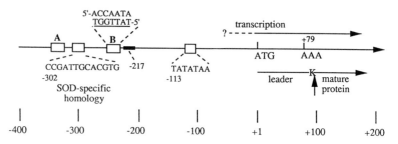

FIG. 2. Map positions and structure of the *SOD1* (Cu,ZnSOD) and *SOD2* (MnSOD) gene loci in *Saccharomyces cerevisiae*. (A) Genetic map positions are indicated relative to other well-mapped genetic loci. Both genes have been physically mapped in Maynard Olson's laboratory (Washington University, St. Louis, Missouri) by Southern blotting of *Not*I and *Sfi*I fragments of chromosomal DNA. The restriction fragments to which probes for each gene hybridize are indicated. The alignment of these fragments was afforded by reference to other cloned genes that genetically map to these regions of the two chromosomes. (B) Diagram (to scale) of the promoter regions of the two SOD genes. Except as noted, sequences are from the coding strand, 5′ → 3′. Of the promoter elements indicated, only the Ace1 (Copper-responsive) binding site in *SOD1* is well-characterized. The X distal to the *SOD1* transcription start locates the site of the mutation in a thoroughly studied *sod1⁻* strain. The A in *SOD2* indicates a region (centered at −330) suggested as a *CYC1*-UAS sequence; this suggestion has been discarded. The B indicates a region that includes a sequence homolgous to the CCAAT box found in UAS2_{CYC1} that is specific for the HAP2,3,4 complex. The solid bar centered at −217 indicates another CCAAT-containing sequence; its deletion suppresses HAP2,3,4-dependent transcription. See text for references to and details of these various features.

where there is a deletion ($-$ GG), and at -51, where there is an insertion ($+$ A) (Chang et al., 1991) (Fig. 2). The relationship between these mutations in the DNA to the protein and activity differences observed remains an intriguing question. The physiology of sod1 $^-$ strains is described in Section V.

2. Regulation of Gene Expression: Oxygen

In comparison to MnSOD, Cu,ZnSOD is generally expressed constitutively at a high level in most organisms. It does exhibit some degree of regulation, however. Certainly in fungi and yeasts, the levels of Cu,ZnSOD can vary considerably, depending on the growth conditions. In Table 3, the response of SOD1 to various culture stimuli is compared to the responses of the other antioxidant enzyme genes in S. cerevisiae.

Early work by Gregory et al. (1974) showed oxygen induction of SOD activity in S. cerevisiae. Cells grown in 100% oxygen had 6.5-fold more SOD activity than cells grown anaerobically. Interestingly, the increase in activity was proportionally distributed between the two SODs. A protective mechanism was induced during growth in 100% oxygen, which rendered the cells highly resistant to 20 atm oxygen.

TABLE 3

Stimuli Contributing to Antioxidant Enzyme Levels in Saccharomyces cerevisiae[a]

Stimulus	SOD1	SOD2	CTT1	CTA1
Shift to aerobiosis	+	+	+	+
O_2 (100%, 1 atm)	+ +	+ +	+ +	+ +
Paraquat (1 mM)	+ +	+	−	−
Cu(II) ($\geq 50 \mu M$)	+	−	−	−
Respiration derepression	+	+	+ +	+ +
Lactate[b]	+	+	+ + +	ND
Acetate[b]	+	+	+	ND
Oleic acid[b]	+	+	+ +	+ + +
Diauxic shift[b]	+ +	+ +	+ + +	ND
N starvation	ND	ND	+ +	+ +
Heat shock	−	ND	+	+

[a]Entries indicate degree of increase in respective mRNAs and/or protein or enzyme activity on change in growth condition or media supplement; −, no significant change; ND, not determined. Quantitative comparisons between conditions may not be valid. See text for references to and details of the results summarized in this table. Cytochrome c peroxidase is not included because data are unavailable.

[b]The increases represent effects of specific carbon source or shift in growth from log to stationary phase (diauxic shift) in addition to effects of glucose (respiration) derepression.

Presumably, SOD and catalase (which was induced 2.3-fold) play a major role in this protection, but are probably not the only players. *Escherichia coli* cells have been shown to induce a host of proteins when presented with oxidative stresses (Farr and Kogoma, 1991); it is likely that the more complex yeasts will show at least as complex a protective system. Oxidant stress can be induced by other agents as well. Paraquat, a redox-cycling drug, causes a twofold to threefold induction of both SOD activities in *S. cerevisiae* (Lee and Hassan, 1985).

The molecular basis of the apparent regulation by oxygen and/or oxidative stress is unknown. Unlike the *SOD2* locus, sequences with homology to known cis-acting elements, such as those associated with carbon source-dependent regulation (HAP2,3,4; see below), are not evident in the 5' noncoding region of the *SOD1* gene. The one region with potential as an oxygen-specific cis-acting sequence is the SOD homology centered at -300 (Fig. 2). Promoter-deletion and mobility-shift assays are in progress to quantitate the precise role this region may have on expression of both the *SOD1* (E. Gralla, unpublished results) and *SOD2* genes (J. Pinkham, personal communication). The results of these investigations should be most revealing about the promoter activity of both loci.

3. Regulation of Gene Expression: Copper

Though oxygen appears to be a major regulator of *SOD* expression, other regulatory processes also play a role. Copper is a strong candidate as a regulator of Cu,ZnSOD synthesis because of its presence in the enzyme and because of its possible role in the generation of oxy radicals. This suggestion is indicated, in part, by the observation that in Cu-depleted, aerobic cultures ([Cu] $\leq$ 50 nM), *S. cerevisiae* had 50% (22 units/mg protein) of the Cu,ZnSOD activity found in cultures grown in Cu-supplemented media (45 units/mg) (Greco *et al.*, 1990). There were comparable changes in both Cu,ZnSOD mRNA and polypeptide, suggesting that the effect of Cu depletion on Cu,ZnSOD activity was transcriptional.

Saccharomyces cerevisiae has one copper-regulated system that is quite well characterized. Copper in the medium ($\geq$10 μM) induces transcription of copper metallothionein, the product of the *CUP1* gene. Yeast copper thionein is a 6500-Dal protein that binds 7 equivalents of copper (Butt and Ecker, 1987). The induction occurs through an activator protein, ACE1 (the *ACE1* gene product), which itself binds copper, and in the Cu(I)-bound form (only) binds to specific DNA sequences in the *CUP1* promoter activating transcription (Thiele, 1988;

Furst *et al.*, 1988). The *ACE1* locus has also been referred to as *CUP2* (Welch *et al.*, 1989).

The ACE1 protein involvement in regulation of the Cu,ZnSOD gene has been shown by two groups (Carri *et al.*, 1991; Gralla *et al.*, 1991). Carri *et al.* (1991) showed that *SOD1* mRNA levels were elevated when copper was continuously present during batch growth. The same effect was seen in the presence of silver, the only other metal known to activate the ACE1 protein, and was not observed in an *ace1* mutant.

Gralla *et al.* (1991) used short periods of Cu-dependent induction during growth on synthetic, low-copper medium (SC) to demonstrate a similar pattern of *SOD1* expression. In addition, the binding site for the ACE1 protein in the *SOD1* promoter was located and characterized in detail. A DNase I-protected region about 25 bp long was located about 200 bases upstream of the transcription start (Fig. 2). The core of this region showed a 12- out of 15-base identity with a strong ACE1-binding site in the *CUP1* promoter, including the seven G residues that were shown by methylation interference to be important for binding of ACE1 in both sites (Gralla *et al.*, 1991; Buchman *et al.*, 1990).

Transcription of the chromosomal *SOD1* gene was induced twofold to threefold by copper (50 μM), and this induction was dependent on both the presence of a functional ACE1 protein and on the presence of a functional ACE1-binding site. Thus, the effect was not observed when this binding site was mutated (changing each of the important seven G residues to A) to eliminate ACE1 binding. mRNA levels were measured by primer extension assays using radiolabeled primer complementary to the RNA of interest. The *SOD1* promoter region also was fused to the *lacZ* gene, which served as a reporter of transcriptional activity. The copper-dependent induction of β-galactosidase mRNA synthesis (measured also by primer extension) from this fusion was much more dramatic than that from the native *SOD1* gene and promoter. However, this Cu-dependent expression was strictly dependent on ACE1 and on a functional ACE1-binding site in the *SOD1* promoter (Gralla *et al.*, 1991).

Quantitatively, the effect of ACE1 and copper on the *SOD1* promoter is much smaller than that seen in the *CUP1* promoter. Copper thionein is induced 10- to 50-fold in the same conditions. There are differences in the two promoters that may explain this discrepancy. A single ACE1 site is present in the *SOD1* promoter, whereas the *CUP1* promoter has four. Also, this site in *SOD1* is inverted relative to the direction of transcription and is located further upstream (both features in comparison to the sites in the *CUP1* promoter). The *SOD1* gene may also be subject to other sorts of regulation that limit or modulate the copper effect.

It has been reported that Cu,ZnSOD accumulates as the apo (unmetallated) form in anaerobic conditions (Galiazzo *et al.*, 1991). Up to 60% of the Cu,ZnSOD can be present in an inactive form that is identified by its ability to be reactivated in cellular extracts by addition of copper. It is not clear whether this is a specific regulatory process or a consequence of different copper utilization in anaerobiosis. Excess copper during growth can decrease, but not eliminate, the amount present as apo-SOD. Yeast is somewhat less sensitive to Cu(II) under N_2 (Greco *et al.*, 1990).

In copper-depleted cultures of *S. cerevisiae*, Cu,ZnSOD activity decreased by about 50% (to 22 units/mg protein). Interestingly, MnSOD activity increased from 5 to 23 units/mg protein in these cultures, apparently compensating for the decrease in Cu,ZnSOD activity (Greco *et al.*, 1990). In a similar fashion, Shatzman and Kosman (1979) studied the effect of Cu limitation on Cu,ZnSOD levels in the basidiomycete, *Dactylium dendroides*. As in *S. cerevisiae*, the decline in Cu,ZnSOD activity observed with Cu starvation correlated with a decrease in immunoreactive Cu,ZnSOD protein. Also, the decline in Cu,ZnSOD activity was compensated for by an increase in MnSOD. That is, depending on the availability of copper (and/or manganese), the ratio of Cu,ZnSOD to MnSOD changed dramatically. In normal media, MnSOD contributed only 13% of the total cellular SOD activity. In low-copper concentration media, MnSOD contributed 87% of cell total SOD activity; this additional activity was found in the nonmitochondrial fraction. This kind of flexibility in SOD levels has not been observed in all organisms. Nonetheless, the balance between different antioxidant enzymes appears to be physiologically important. Extending this a bit further, there is some evidence that too much SOD is deleterious but can be compensated for by a corresponding increase in catalase. This behavior is discussed in more detail in Section V.

The work described above suggests that though copper is a factor controlling *SOD1* expression, it is not necessarily a major one. A role for copper is logical and intuitively satisfying. In very low concentrations of copper it would be advantageous to prevent synthesis of any copper-containing proteins above the minimum necessary for survival. This might be associated with a compensatory mechanism involving an increase in MnSOD, as has been observed in *S. cerevisiae* and *D. dendroides*. An excess of Cu,ZnSOD, which normally is a relatively abundant protein with a high affinity for copper, could scavenge copper that is needed to support other cell activities, such as respiration. However, at higher environmental copper concentrations, an organism could afford to synthesize and metallate more Cu,ZnSOD to protect against the oxidative challenges that could be exacerbated by excess free copper. An O_2-dependent induction of Cu,ZnSOD by copper at ≥ 100 μM in *S. cerevisiae* provides some support for this suggestion (Greco *et al.*, 1990). This induction is specific for Cu,ZnSOD; MnSOD levels do not increase with excess copper (Lee and Hassan, 1985; Greco *et al.*, 1990).

4. Regulation of Gene Expression: Carbon Source and Stage of Growth

There are no published systematic studies that have characterized either the changes in Cu,ZnSOD levels as a function of carbon source or culture growth or the underlying mechanisms of such changes. There are data, however, that indicate that such regulation does modulate *SOD1* expression. For example, Lee and Hassan (1985) have shown that SOD1 activity increased threefold at about the time in culture at which the cells were undergoing the shift from log to stationary phase. Chang *et al.* (1991) reported a sixfold increase in Cu,ZnSOD activity in stationary-phase cells in comparison to cells in early log phase. MnSOD and catalase T activities increased 4- and 31-fold in stationary phase, respectively. In neither case was analysis of these changes carried out at the transcriptional level, however.

The shift into stationary phase growth has been called the diauxic shift (Pringle and Hartwell, 1981). Dramatic changes in metabolic activities occur at the diauxic shift due primarily to decreases in cAMP and thus in cAMP-dependent protein kinase A activity (Boorstein and Craig, 1990, and references therein). As discussed further below, with respect to the increase in MnSOD in the diauxic shift, promoter sequences involved in this cAMP-dependent up-regulation (termed PDS, for postdiauxic shift) have been identified in several other genes (e.g., *CTT1, SSA3*). Such sequences are not obvious in the *SOD1* or *SOD2* promoter. Given the appreciably weaker response of Cu,ZnSOD and MnSOD to the shift to stationary-phase growth (Chang *et al.*, 1991) (Table 3), it is possible that PDS-like elements are present in the two promoters but are not similar enough to be readily apparent. The weak transcriptional response of the SOD genes (cf. *CTT1*) would be consistent with this type of weakened promoter element. The attentuation of the efficiency of the *CUP1*-like promoter at the *SOD1* locus noted above is a good example of this model.

With respect to carbon source, data must be assessed carefully to separate effects of carbon source from those due to the diauxic shift. For example, in studies by Galiazzo et al. (1991), cells were taken for study from late log-phase cultures; it is not clear in this report what the growth phase was metabolically (using data on catalase T, for example). Thus, the 2.5-fold increase in SOD1 activity with glucose derepression observed in these cells could be different than the increase seen in log-phase cells. This possibility is indicated by other work in which early log-phase cells exhibited 50% *more* Cu,ZnSOD activity in 10% glucose (repressing) in comparison to growth on nonfermentable carbon sources (Westerbeek-Marres et al., 1988). This contrasted with MnSOD, which was 8- to 10-fold higher in cells grown on nonfermentable carbon sources.

The qualitative comparisons given in Table 3 with respect to carbon source (glucose, lactate, acetate, and oleic acid) are based on Northern analyses of poly(A)$^+$-selected RNA (J. Lee and D. Kosman, unpublished results). These indicate that SOD1 (and SOD2) are weakly and similarly induced by glucose derepression alone and by growth on a carbon source that requires mitochondrial respiration only (lactate), β oxidation in peroxisomes (oleate), or the glyoxylate cycle (acetate) for its utilization. In general, both catalases are more strongly induced by all three nonglycolytic carbon sources, particularly by oleate, which also induces peroxisome biogenesis (Table 3) (Veenhuis et al., 1987; Skoneczny et al., 1988; Simon et al., 1991; Veenhuis and Harder, 1991). In part, the induction of transcription from CTA1 is under control of the ADR1 protein (Simon et al., 1991), which is very likely an important trans-acting factor regulating expression of other peroxisomal proteins (Veenhuis and Harder, 1991). As discussed further below, the SOD1 locus does not contain an obvious consensus ADR1-binding site.

B. MNSOD—SOD2 IN Saccharomyces cerevisiae

1. Gene Structure and Mapping of SOD2

The MnSOD gene of S. cerevisiae was cloned serendipitously by van Loon et al. (1983). In one purification protocol, the enzyme copurifies with the QH_2:cytochrome c oxidoreductase (cytochrome bc_1 complex) of yeast mitochondria; it also has the same subunit size as the Rieske Fe–S protein. Thus, an antiserum thought to recognize the Fe–S protein in reality recognized the MnSOD, and led to the cloning of its gene. The clone was properly identified by its DNA sequence (Marres et al., 1985). The possibility of a functional association of MnSOD, a matrix protein, with the cytochrome bc_1 complex is intriguing because the QH_2:cytochrome c reductase is a likely source of superoxide radicals, which could arise from autooxidation of bound ubisemiquinone (cf. Fig. 1) (de Vries *et al.*, 1981; Boveris and Cadenas, 1982).

SOD2 maps to the right arm of chromosome VIII in *S. cerevisiae* (Fig. 2). Genetically, *SOD2* mapped 4 cM distal from *pet1* (van Loon *et al.*, 1986); physically it maps to *Not*I and *Sfi*I fragments 08.3-1 and 08.3-2, respectively, which include the *PET1* locus (D. Kosman, J. Kaput, and M. Olson, unpublished results). It is immediately proximal to the *ERG11* locus as determined by DNA sequence analysis (Turi *et al.*, 1991). The *SOD2* locus contains the normal elements of transcription initiation and capping, and translation initiation and termination found in eukaryotic genes (Marres *et al.*, 1985). Some of these are indicated in Fig. 2.

Two additional and potentially important types of cis elements might be located in the 5' noncoding region. One type (suggested by Marres *et al.*, 1985) is represented by putative upstream activating sequences (UASs), as are found upstream from the genes encoding a number of heme-containing proteins. These proteins are transcriptionally regulated by oxygen via heme biosynthesis and by carbon source. The other type is a sequence similar to one found in the 5' noncoding region of the *SOD1* gene (Bermingham-McDonogh *et al.*, 1988; and see above). The relationships of these two types of cis elements to transcription from the *SOD2* locus are discussed below. It should be noted also that there are some mistakes in the reported sequence (Marres *et al.*, 1985) in the 5' noncoding region of the *SOD2* gene. For example, the SOD-specific homology box is actually larger than is apparent based on published data (Pinkham, 1992).

Van Loon *et al.* (1986) disrupted the *SOD2* locus by the insertion of the *LEU2* gene into the coding sequence for MnSOD. This construct was introduced into the wild-type genome by homologous recombination in a diploid strain and the insertion was recovered from haploid spores by selection for the Leu$^+$ phenotype. A deletion of the *SOD2* gene has been constructed also (E. Gralla and J. Valentine, unpublished results).

2. Regulation of Gene Expression: Oxygen and Heme

Many mitochondrial electron transport chain proteins are known to be tightly regulated by oxygen, an effect that is mediated, at least in part, by the synthesis and availability of heme (Guarente and Mason, 1983; Zitomer and Lowry, 1992, and references therein). *Saccharomyces cerevisiae* is an ideal organism for study of these mechanisms because respiration is not essential for cell growth of this fermentative yeast (Gancedo and Serrano, 1989). In anaerobic conditions, in conditions of high glucose (2–10%), or in mutants lacking mitochondrial respiratory capacity, *S. cerevisiae* grows quite well by fermentation (Fraenkel, 1982; Gancedo and Serrano, 1989). In anaerobic conditions (or in hemeless mutants), heme is not synthesized and consequently synthesis of respiratory enzymes is repressed (Sherman and Stewart, 1981). Linking the synthesis of the proteins of the respiratory chain to heme levels is advantageous because many of these enzymes are heme-containing proteins, the most well-studied example being cytochrome *c* (*CYC1*) (Lalonde *et al.*, 1986; Pinkham *et al.*, 1987; Pfeifer *et al.*, 1989; Forsberg and Guarente, 1989).

Although MnSOD is not a heme-containing protein, its expression is influenced by the heme status of the host (Autor, 1982). Thus, a mutant carrying the *hem1-3* allele (*HEM1* codes for δ-aminolevulinic acid synthase), and therefore heme deficient, contained about one-third the amount of MnSOD protein as did wild type in normoxic conditions. However, other heme-regulated respiratory proteins are repressed to a much greater degree in the same conditions, indicating that other factors, such as the presence of oxygen (independent of heme synthesis), also play a role. This suggestion is consistent with the fact that respiration is not absolutely required for MnSOD expression (Autor, 1982).

Genes such as *CYC1* (Pfieffer *et al.*, 1987, 1989) and *CTT1* (Winkler *et al.*, 1988) have a binding site(s) for the HAP1 protein, which is the trans-acting mediator of the heme-dependent control of expression of these genes; this site is found in a UAS1 at those loci. Based on early work (Guarente and Mason, 1983), the *SOD2* locus was thought to contain a UAS1$_{CYC1}$-like sequence centered at about -330 (see Fig. 2B, box A) (Marres *et al.*, 1985). However, this region can be discounted as a homologous UAS1$_{CYC1}$. Methylation interference analysis of protein–DNA contacts in the UAS1$_{CYC1}$ reveals that these contacts would be missing in any interaction between HAP1 and the putative UAS1$_{SOD2}$ (data from Pfeifer *et al.*, 1987, and Marres *et al.*, 1985). These comparisons indicate that strong regulation by HAP1 of *SOD2* expression is not likely.

This suggestion has been confirmed by study of MnSOD gene expression using a gene fusion construct of the MnSOD (*SOD2*) promoter region and the β-galactosidase (*lacZ*) structural gene (J. L. Pinkham, unpublished results). On the one hand, heme was clearly an important regulator of MnSOD expression in glucose-grown cells because a 30-fold decrease in MnSOD promoter activity was observed in a heme-deficient mutant. (Note that this reporter construct was more sensitive to the absence of heme than was the chromosomal *SOD2*.) On the other hand, the expression from the *SOD2* promoter in a *hap1*⁻ strain was 50% of that in wild type. This is in contrast to the severalfold effect of a *hap1* allele on *CYC1* expression, for example. Gel-shift assays were consistent with this result. No specific binding of the HAP1 protein to the 5' noncoding region of the *SOD2* locus was detected, as was predicted by the sequence comparisons noted above.

Oxygen regulation of MnSOD is less well characterized, although Westerbeek-Marres *et al.* (1988) have shown that oxygen levels do modulate MnSOD expression. In cells grown in 80% O_2, the MnSOD levels increase relative to those of core 2 or cytochrome c_1, whatever the carbon source (increases not quantitated). This result indicates that MnSOD expression is more responsive to hyperoxia than is the expression of respiratory proteins. On the other hand, the oxygen-induced level of MnSOD in glucose-repressed cultures is lower than the MnSOD level in normoxic cultures grown on nonfermentable carbon sources. This result shows that carbon source has a relatively stronger effect on MnSOD expression than does O_2 and thus appears to be the dominant mediator of MnSOD levels in yeast. Carbon source-dependent regulation of *SOD2* is discussed further below.

3. Regulation of Gene Expression: Carbon Source and Stage of Growth

Yeast can be grown on a variety of carbon sources, which can be classified as to their ability to induce respiratory proteins and the requirement for respiratory activity to support growth. *Saccharomyces cerevisiae* and some other yeasts are notable in that fermentation of glucose to alcohol and CO_2 is the preferred mode of growth (Gancedo and Serrano, 1989). If glucose is available, it is used exclusively, and respiration and respiratory protein synthesis are shut down. Other five- and six-carbon sugars, such as galactose, maltose, or raffinose, can be metabolized by fermentation, but are known to induce and utilize respiration as well. These are known as partially derepressing. Fully derepressing carbon sources are two- and three-carbon compounds, which cannot support growth unless the respiratory apparatus is fully functional. Some examples are ethanol, acetate, lactate, glycerol, and pyruvate (Fraenkel, 1982; Wills, 1990).

As mentioned, Westerbeek-Marres *et al.* (1988) demonstrated that the synthesis of MnSOD is regulated by the carbon source, in parallel with most other respiratory chain proteins. In batch cultures grown to early log phase in fully repressing conditions (10% glucose), MnSOD expression is low, but detectable. In nonfermentable carbon sources, the level increases 10- to 15-fold. Under the same conditions, Cu,ZnSOD expression was 40% higher in glucose as compared to the nonfermentable carbon sources (Westerbeek-Marres et al., 1988).

In the transition from glucose repression to derepression (glucose to ethanol as carbon source), MnSOD content increases in parallel with representative mitochondrial proteins such as core 2 and cytochrome c_1. Oxygen consumption did not increase until several hours later, however, implying that the induction of these three proteins was not dependent on oxygen utilization *per se,* or, by extension, increased levels of oxygen radicals. Further evidence for the noninvolvement of oxygen in this regulation comes from the fact that MnSOD is fully induced in a mutant with the major sources of O_2^- missing, e.g., in a respiratory-deficient strain (Westerbeek-Marres et al., 1988). Unfortunately, the absolute (total) levels of SOD were not measured in these experiments, so it was not possible to determine the contribution each SOD was making to the cellular O_2^- dismutation activity. The Cu,ZnSOD is more highly expressed than the MnSOD in yeast (usually accounting for >90% of the cell total), so that a 10-fold increase in MnSOD activity on derepression could be approximately balanced by the 40% decrease in Cu,ZnSOD that is observed in such conditions.

Carbon source-induced change in cytochrome *c* expression is due primarily to the HAP2,3,4 protein complex that binds to UAS2$_{CYC1}$ (Pinkham et al., 1987; Forsberg and Guarente, 1989). The second putative UAS region in the *SOD2* locus centered at about -240 (Fig. 2B, box B) could be compared to UAS2$_{CYC1}$. It contains a "CCAA(T)" box to which the HAP2,3,4 protein complex binds (Forsberg and Guarente, 1988). This sequence is found on the noncoding strand in *CYC1*. Thus in *CYC1* the coding strand is TGGTTGGT; a consensus sequence for this element in several genes is TNATTGGT (Guarente et al., 1984). A putative CCAAT homology in *SOD2* is contained in the sequence TTGTTGGC (Fig. 2B, box B). There is also a strong CCAAT consensus downstream centered at -217 with sequence (T)CCAATAAC, or GTTATTGG(T) on the noncoding strand (Fig. 2B, solid bar; J. L. Pinkham, personal communication).

The role of HAP2,3,4 in the expression of MnSOD also has been investigated using the *SOD2* promoter-driven β-galactosidase reporter gene (J. L. Pinkham, unpublished results). First, expression of this construct in *hap2,3,4*-containing backgrounds was 50% of expression in wild type. Even this relatively weak effect was not due to the putative UAS2 centered at -240 because its deletion was essentially neutral. The CCAAT box centered at −217 (Fig. 2B, solid bar) was also deleted from the *SOD2* promoter in the fusion gene; significantly, this construct was transcriptionally insensitive to glucose derepression (J. L. Pinkham, unpublished results). Extension of these interesting studies will undoubtedly reveal further details about the role of these various cis- and trans-acting elements in *SOD2* expression.

Special consideration must be given to acetate and oleic acid as carbon sources in yeast. In many yeast strains, both induce the proliferation of microbodies and the expression of a variety of peroxisomal/glyoxysomal proteins, e.g., catalase A, acyl-CoA oxidase, and isocitrate lyase (Veenhuis and Harder, 1991). In *S. cerevisiae,* only oleic acid induces all of these various enzymes as well as the appearance of microbodies (Veenhuis *et al.,* 1987). In contrast, the activities of the glyoxylate cycle-specific enzymes, isocitrate lyase and malate synthase, but not microbodies, increase with acetate; these enzymes are essential to assemble the four-carbon intermediates that sustain cell growth on two-carbon units (Duntze *et al.,* 1969). Northern analysis of *SOD1* and *SOD2* transcripts indicates that neither transcript is increased more in the presence of oleic acid or acetate than in the presence of another nonfermentable carbon source (Table 3) (J. Lee and D. Kosman, unpublished results). In any event, the promoter region of neither gene contains sequences comparable to those found in *CTA1,* for example, which have been implicated in the regulation of peroxisomal proteins *via* the ADR1 protein (Simon *et al.,* 1991). Nonetheless, as with the potential for cAMP-dependent regulation of *SOD* expression (see below), whether this expression depends on the ADR1 protein needs to be tested directly in *adr1*-containing mutants and by DNA-binding analysis.

A fourth apparent component of the regulation of MnSOD is by stage in growth. A decrease of about 50% in the immunologically reactive MnSOD protein accompanies the transition from log-phase to stationary-phase growth (diauxic shift). This decrease is not mirrored in the other respiratory proteins (Westerbeek-Marres *et al.,* 1988). In contrast, the levels of Cu,ZnSOD protein increase during growth in the transition from log to stationary phase, although the absolute change is less dramatic than that for SOD2. However, these changes have not been correlated directly with nutrient depletion on the one hand, nor with cell age on the other. Clearly, both aspects of culture condition could modulate the level of either or both SODs.

The mechanism of nutrient-dependent changes in MnSOD (and Cu,ZnSOD) levels is not known. For other proteins susceptible to glucose (catabolite) repression and starvation in general, regulation of transcription is partially and negatively RAS- and cAMP-dependent (Tatchell, 1986). Noncoding sequences that are involved in this cAMP-dependent regulation have been identified 5' to some of these genes, e.g., CTT1 (Belazzi et al., 1991) and SSA3 (Boorstein and Craig, 1990). The cAMP and RAS dependence of this regulation has been determined in part by study of expression of these various proteins in ras, tpk1,2,3, and byc1 mutants; TPK1 and BYC1 code for the catalytic and regulatory subunits of the yeast protein kinase A, respectively. Thus, mutants having high constitutive protein kinase A activity exhibit low levels of the regulated protein whereas mutants that lack this activity exhibit high levels. There is also cAMP-independent regulation of such genes that is less well characterized (Cameron et al., 1988). Studies of MnSOD and Cu,ZnSOD expression in such RAS- and protein kinase A-deficient backgrounds have not been carried out nor have the promoter regions of SOD2 and SOD1 been systematically analyzed with respect to their potential to bind cAMP-dependent trans-acting transcription factors. These regions do not obviously contain sequences homologous to those indicated by the work on CTT1 and SSA3, however.

4. Regulation of SOD Gene Expression: Summary

The work described above indicates clearly that the regulation of the Cu,Zn and MnSOD genes is complex, and although much progress has been made complete understanding remains lacking. Clearly, there are some inconsistencies in the results described by the various groups working on the control of SOD gene expression. Particularly perplexing are the differences reported in heme response. These may be explained by differences in experimental method, culture conditions, and/or strains used. There is the potential for leakiness in heme-negative strains, for example. On the other hand, it is also possible that more than one heme-dependent regulatory mechanism exists in yeast. In addition to its necessary role in controlling expression of heme-containing enzymes, heme could also be involved in regulation of oxygen-related enzymes but by a different mechanism involving different putative cis- and trans-acting elements. Heme is chemically suited to be an independant oxygen sensor. As a prosthetic group of an appropriate protein, it could be an exquisitely sensitive sensor of oxygen tension (consider the binding and release of oxygen by hemoglobin and the accompanying changes in protein conformation). Indeed, transcriptional activation of the erythropoietin gene in hypoxia has been ascribed to a heme protein acting as an oxygen sensor (Goldberg *et al.*, 1988). This concept might be fruitfully explored in yeast.

C. Catalases and Peroxidases

As indicated in Table 1, the SODs are not the only antioxidant enzymes present in yeast. There are two catalases in *S. cerevisiae*: catalase T (*CTT1*) is cytoplasmic and catalase A (*CTA1*) is peroxisomal. *CTT1* has been cloned (Spevak *et al.*, 1983) and sequenced (Hartig and Ruis, 1986), as has *CTA1* (Cohen *et al.*, 1985, 1988). The regulation of both loci has been studied in some detail, as has been made evident in the discussion above. *Saccharomyces cerevisiae* also contains cytochrome *c* peroxidase (*CCP1*), located in the intermembrane space of the mitochondria (Kaput *et al.*, 1982, 1989). The regulation of this gene has been little studied in *S. cerevisiae*. However, *Hansenula polymorpha* exhibits a strong induction of this activity when grown in the presence of H_2O_2. Under conditions that induced peroxisomes (methanol as carbon source), catalase and not cytochrome *c* peroxidase was induced, indicating that the latter is important in the detoxification of environmental H_2O_2, whereas catalase scavenges the peroxide formed *in situ* (Verduyn *et al.*, 1988). Nonetheless, a catalase-negative mutant of *H. polymorpha* was able to metabolize methanol and grow on mixtures of glucose and methanol by inducing the mitochondrial cytochrome *c* peroxidase, even though it could not grow on methanol alone due to lack of catalase. Thus the peroxidase activity present in the active mitochondria apparently compensated for the catalase activity missing from the peroxisomes in this mutant (Verduyn *et al.*, 1988).

The glutathione peroxidases are important antioxidant enzymes in mammalian systems. By analogy, yeasts might be expected to contain such peroxidases. However, it is not clear whether this is the case. Recent reports conflict. Galiazzo *et al.* (1988) reported glutathione peroxidase activity in *S. cerevisiae,* but Verduyn *et al.* (1988) found none in *H. polymorpha*. Furthermore, no H_2O_2 decomposition was observed in strains of *S. cerevisiae* containing mutations in both *CTT1* and *CTA1*, suggesting that the gene products from these two loci were the only peroxide-utilizing enzymes in this yeast (Bilinski *et al.*, 1989; T. Bilinski, personal communication).

D. SOD in Other Yeasts and Fungi

Little work on SOD in yeasts other than *Saccharomyces* has been published. However, another organism for which extensive genetics, and methods for genetic engineering, exist is the filamentous fungus *N. crassa*. The protein sequence of the *Neurospora* Cu,ZnSOD has been determined (Lerch and Schenk, 1985). More recently the cloning and sequencing of the gene for Cu,ZnSOD (*SOD1*) was accomplished (Chary *et al.*, 1990), setting the stage for further investigations of the role and regulation of SOD in this organism. A Cu,ZnSOD-deficient mutant has been produced, and shows a phenotype similar to that of *sod1*⁻ yeast, except there are no conditional nutritional auxotrophies. For example, plating efficiency is reduced, and mutation rates increased in the *sod1 N. crassa* mutant. SOD work has been hampered in *Neurospora* by the fact that MnSOD and Cu,ZnSOD run very close together on native gels, making their identification by differential activity staining difficult. With the *sod1*-carrying mutant it is now possible to identify unequivocally the bands. Interestingly, in *Neurospora*, the mutant lacking Cu,ZnSOD shows increased MnSOD activity, which has not been consistently observed in *Saccharomyces*. The MnSOD gene (*SOD2*) has also been cloned and is being characterized (Natvig, 1991; D. Natvig, personal communication).

V. Cell Physiology of Mutants in Cu,Zn and Mn Superoxide Dismutases

Many targets and mechanisms of action have been proposed for oxy radicals in general and for superoxide in particular. Most of the major types of biomolecules are potential targets for oxy radicals; DNA, lipids, and proteins have all been suggested as important targets of superoxide-mediated damage (cf. Halliwell and Gutteridge, 1989, for review). In some cases, of course, the phenotype(s) of oxidant stress will undoubtedly arise from damage to more than a single target. Even in the same organism, different growth conditions may expose different vulnerabilities, and across species, targets and sensitivities are likely to be quite diverse.

In *S. cerevisiae*, the phenotype(s) that results from deletion of a SOD activity should be explicable with respect to the cellular locale of that activity. That is, *sod*⁻ yeast should be phenotypically different from *E. coli* mutants, in which the two SODs do not appear differentially localized. Consequently, in yeast, the phenotype of a *sod1*⁻ strain should be different from that of an *sod2*⁻ strain, and it is. The differences observed are understandable with respect to the types of metabolic functions carried out in the cytosol (and compartments in equilibrium with the cytosol, e.g., the nucleus) *versus* the mitochrondrion (or peroxisome).

In this section the phenotypes of strains of *S. cerevisiae* that lack one or both of the SODs common to all eukaryotes are described. Many of these phenotypes are listed in Table 4. One of the things that will become clear from this summary is that superoxide-dependent damage does occur and that specific enzymes, activities, and cellular locales can be targeted. What should be appreciated, however, is that the actual chemical mechanisms underlying this apparent damage remain essentially uncharacterized. The unraveling of these mechanisms remains a critical area of further investigation.

A. OXIDANT SENSITIVITY

The hallmark of *sod1*⁻ (Cu,ZnSOD) strains in many species has been their sensitivity to hyperoxia and to redox cycling drugs such as paraquat or menadione. The *sod1*⁻ yeast strains show the same exquisite sensitivity to both stressors (Bilinski *et al.*, 1985; Gralla and Valentine, 1991). The *sod1*⁻ strains do not grow on rich medium under 100% O_2 (Table 4). Strains lacking MnSOD (*sod2*⁻) are also more sensitive to paraquat than are the wild type, but are more resistant than *sod1*⁻ strains when growth is on glucose. When nonfermentable carbon sources are used, *sod2*⁻ strains are equally sensitive (E. Gralla and J. Valentine, unpublished results). The *sod1*⁻ strains have been tested for resistance to other drugs, and usually appear slightly more sensitive than the wild type. However, in light of the poor health of these strains, this difference could be ascribed to the nonspecific and cumulative effect of stress in general, rather than to some specific sensitivity. Agents tested in this way include methyl methane sulfonate, *tert*-butyl hydroperoxide, hydrogen peroxide, and canavanine (E. Gralla and J. Valentine, unpublished results).

B. METABOLIC CONSEQUENCES

1. Aerobic Auxotrophies

Mutants in *SOD1* in *S. cerevisiae* exhibit an absolute requirement for cysteine or methionine and lysine when grown in synthetic medium under air (20% O_2). When grown under N_2, neither amino acid is required (Bilinski *et al.*, 1985; Chang *et al.*, 1991). The basis of the methionine/cysteine requirement was subsequently investigated by supplementation analysis (Chang and Kosman, 1990). This is shown in Fig. 3, which illustrates the biosynthesis of methionine in *S. cerevisiae* (Jones and Fink, 1982). The precursors and intermediates that were

TABLE 4

Selected Growth Characters of sod1, sod2, and sod1 sod2 Mutants in *Saccharomyces cerevisiae*

Growth character	Wild type	sod1	sod2	sod1 sod2	Comments	Ref.[a]
Growth without Lys or Met under air	+ +	—	+ +	—	Growth character exhibited only in aerobic growth	a, b
Growth in 100% O_2 on rich medium (YPD)	+ +	—	—	—	—	a–g
Anaerobic growth, any medium	+ +	+ +	+ +	+ +	—	f, h
Growth in air on plates, defined medium						
Glucose	+ +	+	+ +	+	Relatively weak growth for sod1 and sod1 sod2	d, f
Glycerol	+ +	—	±	—	—	
Ethanol	+ +	—	—	—	—	
Growth in liquid media (air)						
Glucose	+ +(+ +)	+ +(+)	+ +(+ +)	+ +(+)	In liquid culture, growth of sod1 is highly dependent on vigorous aeration (values in parentheses), a condition that may account for differences in reported growth; better glycerol growth may be due to glycerol's radical-scavenging ability, or to entry into gluconeogenesis as triose	c–f, h
Glycerol	+ +(+ +)	+(—)	+(±)	—(—)		
Ethanol	+ +(+ +)	+(—)	+(—)	—(—)		
Acetate, pH 5.5	+ +	+	—	ND	Difference between 5.5 and 6.5 may be due to greater influx of acetic acid closer to pK_a	c, d, h
Acetate, pH 6.5	+ +	+ +	+ +	ND		

					Comments	
Rescue by metal supplementation: 200 μM Cu^{2+} or 2mM Mn^{2+}						
Glucose, 100% O_2 (Mn^{2+})	++	++	ND	ND	—	f, g
Glycerol (Cu^{2+}), air	++	+	++	±	—	f, i
Ethanol (Cu^{2+}), air	++	+	++	±	—	f, i
Spontaneous mutation rate (aerobic)	Normal	3 × up	Normal	ND	*sod1* rate from reversion of amber codon in *trp1*, difference in *sod1* lost under N_2	e
					sod2 by reversion of canavanine resistance	c
Paraquat resistance (aerobic)						
2% glucose, plates	2 mM	<10 μM	<1 mM	<10 μM	Concentration to which strain is resistant; similar pattern seen with other redox cycling drugs	f
3% glycerol, plates	1 mM	<10 μM	<10 μM	<10 μM		f
Paraquat toxicity: 2.5 mM paraquat; % survival, 30 min	125	6.5	115	6.0	Survival by colony formation on YPD in microaerobic atmosphere; wild type exhibits killing at >10 mM paraquat	f
Sulfite resistance (YPD, liquid media)						
Air	≥2 mM	≥10 μM	≥1 mM	ND	Sensitivity related to Met requirement of *sod1* strains when grown aerobically and sulfite redox chemistry	b
N_2	≥2 mM	≥1 mM	≥1 mM	ND		

[a] References: (a) Bilinski *et al.*, (1985); (b) Chang and Kosman, 1990; (c) van Loon *et al.*, 1986; Westerbeek-Marres *et al.*, 1988; (e) Gralla and Valentine, 1991; (f) E. Gralla and J. Valentine, unpublished results; (g) Chang and Kosman, 1989; (h) J. Lee and D. Kosman, unpublished results; (i) K. Tamai and D. Thiele, unpublished results.

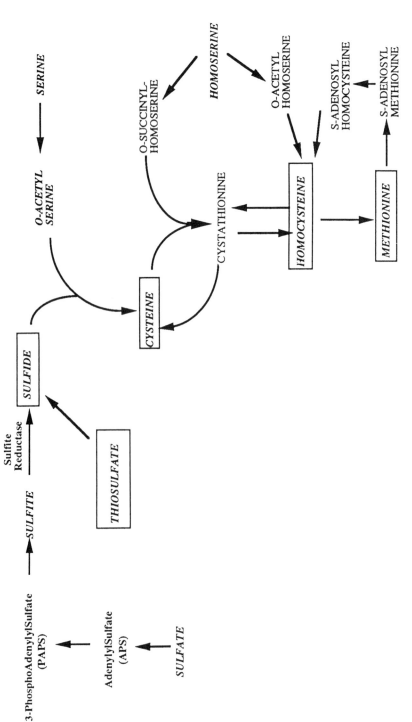

Fig. 3. Methionine biosynthetic pathway indicating precursors and intermediates used in a supplementation anaysis of the metabolic defect in *sod1*⁻ strains of *S. cerevisiae* that causes aerobic methionine auxotrophy. Compounds used as supplements are italicized; those that supported aerobic growth are boxed. Data from Chang and Kosman (1990).

used as supplements are given in italics; of these, those which supported aerobic growth are boxed. The data suggested that the failure
to reductively assimilate SO_4^{2-} was due to a block in converting SO_4^{2-}
to S^{2-} and not in the incorporation of S^{2-} into serine or homoserine, or
the interconversions of Cys, homoCys, and Met. The mechanism of this
block was suggested by the fact that SO_3^{2-} not only did not support
growth, but was, in fact, extremely cytotoxic to $sod1^-$ strains when
added exogenously. This cytotoxicity was seen only under air, and only
in a $sod1^-$ strain, not a $sod2^-$ strain or in the wild type (Table 4).
Enzyme activity measurements showed that the $sod1$ mutants had the
ability to reduce SO_3^{2-} to S^{2-}, e.g., they had wild-type levels of sulfite
reductase activity (Chang and Kosman, 1990). That is, the loss of sulfate assimilation was not due to the inactivation of an enzyme involved, but to the toxicity of the reduction intermediate SO_3^{2-}.

The question of why sulfite is so toxic in the absence of the cytosolic
dismutation activity contributed by Cu,ZnSOD can be addressed by
reference to the well-characterized chemistry of this species (Neta and
Huie, 1985). Sulfite can be reduced by O_2^- to the sulfur trioxy radical,
SO_3^-, which is as reactive as OH·. In this model, the *protection* by Met
(or Cys) was associated with the repression of the sulfate assimilatory
pathway by its end product, S-adenosylmethionine (Jones and Fink,
1982; Thomas *et al.*, 1989). As expected, growth of wild type under air
or mutant under N_2 in the presence of Met resulted in a 10-fold decrease in metabolite flux through sulfite reductase (Chang and Kosman, 1990). Thus, a $sod1^-$ strain requires Met, not because it lacks
specific synthetic machinery, but because Met turns off assimilation of
SO_4^{2-}. This, in turn, prevents the accelerated accumulation of the toxic
sulfur trioxy radical.

The molecular basis of the aerobic Lys auxotrophy has not been
established. However, Liu *et al.* (1992) have carried out supplementation analysis in an $sod1^-$ strain which suggests that the transaminase
responsible for the conversion of α-ketoadipic acid (KA) to α-amino
adipic acid (AA) may be impaired in this strain under air; AA, but
not KA supported growth under air. Interestingly, spermine can fulfill
the aerobic Lys requirement in $sod1^-$ strains, also (J. Lee and D. Kosman, unpublished observations). This result cannot be explained on
the basis of the interconversion of the added spermine to lysine, which
does not occur in yeast (Jones and Fink, 1982). The precise roles of
polyamines are essentially uncharacterized (Tabor and Tabor, 1984).
However, *S. cerevisiae* exhibits an absolute requirement for spermidine or spermine for aerobic growth (Balasundaram *et al.*, 1991). This

has been established by study of a deletion mutant in *SPE2*, which encodes the essential enzyme in spermidine and spermine biosynthesis, *S*-adenosylmethionine decarboxylase (Balasundaram *et al.*, 1991). Spermidine-deficient *E. coli* is also oxidant sensitive, exhibiting a significantly greater sensitivity to paraquat than spermine-sufficient cultures (Minton *et al.*, 1990). In addition, spermine appears to be a natural antioxidant, functioning, in part, perhaps, by chelating metal ions, such as Fe^{2+}, which can catalyze the Fenton reaction [Eq. (3)] (Løvaas and Carlin, 1991). These results indicate that further study of the role of polyamines in oxidant stress are clearly warranted.

The above phenotypes of *sod1*⁻ strains raise the possibility that mutants previously identified as auxotrophs may actually be mutant in one of the antioxidant enzymes. Thus, one of the *MET* loci could be at *SOD1* because most mutant screening is carried out under air on minimal, selective media. This particular possibility is under investigation (Y. Surdin-Kerjan, personal communication).

2. Deficiencies in Growth and Carbon Utilization

Doubling times have been reported for *sod1* (Cu,ZnSOD) mutants grown on glucose (2%) or glycerol (3%) (Gralla and Valentine, 1991). In glucose, *sod1*⁻ cells grew more slowly, and the effect was especially pronounced in 10% glucose. In glycerol, the growth rates were more nearly equal, although the *sod1* mutants were slightly slower than wild type. An explanation for these results can be found in the cellular compartmentalization of the SODs, and the glucose repression of the MnSOD. That is, in glucose, especially 10% glucose, the MnSOD is strongly repressed; the Cu,ZnSOD accounts for essentially all the SOD activity in the cell. When functional *SOD1* is lost by mutation, SOD activity in glucose-repressed cells is eliminated and the cells suffer accordingly. On the other hand, when the cells are growing on the nonfermentable carbon source glycerol, MnSOD is induced and may be able to compensate partially for the loss of the cytoplasmic Cu,ZnSOD (Gralla and Valentine, 1991). One mechanism of this compensation would be the scavenging of $O_2^{.-}$ generated in the mitochondria such that this oxy radical could not leak out into the Cu,ZnSOD-deficient cytoplasm.

However, other results indicate that whether growth of *sod*⁻ strains occurs on nonfermentable carbon sources must depend on other factors as well. A critical factor appears to be the state of aeration. Others include the strain being tested, the status of other nutrients, whether the carbon source is capable of scavenging radicals, and how and where the carbon source is metabolized. Published growth rates of the *sod1*⁻ strains were carried out on strains that had been backcrossed twice, and therefore grew up from spores twice in air (Gralla and Valentine, 1991). This potentially selective treatment may have allowed the recovery of some compensatory mechanism(s) that allowed better growth on glycerol. Recent results with strains that were not backcrossed and have been protected from highly aerobic conditions indicate that growth on many nonfermentable carbon sources can be significantly impaired (although to different degrees) in *sod1*⁻, *sod2*⁻, and double mutant strains (E. Gralla, D. Guidot, and J. Valentine, unpublished results) (Table 4). These experiments indicate that some carbon sources support SOD⁻ cell growth less well than others. Some of these differences may be due to the superoxide-scavenging ability of the carbon source; for example, ethanol has no such ability, and is relatively poor as a carbon source.

The *sod2*⁻ strains exhibit no aerobic "nutritional" auxotrophies as above, exhibit normal growth on glucose (rich medium) at 80% O_2 (Westerbeek-Marres *et al.*, 1988), and even some growth at 100% O_2 on such media (van Loon *et al.*, 1986). However, *sod2*-carrying strains differ from wild type in their ability to utilize two-carbon units (ethanol, acetate) under air, or two- or three-carbon units under 80% O_2. Thus, at pH 5.5, these mutants grow weakly on ethanol and fail to grow on acetate under air, but exhibit close to wild-type growth when supplied with three-carbon units (glycerol, pyruvate, lactate) (Westerbeek-Marres *et al.*, 1988; van Loon *et al.*, 1986). At pH 6.5, growth on acetate by the *sod2*⁻ strain is at wild-type levels, suggesting that the lack of MnSOD activity limits the strain's ability to pump out the protons brought in by acetic acid (J. Lee and D. Kosman, unpublished results).[2]

Under 80% O_2, *sod2*⁻ strains will grow on 10% glucose (Westerbeek-Marres *et al.*, 1988), which suppresses respiration (cf. Section IV). Despite the high O_2 tension, the rate of electron flow through the respiratory chain is low and the MnSOD-deficient mitochondria are not exposed to an elevated O_2^- flux (Boveris, 1978). In effect, the presence of 10% glucose, like the addition of Met, protects the cell from an oxidative stress by repressing it. Put another way, in 10% glucose, there

---

[2] "Acetate" diffuses into yeast cells as the conjugate acid; at pH values near the pK_a of acetic acid, the velocity of diffusion overwhelms the ability of the cell to pump out protons. This condition appears to obtain at higher pH values (lower concentrations of the conjugate acid, slower rates of acidification) for the mutant than for the wild type.

is little MnSOD expressed in any case, and, not surprisingly, its deletion has no phenotype.

3. Peroxisome-Dependent Metabolism (Microbodies)

Why three- but not two-carbon units can serve as carbon sources for *sod2⁻* strains under all conditions is not clear. A fundamental difference between the two classes is that the latter require extramitochondrial conversion to a tricarboxylic acid (TCA) cycle intermediate, e.g., acetate to malate via the glyoxylate cycle, and then transport of the intermediate into the mitochondria via the malate–aspartate shuttle prior to oxidation. Possibly, one of the enzymes unique to this cycle (isocitrate lyase, malate synthase) and/or the transporter are sensitive to an O_2^--dependent species. The glyoxylate cycle in at least some yeasts and fungi is localized to specialized microbodies called glyoxysomes, which are induced by growth on acetate (Veenhuis and Harder, 1991). However, in *S. cerevisiae* growing on acetate, these microbodies appear absent (Veenhuis *et al.*, 1987). This result suggests that the glyoxylate cycle enzymes are in the cytoplasm in this yeast, although they are induced by acetate nonetheless (Duntze *et al.*, 1969).

Glyoxysomes from plants have been shown to generate O_2^- (see below; Sandalio *et al.*, 1988). Given the potential for leakage of univalent O_2 reduction intermediates from these microbodies, the additional O_2^- burden might exceed the dismutation capacity of a *sod⁻* strain. The fact that a *sod2⁻* strain is unable to grow on acetate suggests that MnSOD may be associated with the glyoxylate cycle whether or not it is confined to microbodies. In fact, MnSOD is thought to be associated with the membrane of some plant peroxisomes/glyoxysomes (Sandalio and Del Río, 1987; see below). Clearly, understanding the lack of acetate-based growth in *sod2⁻* strains at lower pH values or higher O_2 tensions should offer real insight with respect to the anaplerotic and energy metabolism in these strains.

Yeasts are exceptionally useful for the study of microbody biogenesis and metabolism. Peroxisomes/glyoxysomes are induced in essentially all yeasts by one or more nutritional conditions. Some yeasts (e.g., *H. polymorpha, Candida bodinii, Candida utilis,* and *Aspergillis niger*) metabolize methanol (for energy) and alkyl and aryl amines (for nitrogen) via peroxisomal oxidases that produce H_2O_2 as end product (Veenhuis and Harder, 1991). Although *S. cerevisiae* cannot do this, it does catabolize oleic acid as a carbon source via peroxisomal β oxidation; oleic acid induces proliferation of peroxisomes (microbodies) in this yeast (Veenhuis *et al.*, 1987). Not surprisingly, peroxisomes are rich in catalase, and in *S. cerevisiae* catalase A (*CTA1*) is induced by oleic acid (Veenhuis *et al.*, 1987; Table 3). Leakage of O_2^- from H_2O_2-generating enzymes is possible; as noted above, O_2^- is produced by glyoxysomes (found in oil seeds), which also contain β-oxidation enzymes (Sandalio *et al.*, 1988). Therefore, one might expect microbodies to contain significant levels of SOD activity as well. Glyoxysomes from watermelon cotyledons may contain Cu,ZnSOD (Sandalio and Del Río, 1987). Also, a recent report showed that human *SOD1* expressed in *S. cerevisiae* accumulated SOD in peroxisomes in that organism (Keller *et al.*, 1991). However, in general, there is little information about the relationship between regulation of microbody biogenesis and either *SOD1* or *SOD2* expression in yeasts and fungi.

This relationship is ambiguous because growth on oleic acid involves glucose derepression, which, irrespective of the addition and effects of oleic acid, results in strong stimulation of transcription from a variety of genes, including *SOD2* (Westerbeek-Marres *et al.*, 1988), *CTA1*, and *CTT1* (Hörtner *et al.*, 1982). This behavior is suggested by the comparisons illustrated in Table 3 (see above), which show that oleic acid-induced transcription from *SOD1* and *SOD2* increases weakly in comparison to *CTA1* and *CTT1* transcription, only about twofold. Whether this increase is functionally significant is not obvious. Data on *S. cerevisiae* (Chang and Kosman, 1989) suggest that yeast can contain a range of SOD1 activity (10–100 units/mg soluble protein) and still appear to be wild type. As discussed below, data on other organisms suggest that a balance between O_2^- and H_2O_2 activities is necessary for cellular homeostasis. Also, whether the *endogenous* SOD1 is localized in peroxisomes in yeast has not been determined. Clearly, in the absence of oleic acid (i.e., in the absence of peroxisomes), there is a substantial amount of SOD1 in the cell; presumably it is cytosolic. In any event, *SOD1* and *SOD2* expression and peroxisome biogenesis are events that deserve close scrutiny and their study should be encouraged.

C. Genetic Consequences

1. Mutation Rates

Mutants deficient in one or more of the normal complement of superoxide dismutation activities in an organism appear to share one characteristic in common: increased rate of mutation and/or sensitivity to ionizing radiation. For example, genetic selection in *Drosophila* resulted in an isolate that contained only 3.5% of the immunoreactive SOD found in wild-type homozygotes (Peng *et al.*, 1986). The corresponding allele was designated N (null). Wild type was more resistant to ionizing radiation than were N/N homozygotes; LD_{50} values (in kR) were 5.31 and 3.16, respectively. Similar kinds of data have been obtained for *E. coli* (Scott *et al.*, 1989).

In addition to cell killing due (in part) to DNA damage, however, are the more subtle effects of mutagenesis. These may be of minor importance in acute oxidant stress, but are critical to the cell line because the results of mutagenesis persist as heritable traits. In *E. coli* the spontaneous rate of $Rif^S \rightarrow Rif^R$ was compared in *sodB* (FeSOD, noninducible), *sodA* (MnSOD, inducible), and *sodA sodB* mutants (Farr *et al.*, 1986). The relative mutation rates (with wild type as 1.0) were 0.9, 8.7, and 41.0, respectively. Enhanced mutation rates in the *sod* mutants were not seen anaerobically, indicating that the mechanism(s) involved an O_2-dependent process, and were not seen in mutant strains which lacked a functional *xth* gene. The *xth* gene encodes exonuclease III, which has several activities involved in DNA repair. This result suggests that the increased flux of O_2^- in SOD$^-$ strains led to premutagenic DNA damage that was "repaired" by ExoIII, leading to persistent genetic lesions (Farr *et al.*, 1986). In *Drosophila*, chromosomal effects can be inferred from the fact that SOD-null females exhibit elevated levels of germ line mutation (Phillips and Hilliker, 1990).

In *S. cerevisiae*, as a paradigmatic eukaryote, the relation between the compartmentalization of SOD activities and mutagensis of the nuclear and mitochondrial genomes is of additional interest. One might presume that a *sod1*$^-$ strain would exhibit a higher rate of nuclear mutation, because the nuclear membrane could be permeable to Cu,ZnSOD (M_r = 32 kDa). Cu,ZnSOD has been observed in the nucleus in rat hepatocytes (L.-Y. Chang *et al.*, 1988). In contrast, a *sod2*$^-$ strain might be more likely to become a mitochondrial (cytosolic) petite. There are some data to support both suppositions.

A strain carrying a deletion at the *SOD1* locus (*sod1 Δ1*) exhibited a 3.9-fold increase in spontaneous reversion rate of the *trp1-289* allele over wild type (13.2 $\times$ 10^{-8} versus 3.4 $\times$ 10^{-8} reversions/cell/generation) (Gralla and Valentine, 1991). This frequency was O_2 dependent because under N_2 the reversion rate decreased for both strains (to 2.1 $\times$ 10^{-8} and 1.8 $\times$ 10^{-8}, respectively); anaerobiosis essentially eliminated the differential between mutant and wild type.

Quantitative nuclear mutation rate studies have not been completed on a *sod2*$^-$ strain. Unpublished data (mutation to canavanine resistance) are said to show no difference in nuclear mutation frequency in wild type and a *sod2*$^-$ carrying strain (van Loon *et al.*, 1986). Nor have rates of spontaneous petites been determined in a *sod1*$^-$ strain. The *sod2*$^-$ strains, however, do exhibit a greater tendency toward cytoplasmic petite genotype than do wild type in an O_2-dependent manner (E. White and D. Kosman, unpublished observations). No studies have been performed in these mutants using ionizing radiation. In general, then, this area of research remains largely unexplored.

2. *Effects on Gene Expression*

There is no evidence that either *SOD1* or *SOD2* is an allele that regulates the expression of another gene directly. On the other hand, in the context of the hypothesis that the absence of one or both of the encoded enzymes results in an increased flux of O_2^{-}, one might expect that the expression of (some) other genes might be altered as a result. This response could represent a compensatory one: an increase in damage-repair activities, for example. Alternatively, the change could be deleterious in that it could reflect the loss of cellular capacity for renewal and/or adaptation. There is some evidence that *sod1*$^{-}$ strains have impaired ability to adapt at the level of gene expression to altered nutrient conditions under air. Because this failure occurs only under air and not under N_2, it is likely to involve the deleterious effects of O_2-derived free radicals. These results (J. Lee and D. Kosman, unpublished observations) are outlined below.

In *S. cerevisiae,* glucose derepression induces transcription from *SUC2*, the gene for invertase, and an increased synthesis of the secreted form of its enzyme (Carlson and Botstein, 1982). EG1, which carries a deletion in *SOD1* (Gralla and Valentine, 1991), synthesizes and secretes in 1 hour ~50% of the invertase activity of wild type when the strains are switched from glucose to sucrose medium under air. Under N_2, mutant and wild-type strains are not different in their response to nutrient switch whereas under O_2 the mutant fails to secrete any invertase at all (J. Lee and D. Kosman, unpublished results). These effects appear due to a failure to induce fully the 1.9-kb *SUC2* message for the secreted form of invertase; in the mutant under air this *SUC2* mRNA is one-sixth of what is induced under N_2 (which is comparable to wild type under both conditions; Table 5). Similar results have been obtained for transcription from *PHO5* (acid phosphatase) on removal of phosphate from the medium, i.e., under air, a *sod1*$^{-}$ strain synthesizes a fraction of this enzyme in comparison to that synthesized anaerobically or by wild type under any oxygen tension (Lee and Kosman, 1992).

The molecular bases of these transcriptional deficiencies are not known. However, analysis of the transcriptional activity of genes that encode the other antioxidant enzymes shows that the mRNA levels for these proteins, unlike invertase and phosphatase, do not differ in mu-

TABLE 5

Increases in Selected mRNA Levels Associated with Glucose Derepression and pO_2 in *Saccharomyces cerevisiae*

| | Derepression under[a] | | | | | |
| | N$_2$ | | Air | | O$_2$ | |
mRNA	*sod1Δ*	WT	*sod1Δ*	WT	*sod1Δ*	WT
SUC2	11	5	2	10	0.7	5
SOD1	—	2.0	—	2.4	—	3.4
SOD2	2.6	2.5	4.8	3.3	13.1	4.8
CTT1	5.6	2.3	11.9	6.9	5.7	5.8

[a] Values are fold increases in specific mRNA for 1 hour of derepression in each strain when transferred from N$_2$-flushed, 10% glucose to glucose-depleted media flushed with N$_2$, air, or O$_2$ as indicated; WT, wild type. mRNA levels are normalized to actin mRNA (*ACT1*) as an internal control for all samples. Data are obtained from quantitative Northern blot analysis on poly(A)$^+$-selected RNA (J. Lee and D. Kosman, unpublished data).

tant and wild type when the strains are derepressed under N$_2$, air, or O$_2$; indeed, some of these mRNAs may actually be higher in the *sod1$^-$* strain, EG1 (Table 5). These results indicate that decreased transcription from some loci can occur in a *sod1$^-$* strain under air. Possibly, when nutrient adaptation is associated with this transcription, a *sod1$^-$* strain may be compromised in its ability to compete with wild type in the new culture environment. On the other hand, induction of the remaining antioxidant enzymes under the same conditions is not impaired. This result suggests that the transcriptional deficiency observed is somewhat selective. Further study of this phenotype should reveal additional adapative limitations in SOD$^-$ strains.

D. SENESCENCE

Free radicals have long been postulated as agents of cell aging and senescence (Sohal and Allen, 1986). In the model, accumulated defects in proteins, membrane lipids, and/or DNA due to the action of O$_2$-derived free radicals lead to a failure of the cell to thrive or, in the extreme, to an "error catastrophe" in which loss of function leads to death (Orgel, 1963). Evidence for this model could come from analysis of longevity profiles of organisms that differed with respect to the level of one or more of the components of the putative cellular antioxidant defense. Data of this type have been compiled for higher eukaryotes

(Tolmasoff *et al.*, 1980). However, the significance of this latter correlation suffers from the lack of appropriate controls, which often characterizes population studies. Mutants in lower eukaryotes, including yeasts and fungi, provide their own, parental controls and should therefore be excellent model systems for the elucidation of the mechanisms of cell aging in general, and of the role of oxy radicals in this pathology in particular. Thus, null mutants in Cu,ZnSOD in *Drosophila* exhibit a short-lived phenotype, with a mean adult life span of 11 days versus 60 days for parental wild type and for SOD^-/SOD^+ heterozygotes (Phillips *et al.*, 1989). Catalase mutants in *Drosophila* (Makay and Bewley, 1989) and *Neurospora* (Munkres, 1990b) also exhibit significant reduction in life span, indicating the essentiality of both types of dismutation activity (O_2^- and H_2O_2) for organismal advantage under air; that is, deletion of either activity had no apparent effect on organismal viability per se.

There is more information about the phenotype of old or senescent cells than about the mechanisms of aging at this time. For example, in senescence, the filamentous fungus *Podospora anserina* exibits significant loss of mitochondrial DNA and an accumulation of novel plasmid DNA (Cummings *et al.*, 1979). This DNA, when microinjected into juvenile mycelia, causes premature senescence (Cummings *et al.*, 1990). The molecular basis for this effect is not known but is suggested to be related to a general instability of mtDNA in aging *P. anserina,* which correlates to an increase in the amount of DNA sequence encoding a reverse transcriptase. The notion that cell senescence correlates to decreased mitochondrial function, which in turn is due to accumulated defects caused by oxy radicals, is well established (Miguel and Fleming, 1986) although unproved. Lacking information about superoxide dismutases in *P. anserina* or significant genetics in this organism, it does not appear to be a first choice as model system.

Somewhat more detail about the relation between aging and cellular antioxidant enzymes, including the superoxide dismutases, is available in *N. crassa* (Munkres, 1985, 1990b). Work by Munkres has utilized conidial variants that differed in heritable mean life span (abbreviated "mls"). The genes associated with life span have been designated *age*. The *age^-*, *age^0*, and *age^+* variants have mls of 7, 22, and 62 days, respectively, with *age^0* being wild type (Munkres and Furtek, 1984). The *age^-*-associated alleles mapped to 16 genes in linkage group IR and have been collectively designated the *age-1* locus (Munkres, 1985). The fact that the *age^-* phenotype was light dependent and was suppressed in the presence of antioxidants such as tocopherol suggested that this locus included the genes for one or some of the antioxidant components or for elements that acted in trans with respect to expression of these antioxidant enzyme genes (Munkres and Furtek, 1984). The fact that a metabolic signal transducer such as cGMP, when added to the medium, can suppress the age^--encoded phenotype lends some support for the latter possibility (Munkres and Rana, 1984), as does the fact that cellular cGMP phosphodiesterase activity is $\approx 10^4$ times greater in age^- strains in comparison to wild type (K. Munkres, unpublished observations). The correlations between the specific cellular activities of essentially all of the putative antioxidant enzymes in *Neurospora* and life span is strong, with correlation coefficients (for $n = 14$ strains) of 0.97 to 0.99 (Munkres, 1990b). These enzymes include the SODs (cytosolic and extracellular Cu,Zn and Mn forms), catalase, glutathione peroxidase, and cytochrome *c* peroxidase. That all of these enzyme activities correlate to life span also argues for the variant alleles to be at loci that act in trans. The suggestion that in yeast the expression of SODs and other antioxidant enzymes is dependent in some way on signal transduction has been discussed (Section IV) and may be relevant to senescence in *S. cerevisiae* also (*vide infra*).

Despite the substantial classical genetics that have been presented concerning SOD activity and aging in *Neurospora,* little mechanistic information is available concerning the correlations described. In addition, the less developed state of molecular genetics in this fungus (in comparison to *S. cerevisiae*) has slowed the cloning of the relevant genes, construction of deletion mutants, and gratuitous expression of specific genes or chimeras, all of which would facilitate the design of systematic experiments to address mechanistic hypotheses. For these reasons, the etiology of eukaryotic (microbial) gerontology and its relation to cellular O_2^- (and H_2O_2) dismutation can be best elucidated in *S. cerevisiae,* as discussed below.

There are compelling reasons for considering *S. cerevisiae* an excellent experimental system for the molecular analysis of the aging process in general (Jazwinski, 1990a), and of the specific role of SODs in determining cellular longevity. Yeast is a true aging species in that cell mortality increases exponentially with age (Jazwinski *et al.,* 1989). Mother and daughter cells are easily distinguished. The mls for a typical wild-type strain (X2180-1A, available from the Yeast Genetics Stock Center, Berkeley, California) is 24 ± 9 generations; 50% survival also occurs at 24 generations (Egilmez and Jazwinski, 1989).

What is so significant about the aging studies in yeast to date are the possible molecular mechanisms of aging that this work has ruled out. The level of chitin (bud scarring) does not correlate with life span, as has been suggested (Mortimer and Johnston, 1959). More generally, neither loss of mitochondrial function nor accumulated damage to chromosomal DNA or other cellular components (membranes, proteins) can account for the aging phenotype in yeast. Daughter cells exhibited generation times of young cells irrespective of the age of their mother, although daughters of old mothers required two to three generations to achieve this rate of growth (Egilmez and Jazwinski, 1989). Because the daughter cell is a copy of the mother at cytokinesis (both mitochondria and genotype are at equilibrium between the two cells), the nascent senescence of the mother cannot be due to DNA or mitochondrial damage; the daughter would be equally at risk. Instead, it appears that senescence may be associated with a cytosolic factor, accumulating in an aging mother, which is transferred to the daughter (thus initially slowing growth). This factor persists in the mother cell but not in the daughter, due either to differences in generation or degradation of the factor in the two cells (Egilmez and Jazwinski, 1989). In this model, this putative factor represents a determinant of cell death; the level of this factor is the probabilistic trigger of the senescence phenotype. Longevity can be related to the threshold level at which the effector triggers the predetermined response (Jazwinski, 1990b). Because there are clear changes in the pattern of gene expression due to aging per se (Egilmez *et al.*, 1989), it is likely that small differences in the concentrations of crucial gene products could have a synergistic effect on where this threshold is set.

These age-dependent genes in *S. cerevisiae* have not been identified. They could represent one or some of the antioxidant enzymes. There may be a relationship between the expression of these genes and cAMP in that longevity in yeast can be modulated by the expression of a *RAS* gene product (Chen *et al.*, 1990). In this context, the hypothesis that cGMP is a second messenger in antioxidant enzyme gene expression in *S. cerevisiae* (Munkres, 1990d) and *N. Crassa* (Munkres, 1990c) is intriguing, although most other data to date have implicated cAMP, not cGMP, in this regulation in *S. cerevisiae* (Section IV). These possibilities and the potential mechanisms of cell aging that they suggest merit thorough study; the strains and reagents are essentially all available to do so.

E. Suppression of *sod*⁻ Strains

1. Suppression by Exogenous Mn(II) and Cu(II)

Interestingly, certain metal ions are able to "rescue" *sod1*⁻ cells (Table 4). Thus 2 mM manganese in the culture medium improves growth of *sod1*⁻ strains in air and allows growth in 100% oxygen in rich medium. This protection is due to the radical-scavenging abilities of the Mn^{2+} that accumulates in these cells (Chang and Kosman, 1989). Copper in the medium at lower concentrations (100 to 500 μM) is also effective in the rescue of mutants in either SOD gene, and allows growth on nonfermentable carbon sources in highly aerated cultures (K. Tamai, E. Gralla, D. Thiele, and J. Valentine, (in preparation). The basis for this effect is not understood at present, although it is under investigation. What is known is that a functional ACE1/CUP1 copper protection system is necessary for the effect. This suggests one model for the Cu-dependent protection, namely, that the CUP1 protein is serving as an antioxidant. Two other models are that either a high level of copper is required for some other protective process, or that the copper is the antioxidant; in either case, however, the required level of copper is toxic to strains that lack the ability to induce copper thionein.

2. Genetic Suppression—Pseudoreversion

SOD deletion mutants offer attractive systems in which to identify cellular sources of and targets for oxygen-derived free radicals. In such nonrevertable strains, recovery of the wild-type phenotype has to be due to physiologic suppression, i.e., decreased production of O_2^- or protection of potential targets. Such pseudorevertants can be selected for by plating under conditions that prohibit growth of the parental mutant. Indeed, the increased rate of mutagenesis observed in SOD$^-$ organisms might predict that the frequency of pseudoreversion, assuming there to be mechanisms of physiologic suppression, would be relatively high. This prediction has been borne out.

Pseudorevertants in sod^- backgrounds have been isolated in both *E. coli* (Imlay and Fridovich, 1981b, 1992) and *S. cerevisiae* (T. Bilinski and D. Kosman, unpublished results; Liu *et al.*, 1992). In yeast, these arise spontaneously with frequency of $\approx 10^{-5}$ when a $sod1^-$-carrying strain is plated on rich medium under 100% O_2 or SC without Met under air (D. Kosman, unpublished). Cells plated under N_2 and subsequently screened for pseudorevertants "revert" at less than 1/100th the frequency seen in cultures under air. (Stocks of $sod1^-$ strains should be grown in low-oxygen atmospheres to avoid selection of such revertants.) The $sod2^-$ strains exhibit little pseudoreversion (E. White and D. Kosman, unpublished observations).

None of the pseudorevertants in $sod1^-$ studied so far has recovered a cytosolic dismutation activity, e.g., by diverting MnSOD from the mitochondria. In general, none of the antioxidant enzyme activities was markedly or systematically different in one set of these pseudorevertants which carried dominant mutations (n = 24; E. Chang, J.

Rhodes, and D. Kosman, unpublished). These pseudodrevertants, which were selected on the basis of the loss of the aerobic Met auxotrophy characteristic of $sod1^-$ strains, also reverted to aerobic prototrophy for Lys and normal growth on rich media under 100% O_2 (E. Chang and D. Kosman, unpublished). In a much larger group selected under 100% O_2 (n = 490) (T. Bilinski, unpublished), 173 were dominant and 317 were recessive. Of the latter, 314 alleles belonged to the same complementation group. Interestingly, in these pseudorevertants, only the sensitivity to hyperoxia was suppressed; the strains still exhibited a Lys/Met auxotrophy under air. Liu et al. (1992) isolated two bsd (bypassed the SOD defect) suppressor alleles, bsd1-1 and bsd2-1, by selection on synthetic media lacking lysine. Both alleles were recessive and conferred Met, as well as Lys, prototrophy and overall resistance towards O_2. The revertants also were deficient in the utilization of many nonfermentable carbon sources. This suggests limited electron transport and a consequent lowering of O_2^- production as a possible mechanism of the physiologic suppression of the $sod1^-$ phenotypes.

One type of physiologic suppression is indicated by somewhat more detailed studies on sodA sodB double deletion mutants in E. coli (Imlay and Fridovich, 1991b, 1992). The frequency of psuedoreversion in E. coli was comparable to that in yeast, $\sim 10^{-6}$. Two complementation groups were identified; one that was studied in detail was designated ssa-1 (suppressor of superoxide-imposed auxotrophy). This allele, which was dominant, allowed the sod^- strain to grow on minimal medium under air. The mechanism of pseudoreversion by ssa-1 is suggested by the observation that partial suppression of the amino acid auxotrophy exhibited by sod^- E. coli can be achieved by addition of stabilizing osmolytes to the growth medium (Imlay and Fridovich, 1992). The model proposed to explain these several data is that O_2^- dependent damage to both amino acid biosynthetic enzymes and the plasma membrane results in a decreased synthesis and increased leakage of essential amino acids from the cell. Reducing the leakage by osmotic stabilization afforded by medium supplementation or by a mechanism associated with ssa-1 conserves the sharply limited pool of branched-chain amino acids in the SOD^- mutant (Imlay and Fridovich, 1992).

One easily could envision mechanisms of suppression of the "suicide" committed by $sod1^-$ yeast strains as a result of their reductive assimilation of SO_4^{2-} to the level of SO_3^{2-}, toward which they exhibit an extreme sensitivity (see above). For example, decreasing the activity of phosphoadenosine phosphosulfate reductase (PAPS reductase, which produces SO_3^{2-}) or increasing the activity of sulfite reductase, which consumes SO_3^{2-} (Fig. 3) would serve to reduce the steady-state level of sulfite. The latter of these two hypothetical mechanisms—representing a gain of function—would appear consistent with the genetic dominance of the putative *ssa* alleles in *S. cerevisiae,* which restore Met prototrophy. It does not necessarily explain why oxygen sensitivity is eliminated in *ssa-1* mutants, however. Biochemical characterization of these pseudorevertants in yeast promises to provide valuable new insight.

F. ANTIOXIDANT ENZYME COMPLEMENTATION IN *sod⁻* STRAINS

Episomal Cu,ZnSOD does not complement a *sod2⁻* allele, nor does episomal SOD2 complement an *sod1⁻* one, unless it is devoid of a mitochondrial targeting sequence (Bowler *et al.,* 1990). This result really just confirms the strict compartmentalization of these two proteins in the cell and provides strong evidence for the notion that all cellular compartments require some level of localized antioxidant defense. In general, single *sod* mutants in yeast do not exhibit elevated levels of the remaining SOD, at least when the cells are in log phase under air. In contrast, *sod1*-deleted *N. crassa* does exhibit higher MnSOD activity (D. Natvig, unpublished results). On the other hand, when put under severe oxidative stress (100% O_2), *sod1⁻* yeast strains exhibit three- to fivefold higher levels of transcription from *SOD2, CCT1,* and *CTA1* than do the wild type (Table 5). Because the cells die anyway, however, these differences may reflect an exaggerated (oxidative) stress response rather than representing a compensatory mechanism.

Heterologous *SOD* genes are efficiently expressed in *S. cerevisiae* and can complement a missing dismutation activity as long as the heterologous protein is targeted to the correct cellular compartment. Thus, the MnSOD gene from *N. plumbaginifolia* (a plant) was expressed, imported into the mitochondria, and processed correctly in *S. cerevisiae,* and effectively complemented a *sod2⁻* strain (Bowler *et al.,* 1989). Similarly, the *Sod3* gene from maize (MnSOD), complemented a *sod2⁻* yeast strain and this required the mitochondrial targeting sequence of the *Sod3* gene (Zhu and Scandalios, 1992). Two conclusions of significance are suggested: (1) the MnSOD preprotein from plants is recognized, targeted, and processed by *S. cerevisiae* and (2) the MnSOD thus imported into the mitochondrial matrix is functional.

The type of complementing SOD does not appear to matter as long as it goes where it is needed. The MnSOD gene from *B. stearothermophilus* can complement yeast *sod1⁻* strains because it lacks a mitochondrial targeting sequence and is expressed and is active in the cytoplasm (Bowler *et al.,* 1990). It has not been possible to express the yeast

MnSOD in the yeast cytosol, even with the leader sequence deleted; the reasons for this are not clear. The converse of this experiment—complementing MnSOD deficiency with a Cu,ZnSOD fusion protein that contains an N-terminal mitochondrial targeting sequence—has not been reported.

G. Antioxidant Enzymes: Physiologic Interrelationships

The comparative regulation of the antioxidant enzymes in *S. cerevisiae* outlined in Section IV suggests that the levels of these activities, although they can complement one another in specific cases, are maintained at some functionally significant ratios. That is, an excess of SOD activity could conceivably be deleterious. For example, it could result in an increase in the steady-state flux of H_2O_2, which, if not removed by a comparable increase in catalase and/or peroxidase activities, would provide a ready source of OH· via the Fenton reaction [Eq. (3)]. This suggestion has not been tested directly in a yeast or fungus as yet, but has been in both *E. coli* and mammalian systems. The results indicate that an optimum ratio of $O_2^{.-}$ to H_2O_2 dismutation activity is necessary for normal cell and/or organismic function.

The first example of this was provided by studies on Down's syndrome, trisomy 21. The human *SOD1* gene is located in the band 21q22.1, the region associated with the pathophysiology of this inherited disease (Summitt, 1981). Cells from homozygotes contain 1.5 times the normal Cu,ZnSOD activity (Sinet *et al.*, 1975). The lesions in the neuromuscular junctions seen in Down's syndrome have subsequently been reproduced in transgenic mice that were expressing the human *SOD1* (Avraham *et al.*, 1988). Although the transgenics exhibited a spectrum of increases in SOD1 activity (2.3- to 10-fold in brain), even the animals with the smallest increases exhibited statistically significant lesions. A complementary experiment, in which membrane lipid peroxidation and paraquat toxicity were quantitated, was performed in mouse and human cell lines transfected with the human *SOD1*. The 3-fold increases in SOD activity in these cells resulted in 1.5- to 2-fold increases in lipid peroxidation and a marked increase in sensitivity to paraquat (Elroy-Stein *et al.*, 1988).

The hypothesis that these paradoxical cellular manifestations of oxidant stress resulting from the overexpression of a putative antioxidant enzyme are due to the resulting overproduction of H_2O_2 can be tested by cotransfection with both *SOD* and *CAT* (catalase) clones. This has been done (Amstad *et al.*, 1991); the results are consistent with the model. That is, though overexpression of human *SOD1* in mouse epidermal cells caused increases in DNA strand breaks, growth retardation, and enhanced cell killing by exogenous O_2^-, SOD/CAT overproducers were much like the parental cell line in phenotype.

Overproduction of SOD in *E. coli* also potentiates the cytotoxicity of O_2-dependent cellular pathophysiology. A strain that carries the *sodB* gene on a high-copy plasmid had >10 times the parental SOD activity (Scott *et al.*, 1987). Growth was 99% inhibited by 0.9 mM paraquat in comparison to wild type, and cell killing by 100% O_2 was doubled in the transformant. H_2O_2 was doubled in the transformant, also, suggesting that these phenotypes resulted from an uncompensated increase in cellular H_2O_2. These phenotypes were independent of the type of SOD being overexpressed; similar results were obtained with plasmids carrying either *sodA* (MnSOD) or *sodB* (FeSOD) (Siwecki and Brown, 1990). As an alternative to the hypothesis that SOD overproduction in *E. coli* results in a phenotype due to H_2O_2 accumulation, Liochev and Fridovich suggest that such overproduction competes for the elements necessary to support the synthesis of the soxR components (Liochev and Fridovich, 1992). The *soxR* regulon codes for a collection of proteins apparently needed for damage-repair activities associated with an oxidative insult. In this model, failure to initiate soxR protein synthesis in a timely fashion leads to the growth retardation characteristic of strains which are overproducing a SOD. Because O_2^- is readily formed from O_2 by ionizing radiation, overproduction of SOD might exacerbate the effects of γ radiation, as well. In fact, even wild-type levels of SOD were positively correlated to the extent of cell killing as a function of radiation dose (Scott *et al.*, 1989). That is, MnSOD- and FeSOD-deficient strains were more resistant than wild type, which was itself more resistant than an FeSOD overproducer. Although the precise explanation of these results is not available, they do indicate that the apparent selective advantage to an aerotolerant organism of an enzymatic antioxidant defense requires a tight coupling of the cellular elements that control the expression of these activities. The molecular details of this coordination must be viewed as an important focus of future research in yeasts and fungi.

ACKNOWLEDGMENTS

We would like to express our appreciation to the following people for sharing with us unpublished data that have been included in this review: Tomasz Bilinski, Chris Bowler, Irwin Fridovich, Kenneth Munkres, Donald Natvig, Jennifer Pinkham, John Scandalios, Yolande Surdan-Kerjan, and Daniele Touati. Several others also supplied us with reprints, which were most useful; our thanks to them as well, particularly Luis Del Río. We hope that we have presented their work faithfully. Work in our labs on the superoxide dismutases and oxidant stress in yeast has been supported by the NIH (EBG, GM28222) and the American Heart Association (DJK), support that is gratefully acknowledged. The critical reading of sections of this review by Joan Selverstone Valentine and her excellent suggestions are much appreciated.

References

Allen, A. O., and Bielski, B. H. (1982). Formation and disappearance of superoxide radicals in aqueous solutions. *In* "Superoxide Dismutases" (L. W. Oberley, ed.), Vol. 1, pp. 125–141. CRC Press, Boca Raton, FL.

Amstad, P., Peskin, A., Shah, G., Mirault, M.-E., Moret, R., Zbinden, I., and Cerutti, P. (1991). The balance between Cu,Zn-superoxide dismutase and catalase affects sensitivity of mouse epidermal cells to oxidative stress. *Biochemistry* **30**, 9305–9313.

Anbar, M., and Neta, P. (1967). A compilation of specific bimolecular rate constants for the reactions of hydrated electrons, hydrogen atoms and hydroxyl radicals with inorganic and organic compounds in aqueous solution. *Int. J. Appl. Radiat. Isot.* **18**, 493–523.

Asada, K., Kanematsu, S., Okaka, S., and Hayakawa, T. (1980). Phylogenic distribution of three types of superoxide dismutase in organisms and in cell organelles. *In* "Chemical and Biochemical Aspects of Superoxide and Superoxide Dismutase" (J. V. Bannister and H. A. O. Hill, eds.), pp. 136–153. Elsevier, New York.

Autor, A. P. (1982). Biosynthesis of mitochondrial manganese superoxide dismutase in *Saccharomyces cerevisiae*. Precursor form of mitochondrial superoxide dismutase made in the cytoplasm. *J. Biol. Chem.* **257**, 2713–2718.

Avraham, K. B., Schickler, M., Sapoznikov, D., Yarom, R., and Groner, Y. (1988). Down's syndrome: Abnormal neuromuscular junction in tongue of transgenic mice with elevated levels of human Cu/Zn-superoxide dismutase. *Cell (Cambridge, Mass.)* **54**, 823–829.

Balasundaram, D., Tabor, C. W., and Tabor, H. (1991). Spermidine or spermine is essential for the aerobic growth of *Saccharomyces cerevisiae*. *Proc. Natl. Acad. Sci. U.S.A.* **88**, 5872–5876.

Bannister, J. V., Bannister, W. H., and Rotilio, G. (1987). Aspects of the structure, function, and applications of superoxide dismutase. *CRC Crit. Rev. Biochem.* **22**, 111–180.

Barra, D., Bossa, F., Marmocchi, F., Martini, F., Rigo, A., and Rotilio, G. (1979). Differential effects of urea on yeast and bovine copper, zinc superoxide dismutases, in relation to the extent of analogy of primary structure. *Biochem. Biophys. Res. Commun.* **86**, 1199–1205.

Belazzi, T., Wagner, A., Wieser, R., Schanz, M., Adam, G., Hartig, A., and Ruis, H. (1991). Negative regulation of transcription of the *Saccharomyces cerevisiae* catalase T (*CTT1*) gene by cAMP is mediated by a positive control element. *EMBO J.* **10**, 585–592.

Berkner, L. V., and Marshall, L. C. (1965). On the origin and rise of oxygen concentration in the Earth's atmosphere. *J. Atmos. Sci.* **22**, 225–261.

Bermingham-McDonogh, O., Gralla, E. B., and Valentine, J. S. (1988). The copper,zinc-superoxide dismutase gene of *Saccharomyces cerevisiae*: Cloning, sequencing, and biological activity. *Proc. Natl. Acad. Sci. U.S.A.* **85**, 4789–4793.

Bertini, I., Banci, L., Luchinat, C., and Hallewell, R. A. (1988). The exploration of the active-site cavity of copper-zinc superoxide dismutase. *Ann. N.Y. Acad. Sci.* **542**, 37–52.

Beyer, W., Imlay, J., and Fridovich, I. (1991). Superoxide dismutases. *Prog. Nucleic Acid Res. Mol. Biol.* **40**, 221–253.

Bilinski, T. (1991). Oxygen toxicity and microbial evolution. *BioSystems* **24**, 305–312.

Bilinski, T., Krawiec, Z., Liczmanski, A., and Litwinska, J. (1985). Is hydroxyl radical generated by the Fenton reaction *in vivo*? *Biochem. Biophys. Res. Commun.* **130**, 533–539.

Bilinski, T., Litwinska, J., Blaszczynski, M., and Bajus, A. (1989). Superoxide dismutase deficiency and the toxicity of the products of autooxidation of polyunsaturated fatty acids in yeast. *Biochim. Biophys. Acta* **1001**, 102–106.

Bjerrum, M. J. (1987). Structural and spectroscopic comparison of manganese-containing superoxide dismutases. *Biochim. Biophys. Acta* **915**, 225–237.

Boorstein, W. W., and Craig, E. A. (1990). Regulation of a yeast HSP70 gene by a cAMP responsive transcriptional control element. *EMBO J.* **9**, 2543–2553.

Borders, C. L., Jr., Saunders, J. E., Blech, D. M., and Fridovich, I. (1985). Essentiality of the active-site arginine residue for the normal catalytic activity of Cu,Zn superoxide dismutase. *Biochem. J.* **230**, 771–776.

Borders, C. L., Jr., Horton, P. J., and Beyer, W. F., Jr. (1989). Chemical modification of iron- and manganese-containing superoxide dismutases from *Escherichia coli. Arch. Biochem. Biophys.* **268**, 74–80.

Boveris, A. (1978). Production of superoxide anion and hydrogen peroxide in yeast mitochondria. *In* "Biochemistry and Genetics of Yeasts" (M. Bacila, B. L. Horecker, and A. O. M. Stoppani, eds.), pp. 65–80. Academic Press, New York.

Boveris, A., and Cadenas, E. (1982). Production of superoxide radicals and hydrogen peroxide in mitochondria. *In* "Superoxide Dismutases" (L. W. Oberley, ed.), Vol. 2, pp. 15–30. CRC Press, Boca Raton, FL.

Bowler, C., Alliotte, T., Van den Bulcke, M., Bauw, G., Vandekerckhove, J., Van Montagu, M., and Inzé, D. (1989). A plant manganese superoxide dismutase is efficiently imported and correctly processed by yeast mitochondria. *Proc. Natl. Acad. Sci. U.S.A.* **86**, 3237–3241.

Bowler, C., van Kaer, L., van Camp, W., Van Montagu, M., Inzé, D., and Dhaese, P. (1990). Characterization of the *Bacillus stearothermophilus* manganese superoxide dismutase gene and its ability to complement copper/zinc superoxide dismutase deficiency in *Saccharomyces cerevisiae. J. Bacteriol.* **172**, 1539–1546.

Brown, O. R., and Yein, F. (1978). Dihydroxyacid dehydratase: The site of hyperbaric oxygen poisoning in branched-chain amino acid biosynthesis. *Biochem. Biophys. Res. Commun.* **85**, 1219–1224.

Brunori, M., and Rotilio, G. (1984). Biochemistry of oxygen radical species. *In* "Methods in Enzymology" (L. Packer, ed.), Vol. 105, pp. 22–35. Academic Press, Orlando, FL.

Buchman, C., Skroch, P., Dixon, W., Tullius, T., and Karin, M. (1990). A single amino acid change in CUP2 alters its mode of DNA binding. *Mol. Cell. Biol.* **10**, 4778–4787.

Butt, T. R., and Ecker, D. J. (1987). Yeast metallothionein and applications in biotechnology. *Microbiol. Rev.* **51**, 351–364.

Buxton, G. V., Greenstock, C. V., Helman, W. P., and Ross, A. B. (1988). Critical review of rate constants for reactions of hydrated electrons, hydrogen atoms and hydroxyl radicals ($\cdot$OH/$\cdot$O$^-$). *J. Phys. Chem. Ref. Data* **17**, 513–886.

Cadenas, E. (1989). Biochemistry of oxygen toxicity. *Annu. Rev. Biochem.* **58**, 79–110.

Cameron, S., Levin, L., Zoller, M., and Wigler, M. (1988). cAMP-independent control of sporulation, glycogen metabolism, and heat shock resistance in *S. cerevisiae. Cell (Cambridge, Mass.)* **53**, 555–566.

Carlson, M., and Botstein, D. (1982). Two differentially regulated mRNAs with different 5' ends encode secreted and intracellular forms of yeast invertase. *Cell (Cambridge, Mass.)* **28,** 145–154.

Carri, M. T., Galiazzo, F., Ciriolo, M. R., and Rotilio, G. (1991). Evidence for co-regulation of Cu,Zn superoxide dismutases and metallothionein gene expression in yeast through transcriptional control by copper via the ACE 1 factor. *FEBS Lett.* **278,** 263–266.

Cashel, M. (1975). Regulation of bacterial ppGpp and pppGpp. *Annu. Rev. Microbiol.* **29,** 301–318.

Chang, E. C., and Kosman, D. J. (1989). Intracellular Mn(II)-associated superoxide scavenging activity protects Cu,Zn superoxide dismutase-deficient *Saccharomyces cerevisiae* against dioxygen stress. *J. Biol. Chem.* **264,** 12172–12178.

Chang, E. C., and Kosman, D. J. (1990). O_2-dependent methionine auxotrophy in Cu, Zn superoxide dismutase-deficient mutants of *Saccharomyces cerevisiae. J. Bacteriol.* **172,** 1840–1845.

Chang, E. C., Crawford, B. F., Hong, Z., Bilinski, T., and Kosman, D. J. (1991). Genetic and biochemical characterization of Cu,Zn superoxide dismutase mutants in *Saccharomyces cerevisiae. J. Biol. Chem.* **266,** 4417–4424.

Chang, L.-Y., Slot, J. W., Geuze, H. J., and Crapo, J. D. (1988). Molecular immunocytochemistry of the Cu,Zn superoxide dismutase in rat hepatocytes. *J. Cell Biol.* **107,** 2169–2179.

Chary, P., Hallewell, R. A., and Natvig, D. (1990). Structure, exon pattern, and chromosome mapping of the gene for cytosolic copper-zinc superoxide dismutase (*sod-1*) from *Neurospora crassa. J. Biol. Chem.* **265,** 18961–18967.

Chen, J. B., Sun, J., and Jazwinski, S. M. (1990). Prolongation of the yeast life span by the v-Ha-*RAS* oncogene. *Mol. Microbiol.* **4,** 2081–2086.

Chevion, M. (1988). A site-specific mechanism for free radical induced biological damage: essential role of redox-active transition metals. *Free Radicals Biol. Med.* **5,** 27–37.

Cloud, P. (1983). The biosphere. *Sci. Am.* **249,** 176–189.

Cohen, G., Fessl, F., Traczyk, A., Rytka, J., and Ruis, H. (1985). Isolation of the catalase A gene of *Saccharomyces cerevisiae* by complementation of the *cta1* mutation. *Mol. Gen. Genet.* **200,** 74–79.

Cohen, G., Rapatz, W., and Ruis, H. (1988). Sequence of the *Saccharomyces cerevisiae CTA1* gene and amino acid sequence of catalase A derived from it. *Eur. J. Biochem.* **176,** 159–163.

Cummings, D. J., Belcour, L., and Grandchamp, C. (1979). Mitochondrial DNA from *Podospora anserina.* II. Properties of mutant DNA and multimeric circular DNA from senescent cultures. *Mol. Gen. Genet.* **171,** 2139–2146.

Cummings, D. J., Domenico, J. M., and Sanford, J. C. (1990). Mitochondrial DNA from *Podospora anserina:* Transformation to senescence via projectile injection of plasmids. *In* "Molecular Biology of Aging" (C. E. Finch and T. E. Johnston, eds.), pp. 91–101. Liss, New York.

de Vries, S., Albracht, S. P. A., Berden, J. A., and Slater, E. C. (1981). A new species of bound ubisemiquinone anion in QH_2:cytochrome c oxidoreductase. *J. Biol. Chem.* **256,** 11996–11998.

Ditlow, C., Johansen, J. T., Martin, B. M., and Svendsen, I. B. (1984). The complete amino acid sequence of manganese-superoxide dismutase from *Saccharomyces cerevisiae. Carlsberg Res. Commun.* **47,** 81–91.

Duntze, W., Neumann, D., Gancedo, J. M., Atzpodien, W., and Holzer, H. (1969). Studies on the regulation and localization of the glyoxylate cycle enzymes in *Saccharomyces cerevisiae*. *Eur. J. Biochem.* **10**, 83–89.

Egilmez, N. K., and Jazwinski, S. M. (1989). Evidence for the involvement of a cytoplasmic factor in the aging of the yeast *Saccharomyces cerevisiae*. *J. Bacteriol.* **171**, 37–42.

Egilmez, N. K., Chen, J. B., and Jazwinski, S. M. (1989). Specific alterations in transcript prevalence during the yeast life span. *J. Biol. Chem.* **264**, 14312–14317.

Elroy-Stein, O., Bernstein, Y., and Groner, Y. (1988). Overproduction of human Cu/Zn-superoxide dismutase in transfected cells; extenuation of paraquat-mediated cytoxicity and enhancement of lipid peroxidation. *EMBO J.* **5**, 615–622.

Farr, S. B., and Kogoma, T. (1991). Oxidative stress responses in *Escherichia coli* and *Salmonella typhimurium*. *Microbiol. Rev.* **55**, 561–585.

Farr, S. B., D'Ari, R., and Touati, D. (1986). Oxygen-dependent mutagenesis in *Escherichia coli* lacking superoxide dismutase. *Proc. Natl. Acad. Sci. U.S.A.* **83**, 8268–8272.

Fee, J. A. (1982). Is superoxide important in oxygen poisoning? *Trends Biochem. Sci.* **7**, 84–86.

Fee, J. A. (1991). Regulation of *sod* genes in *Escherichia coli:* Relevance to superoxide dismutase function. *Molec. Microbiol.* **5**, 2599–2610.

Fenton, H. J. H. (1894). Oxidation of tartaric acid in the presence of iron. *J. Chem. Soc.* **65**, 899–903.

Forsberg, S. L., and Guarente, L. (1988). Mutational analysis of upstream activation sequence 2 of the *CYC1* gene of *Saccharomyces cerevisiae:* A HAP2–HAP3-responsive site. *Mol. Cell. Biol.* **8**, 647–654.

Forsberg, S. L., and Guarente, L. (1989). Identification and characterization of HAP4: A third component of the CCAAT-bound HAP2/HAP3 heteromer. *Genes Dev.* **3**, 1166–1178.

Fraenkel, D. (1982). Carbohydrate metabolism. *In* "The Molecular Biology of the Yeast *Saccharomyces:* Metabolism and Gene Expression" (J. N. Strathern, E. W. Jones, and J. R. Broach, eds.), pp. 1–37. Cold Spring Harbor Lab., Cold Spring Harbor, NY.

Fridovich, I. (1983). Superoxide radical: An endogeneous toxicant. *Annu. Rev. Pharmacol. Toxicol.* **23**, 239–257.

Fridovich, I. (1984). Overview: Biological sources of O_2^-. *In* "Methods in Enzymology" (L. Packer, ed.), Vol. 105, pp. 59–61. Academic Press, Orlando, FL.

Fridovich, I. (1986). Superoxide dismutases. *Adv. Enzymol. Relat. Areas Mol. Biol.* **58**, 61–97.

Fridovich, I. (1989). Superoxide dismutases. An adaptation to a paramagnetic gas. *J. Biol. Chem.* **264**, 7761–7764.

Furst, P., Hu, S., Hackett, R., and Hamer, D. (1988). Copper activates metallothionein gene transcription by altering the conformation of a specific DNA binding protein. *Cell (Cambridge, Mass.)* **55**, 705–717.

Gabor, B. P., Brown, R. D., Koenig, S. H., and Fee, J. A. (1972). Nuclear magnetic relaxation dispersion in protein solutions. V. Bovine erythrocyte superoxide dismutase. *Biochim. Biophys. Acta* **271**, 1–10.

Galiazzo, F., Schiesser, A., and Rotilio, G. (1988). Oxygen-independent induction of enzyme activities related to oxygen metabolism in yeast by copper. *Biochim. Biophys. Acta* **965**, 46–51.

Galiazzo, F., Ciriolo, M. R., Carri, M. T., Civitareale, P., Marcocci, L., Marmocchi, F., and Rotilio, G. (1991). Activation and induction by copper of Cu/Zn superoxide dismutase in *Saccharomyces cerevisiae*. Presence of an inactive proenzyme in anaerobic yeast. *Eur. J. Biochem.* **196**, 545–549.

Gancedo, C., and Serrano, R. (1989). Energy-yielding metabolism. *In* "The Yeasts" (A. H. Rose and J. S. Harrison, eds.), Vol. 3, pp. 205–259. Academic Press, San Diego.

Gardner, P. R., and Fridovich, I. (1991a). Superoxide sensitivity of the *Escherichia coli* 6-phosphogluconate dehydratase. *J. Biol. Chem.* **266,** 1478–1483.

Gardner, P. R., and Fridovich, I. (1991b). Superoxide sensitivity of the *Escherichia coli* aconitase. *J. Biol. Chem.* **266,** 19328–19333.

Gerschman, R., Gilbert, D. L., Nye, S. W., Dwyer, P., and Fenn, W. O. (1954). Oxygen poisoning and X-irradiation: A mechanism in common. *Science* **119,** 623–626.

Getzoff, E. D., Tainer, J. A., Weiner, P. K., Kollman, P. A., Richardson, J. S., and Richardson, D. C. (1983). Electrostatic recognition between superoxide and copper, zinc superoxide dismutase. *Nature (London)* **306,** 287–290.

Getzoff, E. D., Tainer, J. A., Stempien, M. M., Bell, G. I., and Hallewell, R. A. (1989). Evolution of copper zinc superoxide dismutase and the Greek key β-barrel structural motif. *Proteins* **5,** 322–336.

Gilbert, D. L., ed. (1981). "Oxygen and Living Processes." Springer-Verlag, Berlin and New York.

Goldberg, M. A., Dunning, S. P., and Bunn, H. F. (1988). Regulation of the erythropoietin gene: Evidence that the oxygen sensor is a heme protein. *Science* **242,** 1412–1415.

Goscin, S. A., and Fridovich, I. (1972). The purification and properties of superoxide dismutase from *Saccharomyces cerevisiae*. *Biochim. Biophys. Acta* **289,** 276–283.

Gralla, E. B., and Valentine, J. S. (1991). Null mutants of *Saccharomyces cerevisiae* Cu,Zn superoxide dismutase: Characterization and spontaneous mutation rates. *J. Bacteriol.* **173,** 5918–5920.

Gralla, E. B., Thiele, D. J., Silar, P., and Valentine, J. S. (1991). ACE1, a copper-dependent transcription factor, activates expression of the yeast copper, zinc superoxide dismutase gene. *Proc. Natl. Acad. Sci. U.S.A.* **88,** 8558–8562.

Greco, M. A., Hrab, D. I., Magner, W., and Kosman, D. J. (1990). Cu,Zn superoxide dismutase and copper deprivation and toxicity in *Saccharomyces cerevisiae*. *J. Bacteriol.* **172,** 317–325.

Gregory, E. M., Goscin, S. A., and Fridovich, I. (1974). Superoxide dismutase and oxygen toxicity in a eucaryote. *J. Bacteriol.* **117,** 456–460.

Guarente, L., and Mason, T. (1983). Heme regulates transcription of the *CYC1* gene of *Saccharomyces cerevisiae* via an upstream activation site. *Cell (Cambridge, Mass.)* **32,** 1279–1286.

Guarente, L., Lalonde, B., Gifford, P., and Alani, A. (1984). Distinctly regulated tandem upstream activation sites mediate catabolite repression of the *CYC1* gene of *S. cerevisiae*. *Cell (Cambridge, Mass.)* **36,** 503–511.

Haber, F., and Weiss, J. (1934). The catalytic decomposition of hydrogen peroxide by iron salts. *Proc. R. Soc. London, Ser. A* **147,** 332–351.

Haber, F., and Willstätter, R. (1931). Unpaarigheit und Radikalketten im Reaktionmechanismus organischer und enzymatischer Voränge. *Chem. Ber.* **64,** 2844–2856.

Halliwell, B. (1984). Superoxide dismutase and the superoxide theory of oxygen toxicity: A critical appraisal. *In* "Copper Proteins and Copper Enzymes" (R. Lontie, ed.), Vol. 2, pp. 63–102. CRC Press, Boca Raton, FL.

Halliwell, B., and Gutteridge, J. M. C. (1989). "Free Radicals in Biology and Medicine," 2nd ed. Oxford Univ. Press (Clarendon), London and New York.

Hartig, A., and Ruis, H. (1986). Nucleotide sequence of the *Saccharomyces cerevisiae* *CTT1* gene and deduced amino-acid sequence of yeast catalase T. *Eur. J. Biochem.* **160,** 487–490.

Henry, L. E., Cammack, R., Schwitzguebel, J. P., Palmer, J. M., and Hall, D. O. (1980). Intracellular localization, isolation and characterization of two distinct varieties of superoxide dismutase from *Neurospora crassa. Biochem. J.* **187**, 321–328.

Hjalmarsson, K., Marklund, S. L., Engström, Ä., and Edlund, T. (1987). Isolation and sequence of complementary DNA encoding human extracellular superoxide dismutase. *Proc. Natl. Acad. Sci. U.S.A.* **84**, 6340–6344.

Hörtner, H., Ammerer, G., Hartter, E., Hamilton, B., Rytka, J., Bilinski, T., and Ruis, H. (1982). Regulation of synthesis of catalases and *iso*-1 cytochrome *c* in *Saccharomyces cerevisiae* by glucose, oxygen and heme. *Eur. J. Biochem.* **128**, 179–184.

Imlay, J. A., and Fridovich, I. (1991a). Assay of metabolic superoxide production in *Escherichia coli. J. Biol. Chem.* **266**, 6957–6965.

Imlay, J. A., and Fridovich, I. (1991b). Isolation and genetic analysis of a mutation that suppresses the auxotrophies of superoxide dismutase deficient *Escherichia coli* K12. *Mol. Gen. Genet.* **228**, 410–416.

Imlay, J. A., and Fridovich, I. (1992). Suppression of oxidative envelope damage by pseudoreversion of a superoxide dismutase-deficient mutant of *Escherichia coli. J. Bacteriol.* **174**, 953–961.

Jazwinski, S. M. (1990a). An experimental system for the molecular analysis of the aging process: The budding yeast *Saccharomyces cerevisiae. J. Gerontol.* **54**, B68–B74.

Jazwinski, S. M. (1990b). Aging and senescence of the budding yeast *Saccharomyces cerevisiae. Mol. Microbiol.* **4**, 337–343.

Jazwinski, S. M., Egilmez, N. K., and Chen, J. B. (1989). Replication control and cellular life span. *Exp. Gerontol.* **24**, 423–436.

Johansen, J. T., Overballe-Petersen, C., Martin, B., Hasemann, V., and Svendsen, I. (1979). The complete amino acid sequence of copper,zinc superoxide dismutase from *Saccharomyces cerevisiae. Carlsberg Res. Commun.* **44**, 201–217.

Johnson, K. J., and Ward, P. A. (1985). Inflammation and active oxygen species. *In* "Superoxide Dismutases" (L. W. Oberley, ed.), Vol. 3, pp. 129–142. CRC Press, Boca Raton, FL.

Jones, E. W., and Fink, G. R. (1982). Regulation of amino acid and nucleotide biosynthesis in yeast. *In* "The Molecular Biology of the Yeast *Saccharomyces:* Metabolism and Gene Expression" (J. N. Strathern, E. W. Jones, and J. R. Broach, eds.), pp. 181–299. Cold Spring Harbor Lab., Cold Spring Harbor, NY.

Kaput, J., Goltz, S., and Blobel, G. (1982). Nucleotide sequence of the yeast nuclear gene for cytochrome *c* peroxidase precursor: Implications of the presequence for protein transport into mitochondria. *J. Biol. Chem.* **257**, 15054–15058.

Kaput, J., Brandriss, M. C., and Prussak-Wieckowska, T. (1989). *In vitro* import of cytochrome *c* peroxidase into the intermembrane space: Release of the processed form by intact mitochondria. *J. Cell Biol.* **109**, 101–112.

Keele, B. B., Jr., McCord, J. M., and Fridovich, I. (1970). Superoxide dismutase from *Escherichia coli* B: A new manganese enzyme. *J. Biol. Chem.* **245**, 6176–6181.

Keller, G.-A., Warner, T. G., Steimer, K. S., and Hallewell, R. A. (1991). Cu,Zn superoxide dismutase is a peroxisomal enzyme in human fibroblasts and hepatoma cells. *Proc. Natl. Acad. Sci. U.S.A.* **88**, 7381–7385.

Koppenol, W. H. (1981). The physiological role of the charge distribution of superoxide dismutase. *In* "Oxygen and Oxy-Radicals in Chemistry and Biology" (M. A. J. Rogers and E. L. Powers, eds.), pp. 671–674. Academic Press, New York.

Koppenol, W. H., and Butler, J. (1985). Energetics of interconversion reactions of oxyradicals. *Adv. Free Radicals Biol. Med.* **1**, 91–131.

Koppenol, W. H., Butler, J., and van Leeuwen, J. W. (1978). The Haber–Weiss cycle. *Photochem. Photobiol.* **28**, 655–660.

Kroll, J. S., Langford, P. R., and Loynds, B. M. (1991). Copper-zinc superoxide dismutase of *Haemophilus influenzae* and *H. parainfluenzae. J. Bacteriol.* **173,** 7449–7457.

Kuo, C. F., Mashino, T., and Fridovich, I. (1987). α,β-Dihydroxyisovalerate dehydratase: A superoxide-sensitive enzyme. *J. Biol. Chem.* **262,** 4724–4727.

Kusel, J. P., Boveris, A., and Storey, B. T. (1973). H_2O_2 production and cytochrome *c* peroxidase activity in mitochondria isolated from the trypanosomatid hemoflagellate *Crithidia fasciculata. Arch. Biochem. Biophys.* **158,** 799–805.

Lalonde, B., Arcangioli, B., and Guarente, L. (1986). A single *Saccharomyces cerevisiae* upstream activation site (UAS1) has two distinct regions essential for its activity. *Mol. Cell. Biol.* **6,** 4690–4696.

Lee, F. J., and Hassan, H. M. (1985). Biosynthesis of superoxide dismutase in *Saccharomyces cerevisiae:* Effects of paraquat and copper. *J. Free Radicals Biol. Med.* **1,** 319–325.

Lee, J., and Kosman, D. (1992). In preparation.

Lerch, K., and Schenk, E. (1985). Primary structure of copper-zinc superoxide dismutase from *Neurospora crassa. J. Biol. Chem.* **260,** 9559–9566.

Lesuisse, E., Crichton, R. R., and Labbé, P. (1990). Iron-reductases in the yeast *Saccharomyces cerevisiae. Biochim. Biophys. Acta* **1038,** 253–259.

Levine, J. S. (1982). The photochemistry of the paleoatmosphere. *J. Mol. Evol.* **18,** 161–172.

Levine, J. S. (1988). The origin and evolution of atmospheric oxygen. In "Oxidases and Related Redox Systems" (T. E. King, H. S. Mason, and M. Morrison, eds.), pp. 111–126. Liss, New York.

Liochev, S. I., and Fridovich, I. (1992). Effects of overproduction of superoxide dismutases in *Escherichia coli* on inhibition of growth and on induction of glucose-6-phosphate dehydrogenase by paraquat. *Arch. Biochem. Biophys.* **294,** 138–143.

Liu, X. F., Elashvili, I., Gralla, E. B., Valentine, J. S., Lapinskas, P., and Culotta, V. C. (1992). Yeast lacking superoxide dismutase: Isolation of genetic suppressors. *J. Biol Chem.* **267,** in press.

Løvaas, E., and Carlin, G. (1991). Spermine: An anti-oxidant and anti-inflammatory agent. *Free Radicals Biol. Med.* **11,** 455–461.

Macquer, P. (1777). In "Dictionnaire de Chymie," 2nd ed. Libraire de la Facultie de Medicine, Paris; quote taken from Gilbert (1981).

Makay, W. J., and Bewley, G. C. (1989). The genetics of catalase in *Drosophila melanogaster:* Isolation and characterization of acatalasemic mutants. *Genetics* **122,** 643–652.

Marres, C. A., Van Loon, A. P., Oudshoorn, P., Van Steeg, H., Grivell, L. A., and Slater, E. C. (1985). Nucleotide sequence analysis of the nuclear gene coding for manganese superoxide dismutase of yeast mitochondria, a gene previously assumed to code for the Rieske iron–sulphur protein. *Eur. J. Biochem.* **147,** 153–161.

McAdam, M. E., Fox, R. A., Lavelle, F., and Fielden, E. M. (1977). A pulse-radiolysis study of the manganese-containing superoxide dismutase from *Bacillus stearothermophilus.* A kinetic model for the enzyme action. *Biochem. J.* **165,** 71–80.

McCord, J. M., (1974). Free radicals and inflammation: Protection of synovial fluid by superoxide dismutase. *Science* **185,** 529–531.

McCord, J. M., and Fridovich, I. (1969). Superoxide dismutase—An enzymic function for erythrocuprein (hemocuprein). *J. Biol. Chem.* **244,** 6049–6055.

Miguel, J., and Fleming, J. (1986). Theoretical and experimental support for an oxygen radical-mitochondrial injury hypothesis of cell aging. In "Free Radicals, Aging and Degenerative Diseases" (J. E. Johnson, D. Harman, R. Walford, and J. Migeul, eds.), pp. 51–74. Liss, New York.

Minton, K. W., Tabor, H., and Tabor, C. W. (1990). Paraquat toxicity is increased in *Escherichia coli* defective in the synthesis of polyamines. *Proc. Natl. Acad. Sci. U.S.A.* **87,** 2851–2855.

Misra, H. P., and Fridovich, I. (1972). The purification and properties of superoxide dismutase from *Neurospora crassa*. *J. Biol. Chem.* **247,** 3410–3414.

Mortimer, R. K., and Johnston, J. R. (1959). Life span of individual yeast cells. *Nature (London)* **183,** 1751–1752.

Munkres, K. D. (1985). The role of genes, antioxidants, and antioxygenic enzymes in aging in *Neurospora:* A review. *In* "Superoxide Dismutases" (L. W. Oberley, ed.), Vol. 3, pp. 237–248. CRC Press, Boca Raton, FL.

Munkres, K. D. (1990a). Purification of exocellular superoxide dismutases. *In* "Methods in Enzymology" (L. Packer and A. N. Glazer, eds.), Vol. **186,** pp. 249–260. Academic Press, San Diego.

Munkres, K. D. (1990b). Genetic coregulation of longevity and antioxienzymes in *Neurospora crassa*. *Free Radicals Biol. Med.* **8,** 355–361.

Munkres, K. D. (1990c). Pharmacogenetics of cyclic guanylate, antioxidants and antioxidant enzymes in *Neurospora*. *Free Radicals Biol. Med.* **9,** 29–38.

Munkres, K. D. (1990d). Pharmacogenetics of cyclic guanylate, antioxidants and antioxidant enzymes in *Saccharomyces*. *Free Radicals Biol. Med.* **9,** 39–50.

Munkres, K. D., and Furtek, C. A. (1984). Selection of conidial longevity mutants of *Neurospora crassa*. *Mech. Age. Dev.* **25,** 47–62.

Munkres, K. D., and Rana, R. S. (1984). Genetic control of cellular longevity in *Neurospora crassa:* A relationship between cyclic nucleotides, antioxidants, and antioxygenic enzymes. *Age* **7,** 30–35.

Natvig, D. O. (1991). Superoxide dismutases and catalases in *Neurospora crassa*. *Fungal Genet. Newsl.* **38,** 14 (abstr.).

Neta, P., and Huie, R. E. (1985). Free radical chemistry of sulfite. *Environ. Health Perspect.* **64,** 209–217.

Orgel, L. E. (1963). The maintenance of the accuracy of protein synthesis and its relevance to ageing. *Proc. Natl. Acad. Sci. U.S.A.* **49,** 517–521.

Palma, J. M., Garrido, M., Rodríguez-García, M. I., and del Río, L. A. (1991). Peroxisome proliferation and oxidative stress mediated by activated oxygen species in plant peroxisomes. *Arch. Biochem. Biophys.* **287,** 68–74.

Parker, M. W., and Blake, C. C. F. (1988). Crystal structure of manganese superoxide dismutase from *Bacillus stearothermophilus* at 2.4 Å resolution. *J. Mol Biol.* **199,** 649–661.

Peng, T. X., Moya, A., and Ayala, F. J. (1986). Irradiation-resistance conferred by superoxide dismutase: Possible adaptive role of a natural polymorphism in *Drosophila melanogaster*. *Proc. Natl. Acad. Sci. U.S.A.* **83,** 684–687.

Pfeifer, K., Prezant, R., and Guarente, L. (1987). Yeast HAP1 activator binds to two upstream activation sites of different sequence. *Cell (Cambridge, Mass.)* **49,** 19–27.

Pfeifer, K., Kim, K.-S., Kogan, S., and Guarente, L. (1989). Functional dissection and sequence of yeast HAP1 activator. *Cell (Cambridge, Mass.)* **56,** 291–301.

Phillips, J. P., and Hilliker, A. J. (1990). Genetic analysis of oxygen defense mechanisms in *Drosophila melanogaster*. *Adv. Genet.* **27,** 43–71.

Phillips, J. P., Campbell, S. D., Michaud, D., Charbonneau, M., and Hilliker, A. J. (1989). Null mutation of copper/zinc superoxide dismutase in *Drosophila* confers hypersensitivity to paraquat and reduced longevity. *Proc. Natl. Acad. Sci. U.S.A.* **86,** 2761–2765.

Pinkham, J. L. (1992). In preparation.

Pinkham, J. L., Olesen, J. T., and Guarente, L. (1987). Sequence and nuclear localization of the *Saccharomyces cerevisiae* HAP2 protein, a transcriptional activator. *Mol. Cell. Biol.* **7**, 578–785.

Pringle, J. R., and Hartwell, L. H. (1981). The *Saccharomyces cerevisiae* cell cycle. In "The Molecular Biology of the Yeast *Saccharomyces:* Life Cycle and Inheritance" (J. N. Strathern, E. W. Jones, and J. R. Broach, eds.), pp. 97–142. Cold Spring Harbor Lab., Cold Spring Harbor, NY.

Pryor, W. A. (1986). Oxy-radicals and related species: Their formation, lifetimes, and reactions. *Annu. Rev. Physiol.* **48**, 657–667.

Ravindranath, S. D., and Fridovich, I. (1975). Isolation and characterization of a manganese-containing superoxide dismutase from yeast. *J. Biol. Chem.* **250**, 6107–6112.

Rose, M. D., Winston, F., and Hieter, P. (1990). "Methods in Yeast Genetics: A Laboratory Manual," pp. 5–7. Cold Spring Harbor Lab. Press, Cold Spring Harbor, NY.

Saltman, P. (1989). Oxidative stress: A radical view. *Semin. Hematol.* **26**, 249–256.

Sandalio, L. M., and Del Río, L. A. (1987). Localization of superoxide dismutase in glyoxysomes from *Citrullus vulgaris.* Functional implications in cellular metabolism. *J. Plant Physiol.* **127**, 395–409.

Sandalio, L. M., and Del Río, L. A. (1988). Intraorganellar distribution of superoxide dismutase in plant peroxisomes (glyoxysomes and leaf peroxisomes). *Plant Physiol.* **88**, 1215–1218.

Sandalio, L. M., Fernández, V. M., Rupérez, F. L., and Del Río, L. A. (1988). Superoxide free radicals are produced in glyoxysomes. *Plant Physiol.* **87**, 1–4.

Scandalios, J. G. (1990). Response of plant antioxidant defense genes to environmental stress. *Adv. Genet.* **27**, 1–41.

Scott, M. D., Meshnick, S. R., and Eaton, J. W. (1987). Superoxide dismutase-rich bacteria. Paradoxical increase in oxidant toxicity. *J. Biol. Chem.* **262**, 3640–3645.

Scott, M. D., Meshnick, S. R., and Eaton, J. W. (1989). Superoxide dismutase amplifies organismal sensitivity to ionizing radiation. *J. Biol. Chem.* **264**, 2498–2501.

Shatzman, A. R., and Kosman, D. J. (1979). Biosynthesis and cellular distribution of the two superoxide dismutases of *Dactylium dendroides. J. Bacteriol.* **137**, 313–320.

Sherman, F., and Stewart, J. W. (1981). Genetics and biosynthesis of cytochrome *c. Annu. Rev. Genet.* **5**, 257–296.

Sherman, L., Dafni, N., Lieman-Hurwitz, J., and Groner, Y. (1983). Nucleotide sequence and expression of human chromosome 21-encoded superoxide dismutase mRNA. *Proc. Natl. Acad. Sci. U.S.A.* **80**, 5465–5469.

Sies, H., and Cadenas, E. (1983). Biological basis of detoxification of oxygen free radicals. *In* "Biological Basis of Detoxification" (J. Caldwell and W. Jacoby, eds.), pp. 181–211. Academic Press, New York.

Simon, M., Adam, G., Rapatz, W., Spevak, W., and Ruis, H. (1991). The *Saccharomyces cerevisiae ADR1* gene is a positive regulator of transcription of genes encoding peroxisomal proteins. *Mol. Cell. Biol.* **11**, 699–704.

Sinet, P. M., Lavelle, F., Michelson, R. M., and Jerome, H. (1975). Superoxide dismutase activities of blood platelets in trisomy 21. *Biochem. Biophys. Res. Commun.* **67**, 904–909.

Siwecki, G., and Brown, O. R. (1990). Overproduction of superoxide dismutase does not protect *Escherichia coli* from stringency-induced growth inhibition by 1 m*M* paraquat. *Biochem. Int.* **20**, 191–200.

Skoneczny, M., Chelstowska, A., and Rytka, J. (1988). Study of the coinduction by fatty acids of catalase A and acyl-CoA oxidase in standard and mutant *Saccharomyces cerevisiae* strains. *Eur. J. Biochem.* **174**, 297–302.

Sohal, R. S., and Allen, R. G. (1986). Relationship between oxygen and metabolism, aging and development. *Adv. Free Radicals Biol. Med.* **2**, 117–160.

Spevak, W., Fessl, F., Rytka, J., Traczyk, A., Skoneczny, M., and Ruis, H. (1983). Isolation of the catalase T structural gene of *Saccharomyces cerevisiae* by functional complementation. *Mol. Cell. Biol.* **3**, 1545–1551.

Stallings, W. C., Pattridge, K. A., Strong, R. K., and Ludwig, M. L. (1985). The structure of manganese superoxide dismutase from *Thermus thermophilus HB8* at 2.4-Å resolution. *J. Biol. Chem.* **260**, 16424–16232.

Steinman, H. M. (1980). The amino acid sequence of copper-zinc superoxide dismutase from bakers' yeast. *J. Biol. Chem.* **255**, 6758–65.

Steinman, H. M. (1982). Superoxide dismutases: Protein chemistry and structure-function relationships. *In* "Superoxide Dismutases" (L. W. Oberley, ed.), Vol. 1, pp. 11–68. CRC Press, Boca Raton, FL.

Steinman, H. M., and Ely, B. (1990). Copper-zinc superoxide dismutase of *Caulobacter crescentus:* Cloning, sequencing, and mapping of the gene and periplasmic location of the enzyme. *J. Bacteriol.* **172**, 2901–2910.

Summitt, R. L. (1981). Chromosome 21: Specific segments that cause the phenotype of Down's syndrome. *In* "Trisomy 21 (Down's Syndrome) Research Perspective" (F. F. de la Cruz and P. S. Gerard, eds.), pp. 225–235. University Park Press, Baltimore.

Tabor, C. W., and Tabor, H. (1984). Polyamines. *Annu. Rev. Biochem.* **53**, 749–790.

Tainer, J. A., Getzoff, E. D., Beem, K. M., Richardson, J. S., and Richardson, D. C. (1982). Determination and analysis of 2 Å structure of copper, zinc superoxide dismutase. *J. Mol. Biol.* **160**, 181–217.

Tainer, J. A., Getzoff, E. D., Richardson, J. S., and Richardson, D. C. (1983). Structure and mechanism of copper, zinc superoxide dismutase. *Nature (London)* **306**, 284–287.

Tan, Y. H., Tischfield, J., and Ruddle, F. H. (1973). The linkage of genes for the human interferon-induced antiviral protein and indophenol oxidase-B traits to chromosome G-21. *J. Exp. Med.* **137**, 317–330.

Tatchell, K. (1986). *RAS* genes and growth control in *Saccharomyces cerevisiae. J. Bacteriol.* **166**, 364–367.

Taube, H. (1965). Mechanisms of oxidation with oxygen. *J. Gen. Physiol.* **49**, 29–52.

Thiele, D. J. (1988). *ACE1* regulates expression of the *Saccharomyces cerevisiae* metallothionein gene. *Mol. Cell. Biol.* **8**, 2745–2752.

Thomas, D., Cherest, H., and Surdin-Kerjan, Y. (1989). Elements involved in *S*-adenosylmethionine-mediated regulation of the *Saccharomyces cerevisiae MET25* gene. *Mol. Cell. Biol.* **9**, 3292–3298.

Tolmasoff, J. M., Ono, T., and Cutler, R. G. (1980). Superoxide dismutase: Correlation with life-span and specific metabolic rate in primate species. *Proc. Natl. Acad. Sci. U.S.A.* **77**, 2777–2781.

Tsang, E. W. T., Bowler, C., Hérouart, D., Van Camp, W., Villarroel, R., Genetello, C., Van Montagu, M., and Inzé, D. (1991). Differential regulation of superoxide dismutases in plants exposed to environmental stress. *Plant Cell* **3**, 783–792.

Turi, T. G., Kalb, V. F., and Loper, J. C. (1991). Cytochrome P450 lanosterol 14α-demethylase (*ERG1*) and manganese superoxide dismutase (*SOD1*) are adjacent genes in *Saccharomyces cerevisiae. Yeast* **7**, 627–630.

van Loon, A. P. G. M., Maarse, A. C., Riezman, H., and Grivell, L. A. (1983). Isolation, characterization and regulation of expression of the nuclear genes for the Core II and Rieske iron–sulfur proteins of the yeast ubiquinol-cytochrome *c* reductase. *Gene* **26**, 261–272.

van Loon, A. P. G. M., Pesold-Hurt, B., and Schatz, G. (1986). A yeast mutant lacking mitochondrial manganese-superoxide dismutase is hypersensitive to oxygen. *Proc. Natl. Acad. Sci. U.S.A.* **83**, 3820–3824.

Veenhuis, M., and Harder, W. (1991). Microbodies. *In* "The Yeasts" (A. H. Rose and J. S. Harrison, eds.), Vol. 4, pp. 601–653. Academic Press, San Diego.

Veenhuis, M., Materblowski, M., Kunau, W. H., and Harder, W. (1987). Proliferation of microbodies in *Saccharomyces cerevisiae*. *Yeast* **3**, 77–84.

Verduyn, C., Giuseppin, M. L. F., Scheffers, W. A., and van Dijken, J. P. (1988). Hydrogen peroxide metabolism in yeasts. *Appl. Environ. Microbiol.* **54**, 2086–2090.

Welch, J., Fogel, S., Buchman, C., and Karin, M. (1989). The *CUP2* gene product regulates the expression of the *CUP1* gene, coding for yeast metallothionein. *EMBO J.* **8**, 255–260.

Westerbeek-Marres, C. A., Moore, M. M., and Autor, A. (1988). Regulation of manganese superoxide dismutase in *Saccharomyces cerevisiae*. The role of respiratory chain activity. *Eur. J. Biochem.* **174**, 611–620.

Wills, C. (1990). Regulation of sugar and ethanol metabolism in *Saccharomyces cerevisiae*. *CRC Crit. Rev. Biochem. Mol. Biol.* **25**, 245–280.

Winkler, H., Adam, G., Mattes, E., Schantz, M., Hartig, A., and Ruis, H. (1988). Coordinate control of synthesis of mitochondrial and non-mitochondrial hemoproteins: A binding site for the HAP1 (CYP1) protein in the UAS region of the yeast catalase T gene (*CTT1*). *EMBO J.* **7**, 1799–1804.

Yamashoji, S., Ikeda, T., and Yamashoji, K. (1991). Extracellular generation of active oxygen species catalyzed by exogenous menadione in yeast cell suspension. *Biochim. Biophys. Acta* **1059**, 99–105.

Zhu, D., and Scandalios, J. G. (1992). Expression of the maize Mn SOD (*Sod3*) gene in MnSOD-deficient yeast rescues the mutant yeast under oxidative stress. *Genetics* (in press).

Zitomer, R. S., and Lowry, C. V. (1992). Regulation of gene expression by oxygen in *Saccharomyces cerevisiae*. *Microbiol. Rev.* **56**, 1–11.

GENETICS OF THE MAMMALIAN CARBONIC ANHYDRASES

Richard E. Tashian

Department of Human Genetics, University of Michigan Medical School,
Ann Arbor, Michigan 48109

I. Introduction

There are probably few evolutionarily related genes coding for enzymes that can surpass the carbonic anhydrase (CA) multigene family with respect to the distributional and functional diversity of their gene products. The CA gene family in mammals is now known to consist of eight genes coding for seven CA isozymes (CA I–CA VII) and a CA-related protein (CARP). These genes may be expressed in certain cells of virtually all tissues (i.e., CA II), or limited in expression to a single tissue (i.e., CA VI and CA VII, salivary glands), with the other CA genes ranging in their expression between these two extremes. In the cell, they may be membrane associated (CA IV), cytosolic (CA I, II, III, and VII), mitochondrial (CA V), or located in secretory granules (CA VI), where they participate in a variety of biological activities that involve acid–base balance, CO_2 transfer, and ion exchange. These include functions such as respiration; H^+ secretion; HCO_3^- reabsorption; the production of aqueous humor, cerebrospinal fluid, gastric acid, and

ADVANCES IN GENETICS, Vol. 30

Copyright © 1992 by Academic Press, Inc.
All rights of reproduction in any form reserved.

pancreatic juice; bone resorption; and biosynthesis (ureagenesis, gluconeogenesis, and lipogenesis). The catalytic activity of CA in all of these processes is the simple reversible hydration of CO_2 (CO_2 + $H_2O \rightleftharpoons HCO_3^- + H^+$), and for the CA II isozyme, this reaction is one of the fastest known for any enzyme, with a turnover number in excess of 1×10^6 sec^{-1}.

Considerable chemical and biological information has now accumulated on the CAs regarding their primary and tertiary structures, enzyme kinetics, active site mechanisms, physiological roles, cellular and subcellular locations, genetics, and evolution. Thus, the related CA genes appear to be an especially attractive model for the study of the molecular genetics and evolution of a multigene family. Over the past 5 years, our knowledge of the structures, linkage relationships, and regulation of normal and mutated CA isozyme genes has increased greatly, and it is the main objective of this review to discuss some of these developments. [For reviews and volumes devoted to CA studies, see Tashian and Hewett-Emmet (1984), Carter and Jeffery (1985), Deutsch (1987), Fernley (1988), Forster (1988), Tashian (1989), Edwards (1990), Dodgson *et al.* (1991), Botrè *et al.* (1991), Hewett-Emmett and Tashian (1991), and Silverman (1991).]

II. Evolutionary Origins of the Carbonic Anhydrase Genes

As an increasing number of amino acid sequences of the CA I, CA II, and CA III isozymes became available, primarily from mammalian sources, phylogenetic trees based on these sequences were constructed that provided the initial branching pattern for the evolutionary relationships of these three carbonic anhydrases (Hewett-Emmett and Tashian 1984). However, not until more CA sequences (nucleotide as well as amino acid) were determined from nonmammalian and viral sources (e.g., shark, algae, and vaccinia virus), and from newly discovered mammalian CA isozymes (i.e., CA IV, CA V, CA VI, CA VII, and CARP) was it possible to define more accurately the evolutionary patterns of descent (Hewett-Emmett and Tashian, 1991). A generalized branching pattern of the CA gene family incorporating the recently reported sequences is shown in Fig. 1. It appears that the oldest mammalian CA, or CA-like, genes are the CARP, CA VI, and CA IV genes, and those that diverged most recently (i.e., between 300 and 400 million years ago) are the CA I, II, and III genes, with the CA VII and CA Y genes occupying intermediate positions. It is possible that CA Y is the mitochondrial CA V gene, but this has not as yet been deter-

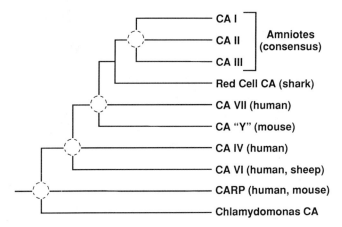

FIG. 1. General branching scheme for the carbonic anhydrase isozymes of amniotes and related shark and algal carbonic anhydrases. The amino acid sequences are from Hewett-Emmett and Tashian (1991), Kato (1990), Fuzukawa *et al.* (1990), Okuyama *et al.* (1992), Aldred *et al.* (1991), and R. E. Tashian (unpublished).

mined (Tashian *et al.*, 1991). The relatively recent formation of the CA I, II, and III genes is supported by the fact that they have remained closely linked on chromosomes 8 and 3 in humans and mice, respectively (see Section IV).

Recently, another type of carbonic anhydrase has been characterized from chloroplasts of higher plants (i.e., pea, spinach) and certain bacteria. Because the amino acid sequences of these CAs show no sequence identity to the previously described animal and algal CAs, they represent an evolutionarily distinct CA gene family that we have tentatively designated the β-carbonic anhydrases. These β-CAs are found in chloroplasts of higher plants (e.g., pea, spinach) and in the cell membranes or carboxysomes of bacteria, e.g., *Escherichia coli* and *Synechoccus* sp. PCC7942 (Chin *et al.*, 1983; Majeau and Coleman, 1990; Burnell *et al.*, 1989; Fawcett *et al.*, 1990; Roeske and Ogren, 1990; Suzuki *et al.*, 1991; Guilloton *et al.*, 1992). Despite the fact that the β-CAs are not homologous to the animal and algal CAs, they nevertheless catalyze the CO_2 hydrase reaction at high rates (cf. Silverman, 1991), making this an especially interesting example of the evolutionary convergence of enzyme activities.

It is possible that both types of carbonic anhydrase may be present in bacteria. The CA found in *Neisseria sicca,* and probably other bacteria (cf. Karrasch *et al.*, 1989; Nafi *et al.*, 1990; Tashian *et al.*, 1991), may be related to the algal and animal CAs, whereas the CA products

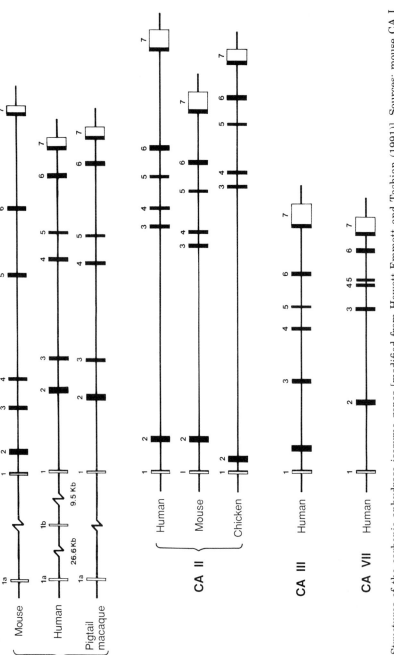

FIG. 2. Structures of the carbonic anhydrase isozyme genes [modified from Hewett-Emmett and Tashian (1991)]. Sources: mouse CA I (Fraser et al., 1989); human CA I (Lowe et al., 1990); pigtail macaque CA I (Nicewander, 1990); human CA II (Venta et al., 1991); mouse CA II (Venta et al., 1985); chicken CA II (Yoshihara et al., 1987); human CA III (Lloyd et al., 1987); human CA VII (Montgomery et al., 1991).

of the *cynT* gene of *E. coli* (Guilloton *et al.*, 1992) and the *IcfA* gene of the cyanobacterium, *Synechoccus* sp. PGG7942 (Suzuki *et al.*, 1991), belong to the β-CA family. Because chloroplasts are thought to have arisen by internalization of cyanobacteria by a eukaryotic cell, the homology of β-CAs from a cynobacterium and chloroplasts would support this hypothesis on the origin of chloroplasts. It would obviously also be of great interest to determine to what extent the β-CAs are present in animal cells, or animal CAs in plant cells.

III. The Carbonic Anhydrase Genes

The structures of the mammalian genes for the CA I, CA II, CA III, and CA VII isozymes have been characterized from one or more species. As shown in Fig. 2, these genes are structurally similar in possessing six introns that separate exons 1–7 at identical positions. However, the CA I genes are unique in having an additional (noncoding) exon, exon 1a, at the 5′ end of the gene that is separated by about 36 kb (human CA I gene) from exon 1 (Fraser *et al.*, 1989; Brady *et al.*, 1989, 1991; Lowe *et al.*, 1990). As yet, no pseudogenes have been reported for any of the CA isozyme genes.

Because of the considerable diversity in expression of the CA genes, studies of their regulatory elements (tissue specific or developmental) should be particularly rewarding. As yet, little is known about specific transcription factors associated with their regulation; nevertheless, certain aspects of the promoter regions of these four CA isozymes have been characterized and limited analyses of these regions have been reported. Some possibly pertinent features of the promoter regions of the CA I, CA II, CA III, and CA VII genes and their expression are presented below. Also, where relevant, certain aspects of their expression, hormonal control, and evolution are discussed. Obviously, future studies on the regulation of the genes encoding the noncytoplasmic membrane-bound CA IV, mitochrondrial CA V, and secreted CA VI isozymes will be of considerable interest.

A. CARBONIC ANHYDRASE I

CA I is primarily expressed in erythrocytes, but is also found in gastrointestinal mucosa, vascular endothelium, myoepithelial cells and, at low levels, in the cells of many other tissues (cf. Tashian, 1989). The most striking feature of this CA isozyme, however, is that although it is the second most abundant nonglobin protein in most mammalian red cells, no hematological abnormalities are evident in its absence as the result of a mutation. Furthermore, CA I is not expressed in the red cells of certain species, e.g., ruminants and felids (cf. Tashian *et al.*, 1971, 1980; Kendall and Tashian, 1977; Tashian, 1977), making the assignment of a physiological role in erythrocytes problematic.

1. Transcriptional Control

The expression of the mouse, human, and pigtail macaque CA I genes has been demonstrated to be under the control of two promoters (Fraser *et al.*, 1989; Brady *et al.*, 1991; Bergenhem *et al.*, 1992b). The promoter region upstream of exon 1a controls the expression of CA I in erythroid cells, and the region 5' to exon 1 appears to control expression in nonerythroid tissue (Figs. 2 and 3). Sequence analysis of the 5' flanking regions of the CA I genes reveals TATA (CATA in human and macaque) and CAAT boxes at expected positions upstream of the start sites for the erythroid RNA transcripts. In addition, several possible binding sites for transcriptional factors, i.e., Oct-1, CCACACCC, AP-1, and GATA-1 (A/T GATA A/G, erythroid), can be identified. In the nonerythroid 5' promoter region, only a TATA box and one potential GATA-1-binding site are evident (Lowe *et al.*, 1990; Brady *et al.*, 1991). Four DNase-hypersensitive sites have been located in the 5' region of the mouse CA I gene (*Car-1*). Two sites are immediately upstream of exon 1a, and two are at the 3' end of intron 1 (Thierfelder *et al.*, 1991). A comparison of these 5' flanking sequences of the human and mouse CA I genes is shown in Fig. 3.

Several potential transcription factor binding sites have also been described in the 3' flanking region of the human CA I gene between the first AATAAA polyadenylation signal site (86 bp from the translation termination codon) and extending 820 bp downstream (Brady *et al.*, 1989). One of these is a GATA-1 site between the first poly(A) site at position 109 and the second AATAAA signal site at 312. Two other GATA-1 sites and one AP-1 site are found after the second poly(A) site, between positions 329 and 820.

2. Developmental Regulation

a. Erythrogenesis. CA I is expressed early during erythroid differentiation in humans (Villeval *et al.*, 1985), and later (in a stage-specific manner) during human fetal erythroid development (Boyer *et al.*, 1983; Weil *et al.*, 1984). Only low levels of CA I are present in fetal red cells; however, at the time of birth, CA I production is induced and reaches adult levels during the first year of life. It has recently been

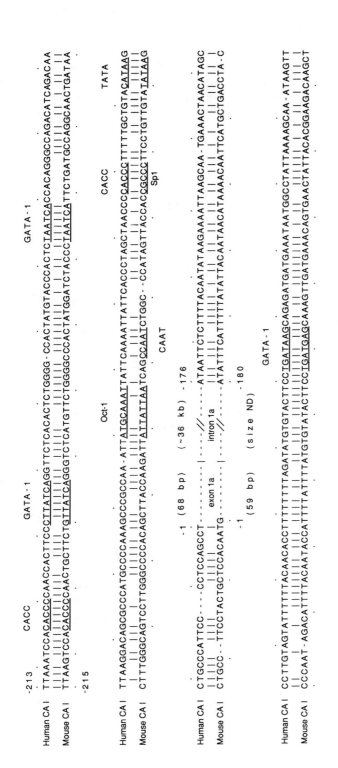

Fig. 3. Comparison of the erythroid and nonerythroid promoter regions 5′ to exon 1a and exon 1, respectively, in the human and mouse CA I genes. Sequence sources: human CA I (Lowe *et al.*, 1990; Brady *et al.*, 1991); mouse CA I (Fraser *et al.*, 1989). TATA and CAAT boxes and proposed potential binding sites for transcription factors are designated and underlined.

shown that this type of development is regulated by trans-acting factors (Brady *et al.*, 1989), and a GATA-1 element located at position −190 from the cap site of exon 1a has been implicated in the regulation of CA I in the transition from fetal to adult red cells (Butterworth *et al.*, 1991).

Ma *et al.* (1991) describe a stage-specific factor, UB2, which binds to a sequence element located about 290 bp upstream of exon 1 in the mouse CA I gene (*Car-1*). This element contains two sites, i.e., I (TTGATA) and II (AGTCAATAGTCA), separated by 6 bp. It is possible that this sequence may correspond to one of the two DNase-hypersensitive sites at the 3′ end of intron 1a (Thierfelder *et al.*, 1991). Similar elements (with sites I and II in orientations opposite to those in the CA I gene) are found between 485 and 690 bp downstream of the last exon of the β-globin genes of human, mouse, and chicken, and appear to be important in the timing of gene expression. It is noteworthy that the binding of UB2 to the β-globin gene element results in active transcription of β-globin in uninduced mouse erythroleukemia (MEL) cells and inactive transcription in the induced cells, whereas the opposite effect is seen when UB2 is bound to the CA I gene element. Ma *et al.* (1991) suggest that perhaps the different orientation of the I and II elements in the CA I and β-globin genes might be responsible for their opposite effects.

The induction of CA I in the human erythroleukemic cell line, K562, indicates that these cells are well-suited for studies on the transcrip-

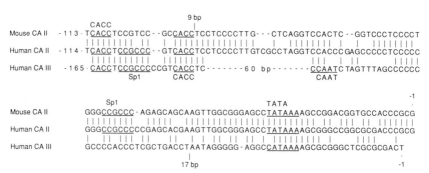

FIG. 4. Comparison of the 5′ flanking regions of human and mouse CA II genes and human CA III gene. Seven additional Sp1 sites in the human CA II gene, seven additional Sp1 sites in the human CA II gene, and two Sp1 sites in the mouse CA II gene, were found between positions −114 and −559, and −113 and −196, respectively, in the human and mouse genes (Venta *et al.*, 1985; Shapiro *et al.*, 1987). Proposed binding sites for transcription factors, including TATA and CAAT boxes, are indicated and underlined. Nucleotides common to mouse and human CA II, or human CA I and CA III are indicated. Insertions are identified by the number of bp they comprise.

tional regulation of the human CA I gene (Kondo *et al.*, 1991). K562 cells do not normally express CA I; however, on treatment with erythropoietin or hemin, CA I was detected after 3 hours of incubation.

b. Hepatogenesis. In the fetal rat, intense CA I immunostaining is observed in hepatocytes at day 12. By day 20, however, CA I immunoreactivity is greatly diminished in the hepatocytes and appears to be partially replaced by CA II and CA III late in prenatal development. During postnatal development, CA I tends to be found in hepatocytes close to the central veins (Laurila *et al.*, 1989).

B. Carbonic Anhydrase II

1. Transcription

CA II is the most widely distributed of all the CA isozymes and is found at variable levels in certain cells of almost all tissues. Although the 5' flanking region of the human CA II gene contains a TATA box and a possible CAAT box, it also contains nine potential Sp1 (CCGCCC/GGGCGG) sequences and is highly G + C rich [about 80% in the region 567 bp upstream of the cap site (Shapiro *et al.*, 1987)], features that are characteristic of tissue-specific and housekeeping genes (Dynan, 1986), respectively. Thus, the CA II promoter region may exhibit unique regulatory features. This region also contains a tandem repeat (TCACCTCC in human and mouse) that is similar to a tandem repeat required for maximal transcription of β-globin genes (Myers *et al.*, 1986), and possibly other genes (Weiher *et al.*, 1983; Weiher and Botchan, 1984; Venta *et al.*, 1985). Deletion analysis of the human 5' promoter region showed a gradual, but differential, increase in promoter activity because constructs (transfected into both murine L cells and human HeLa cells) contained more Sp1-binding sites (Shapiro *et al.*, 1987). When the 5' flanking regions for mouse and human CA II genes are compared (Fig. 4), several regions of high sequence identity are evident between the two genes. Four DNase-hypersensitive sites have been located in the mouse CA II gene, one immediately upstream of exon 1, two in intron 1, and one at the 3' end of intron 2 (Thierfelder *et al.*, 1991). Donehower *et al.* (1989) found a conserved sequence at the 3' end of intron 1 in the human CA II gene that matches 14 of 15 bp of a consensus sequence they describe from the noncoding regions of many human genes. This raises the possibility that a similar sequence may be in the DNase-sensitive region at the 3' end of intron 1 reported in the CA II gene of the mouse (Thierfelder *et*

Human CA II   TTACAAATGTGGAAACTGAGGCACA → 141 bp to 3' end of intron 1
              :::::.::: :::::::::::::::.:
Mouse CA II   TTACAAATAAAAGGACTGAGGCACA → 153 bp to 3' end of intron 1
              :::::.::.: . :::::::::.:
Censensus   TACAGATGAGGAAACTGAGGCTC
              T A GAG A C T   A AT

FIG. 5. Comparison of conserved sequence elements (DNase-hypersensitive sites?) at the 3' end of intron 1 in human and mouse CA II genes. Matches with primary (:) and secondary (.) consensus core sequences are indicated (Donehower *et al.*, 1990).

al., 1991). In Fig. 5, the human CA II sequence is compared to a comparable region in intron 1 of the mouse. As can be seen, eight nucleotides are the same as the consensus sequence and four additional nucleotides show a match with a frequent secondary nucleotide.

A useful cellular model for the study of CA II initiation in mammalian cells would appear to be the human promyelocytic leukemia cell line, HL-60. These cells do not normally express CA II; however, treatment with 1,25-dihydroxyvitamin D_3 or 12-*O*-tetradecanoylphorbol-13-actate results in the marked *de novo* expression of CA II (Shapiro *et al.*, 1989). Identification of the responsive sites for these inducing compounds should provide useful insights into the regulation of CA II expression.

The binding of v-*erbA* oncoprotein to a specific site (T_3RE CA II) 655 bp upstream of the cap site of the CA II gene in chickens suppresses CA II production and contributes to virus-induced erythroblastosis (Disela *et al.*, 1991). Although this is found in an avian gene, it nevertheless suggests that a similar sequence element (GGTGAGTGAAC) may be involved in the control of some mammalian CA II genes.

2. Hormonal Regulation

It was reported in several early studies that progesterone treatment resulted in elevated levels of what was presumably CA II in the endometrial epithelia of human, rabbit, and guinea pig (Nicholis and Board, 1967; McIntosh, 1970; Hodgen and Falk, 1971). CA II in rat liver has also been shown to be regulated by estrogen (Jeffery *et al.*, 1984). More recently, the modulation of CA II by calcitonin and parathyroid hormone was reported (Arlot-Bonnemains *et al.*, 1985). However, one of the most striking examples of the hormonal regulation of CA II is the recent demonstration of the markedly different effect of androgen and estrogen on the levels of CA II in the dorsal and lateral lobes of rat prostate (Härkönen *et al.*, 1991). CA II is the major cytoplasmic protein present in the dorsal and lateral prostate of the rat, where it appears to be expressed exclusively in the epithelial cells (Härkönen *et al.*, 1988). Castration decreased the concentration of CA II in lateral prostate but increased it in the dorsal prostate, and these changes could be reversed in both lobes by testosterone treatment. Estrogen treatment of the castrated rats enhanced CA II concentration in lateral prostate but had no effect in the dorsal prostate. Obviously, identification of the regulatory elements responsible for this differential hormonal control of CA II transcription would contribute greatly to our understanding of an interesting type of transcriptional regulation.

C. CARBONIC ANHYDRASE III

1. Expression in Tissues Other Than Muscle

Even though CA III is expressed primarily in red skeletal muscle (type 1 fibers), varying levels have been reported in several other tissues, including red cells, salivary glands, colon, testis, adipose tissue, lung, and liver. The relatively high levels of CA III in male rat liver (male:female ratio 10–20:1) appear to be tissue specific because a similar increase is not observed in skeletal muscle (Jeffery *et al.*, 1986). This increased expression in male liver has been attributed to the pulsatory effect of growth hormone from the male pituitary gland (Jeffery *et al.*, 1986; 1990). Interestingly, sexual dimorphism of CA III in rat liver is not seen in mouse liver, even though growth hormone is also released in a similar pulsatory manner in male mice. Furthermore, CA III does not appear to be expressed in the liver of certain mammals, e.g., humans and higher primates (Jeffery *et al.*, 1980). The high levels of CA III in adipose tissue (Stanton *et al.*, 1991) also presents an interesting tissue-specific aspect of CA III expression. Obviously, this intriguing variability in CA III expression should prove to be a particularly attractive area of study.

2. Expression in Muscle Development

During human muscle morphogenesis, CA III is expressed at low levels in early fetal muscle, increasing slowly until the last trimester, when a rapid increase takes place that seemingly coincides with the differentiation of the various fiber types, and reaching 50–60% of adult levels at birth (Jeffery *et al.*, 1980; Lloyd *et al.*, 1987). During myogenesis, CA III is expressed in the earliest phases of a rat myo-

CA Forms

Residue Number

CA Forms	7*	29*	61	62	64*	65	66	67*	69	91	92*	94(Z)	96(Z)	106	107*	117	119(Z)	121	131	141	143	145	192	194*	198*	199*	200*	201	202	204	206	207	209	211	244	246*
CA I (consensus)	Y	S	N	[V]	H	S	F	H	N	F	Q	H	H	E	H	V	H	L	L	I	G	W	Y	L	F	T	[H]	P	P	H	S	V	W	I	N	R
CA II (consensus)	Y	S	N	N	H	S	F	[N]	E/V	I	Q	H	H	E	H	V	H	V	L	V	G	W	Y	L	L	T	T	P	P	L	C/S	V	W	V	N	R
CA III (consensus)	Y	S	N	N	L	T	[C]	R	V	F	Q	H	H	E	H	V	H	F	I	V	V	L	W	[F]	T	T	P	P	E	[C]	[I]	W	L	V	N	R
CA IV (human)	Y	S	N	N	H	S	V	[M]	K	[V]	Q	H	H	E	H	V	H	A	F	Y	L	[F]	Y	L	V	T	[D]	[K]	V	W	V	V	N	R	N	R
CA "Y" (mouse)	–	S	N	N	H	S	[F]	F	Q	E	K	Q	H	H	E	H	[F]	V	L	F	Y	L	V	L	T	P	P	A	S	V	W	V	V	N	N	R
CA VI (human, sheep)	Y	S	N	N	H	T	V	Q	S	K/G	Q	H	H	E	H	V	H	[Y]	Y	L	V	A	Y	L	L	T	P	P	T	N	V	W	V	V	N	R
CA VII (human)	Y	S	N	N	H	S	V	Q	D	K	Q	H	H	E	H	V	H	F	L	V	G	W	Y	L	L	T	P	P	S	N	V	W	V	V	N	R
Shark CA	–	S	–	–	–	–	–	–	–	R	Q	H	H	E	H	V	H	F	L	V	G	W	Y	L	L	T	P	P	L	S	W	V	W	V	N	R
Chlamydomonas CA	–	S	T	N	H	T	I	Q	Q	T	Q	H	H	E	H	V	[C]	–	–	I	A	W	Y	L	L	T	[I]	P	S	G	L	W	V	L	N	R
CARP (human, mouse)	Y	S	[T]	N	H	T	I	Q	V	[Y]	[E]	[R]	H	E	H	[I]	I	I	A	W	Y	L	L	T	[I]	P	P	S	G	V	W	L	N	R	N	R

Fig. 6. Amino acids whose side chains either project into, or border, the active site cavity of human CA I and CA II and bovine CA III molecules (cf. Eriksson, 1988; Notstrand *et al.*, 1974). Residues that form part of the hydrogen-binding network associated with the zinc ion are designated with an asterisk, and the three His residues liganded to the zinc ion are designated by a Z. Boxed residues appear to be unique and invariant for that CA form. The CA I, II, and III residues are consensus residues for these isozymes from amniotes (primarily mammals). Sequence sources are the same as those cited in Fig. 1.

genic cell line; however, the levels of CA III (1% of adult) present in undifferentiated myoblasts were not increased during myotube formation, as is characteristic of muscle-specific genes at this stage of muscle development (Tweedie and Edwards, 1989).

In the fetal rat, CA III is not expressed in skeletal muscle until day 17. During postnatal development, however, there appears to be a loss in CA III activity in some muscle fibers. By the third week, the fiber distribution of CA III appears to approach the pattern characteristic of adults (Lauila et al., 1989).

For a review of the effects of thyroid hormone on the expression of CA III in adult skeletal muscle, see Tashian (1989).

3. Transcription

The region 5' to exon 1 of the human CA III gene contains a TATA box and a CAAT box, and as with the CA I and CA II genes the promoter region is also characterized by being 60% G + C rich (Lloyd et al., 1987). In fact, when the promoter regions of CA II and CA III genes are compared (Fig. 4), some similarities are evident. For example, there is a 78% sequence identity in the 41-bp region surrounding the TATA box, and 90% sequence identity in the 21-bp region of the CACC tandem repeats containing an Sp1 site (Edwards, 1991). No such sequence identities are evident with the CA I sequence in this region. These sequence similarities between the CA II and CA III genes, along with their close chromosomal orientation pattern (Fig. 7; discussed in Section IV), suggest that they may have arisen through a duplication event that occurred after the one that gave rise to the CA I gene.

An interesting aspect of a comparison of the CA I, CA II, and III genes is that CA III seems to have acquired its unique tissue specificity and functional differences within a relatively short period of evolutionary time. After these three genes arose by duplications between 300 and 400 million years ago, six unique residues associated with the active site of CA III appear to have initially evolved relatively rapidly (see Figs. 1 and 6). However, since the radiation of the different mammalian orders over the past 80 million years or so, the CA III gene seems to have evolved more slowly than the CA I and CA II genes, indicating possible selection for an important function for CA III other than, or in addition to, CO_2 hydration (Hewett-Emmett and Tashian, 1991; Lowe et al., 1991). Thus, although a considerable amount of information has accumulated on this interesting CA isozyme, its function remains obscure and its regulatory elements await further definition.

D. CARBONIC ANHYDRASE VII

The CA VII gene is the most recent CA gene to be characterized (Montgomery *et al.*, 1991). The CA VII isozyme appears to be primarily expressed in the cytosol of salivary glands where its function is unknown. As shown in Fig. 1, the phylogenetic position of the CA VII gene is located between the CA I/II/III gene cluster, and the evolutionarily older CA IV and CA VI genes.

In contrast to the CA I, CA II, and CA III genes, the CA VII gene does not contain the promoter elements, TATA and CAAT, within 100 bp of the transcriptional start site. It does, however, contain a TTTAA sequence 102 bp upstream of the initiation codon that is located in a G + C-rich region (-243 to $+551$ relative to the start codon), similar to those found in the CA II and CA III genes previously discussed. Calculated over ranges of 100 bases centered around each nucleotide from -243 to $+551$ from the ATG start codon, the G + C content ranges from 65 to 84%, and 80 CpG dinucleotides are found in this region. Several potential Sp1-binding sites are also found in this region at positions -185, -152, -143, $+155$, and $+398$. Two additional cis-acting regulatory elements, both of which match all but one nucleotide of their consensus sequences, are an octomer motif (Oct-1) (ATTGCAG) at -384 and an immunoglobulin NF-κ B box (AGGGGACTTTCCA) at -442. Three *Alu* repeat elements are located in the CA VII gene. Two are close together (in opposite orientations) at the 3' end of intron 2, and one is located at the 3' end of intron 5. A poly(GT) sequence, $(GT)_{14}$, is present in the middle of intron 1, and what appears to be a length polymorphism, $(GT)_{16}$, has been identified (P. J. Venta, unpublished).

IV. Chromosomal Localization

In humans, the CA I, II, and III genes (*CA1, CA2,* and *CA3*) are clustered within about 180 kb on the short arm of chromosome 8 at q22 in the order *CA1, CA3, CA2* (Davis *et al.*, 1987; Nakai *et al.*, 1987; Lowe *et al.*, 1991). As shown in Fig. 7, the CA II and CA III genes are transcribed in the same direction and opposite to that of CA I (Lowe *et al.*, 1991). How these genes are oriented on the chromosome relative to the centromere/telemere has not been determined. Although these three genes are closely linked, no example of gene conversion has been reported.

In mice, the CA I, II, and III genes (*Car-1, Car-2,* and *Car-3*) are all probably closely linked in band 3A2 near the centromere of chromo-

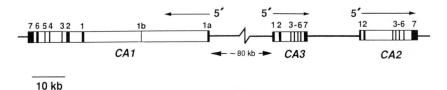

FIG. 7. Order of human carbonic anhydrase genes in the CA I, CA II, and CA III isozyme cluster at q22 on chromosome 8. Transcriptional directions indicated by arrows. Chromosomal orientation of genes in relation to centromere or telemere has not been determined. Data from Davis *et al.* (1987), Nakai *et al.* (1987), Venta *et al.* (1987), and Lowe *et al.* (1991).

some 3 (Beechey *et al.*, 1989). However, based on backcross studies, the order of the genes appears to differ from that in humans (Beechey *et al.*, 1989). No recombinants were observed between the *Car-1* and *Car-2* genes, whereas a recombination frequency of 2.4 ± 1.7% (three-point test) was found between *Car-3* and *Car-1* or *Car-2*. As linkage between the CA I and CA II genes has also been observed in the guinea pig and pigtail macaque (Carter *et al.*, 1972b; DeSimone *et al.*, 1973a), it is likely that these three genes have remained closely linked ever since their formation by gene duplication, which probably occurred before the appearance of reptiles over 300 million years ago.

The finding of the CA II gene in the bovine syntenic group, U23, on chromosome 14, along with the thyroglobulin gene and the *MYC* and *MOS* protooncogenes is of interest because these four genes also span the long arm of human chromosome 8 (Threadgill *et al.*, 1990). These four genes appear on three different chromosomes in mice, i.e., *Car-2*, *MOS*, and *Myc-1* + *Tg* are on chromosomes, 3, 4, and 15, respectively. This indicates a closer chromosomal conservation for the positions of these genes between ox and humans than between ox and mice.

In humans, the gene for CA VI has been assigned to the tip of the short arm of chromosome 1 and the gene for CA VII to chromosome 16 at q22–23 (Sutherland *et al.*, 1989; Montgomery *et al.*, 1991). This may reflect the postulated earlier evolutionary divergence of these genes (Fig. 1). It will be of interest to determine the chromosomal locations of the CA IV, CA V, and CARP genes, which also show early evolutionary branching patterns.

The recent finding that the human calbindin D28k (*CALB1*) and calretinin (*CALB2*) genes are located at 8q21.3–22.1 and 16q22–23, respectively, suggest that these genes may have shared a common evolutionary event with the human *CA1*, *CA2*, and *CA3* gene cluster and the *CA7* gene, which are at similar positions on human chromosomes

336   RICHARD E. TASHIAN

8 and 16 (Parmentier *et al.*, 1991). It will be of interest to determine whether *CALB1* is located in the bovine syntenic group, U23, as discussed above.

V. Genetic Variation

The suitability of the CA isozymes for genetic studies became apparent in the early 1960s, when it was found that the CA I and CA II isozymes were the second most abundant nonglobin proteins in mammalian red cells (cf. Tashian, 1969; Tashian and Carter, 1976). Because of this feature, genetic variants of these CA isozymes could be readily identified electrophoretically, isolated from blood samples, and their amino acid changes determined. Furthermore, as with the hemoglobins, the three-dimensional structures of the human CA I and CA II and bovine CA III isozymes have been characterized (cf. Eriksson and Liljas, 1991; Notstrand *et al.*, 1974). Thus, studies on genetic variation in the CA isozymes, along with the hemoglobin variants, early on provided us with two useful models to study the molecular bases of genetic control and structure–function relationships of both a catalytically active protein and a transport protein.

Because the identification of variants of the electrophoretically separated forms of CA I and CA II in red blood cells was a relatively easy procedure, carbonic anhydrases were among the first inherited variants of human and other mammalian enzymes to be characterized (cf. Tashian *et al.*, 1971; Tashian, and Carter, 1976). Since that time, several accounts and reviews have appeared describing these variants, primarily in human red cells, the majority of which are of rare occurrence (Tashian *et al.*, 1980, 1983; Deutsch, 1984; Jones and Shaw, 1984). Those variants in which the nucleotide changes have been determined are listed in Table 1 together with brief descriptions of restriction fragment-length polymorphisms (RFLPs) associated with the CA genes.

A. ACTIVITY MUTATIONS

Two interesting variants of human CA I have been reported in which the specific activities have been altered, CA II Baniwa (Mohrenweiser *et al.*, 1979, Tashian *et al.*, 1980) and CA I Michigan-1 (Shaw *et al.*, 1962; Tashian and Carter, 1976). CA II Baniwa occurs as a limited polymorphism in a Brazilian Indian tribe, and is characterized by approximately a 10-fold decrease in specific CO_2 hydrase activity. The nucleotide substitution has not been determined.

1. Zinc-Activated CA I Mutant

CA I Michigan-1 was discovered in a Caucasian family from Michigan (Shaw *et al.*, 1962). This mutation is remarkable in that although its CO_2 hydrase activity appears to be normal, its esterase activity toward α- or β-naphthyl acetate is dramatically increased in the presence of free zinc ions. No similar effect is observed when other divalent cations (e.g., Co^{2+}, Cu^{2+}, Mg^{2+}, and Ni^{2+}) or ester substrates (e.g., *p*-nitrophenyl acetate and 4-OH-5-nitro-α-toluenesulfonic acid sultone) are used (Tashian and Carter, 1976). This is a rare example of a mutation that results in the true metal-ion activation of an enzyme. Obviously, it would be extremely useful to determine the exact mechanism responsible for this unusually specific type of enzyme activation, which in turn should contribute to our general understanding of the process of metal-ion activation. The amino acid substitution has been determined for this variant as His-67 → Arg (Chegwidden *et al.*, 1992). Position 67 is also Arg in all CA III isozymes sequences to date (Hewett-Emmett and Tashian, 1991), however, because CA III exhibits no similar Zn^{2+} activation of esterase activity (R. E. Tashian, unpublished results), the basis for this unique, specific type of activation probably involves complex interactions among active site residues that are not readily apparent.

B. CA I DEFICIENCY MUTATIONS

1. Pigtail Macaque CA I Deficiencies

Two types of deficiencies have been reported. In one, the red cell CA I levels of the product of the CA Ia allele are reduced to about one-third of those of the CA Ib and CA Ic alleles (DeSimone *et al.*, 1973b). The CA Ia allele occurs at high frequencies, e.g., 0.56 in Malaysian populations of pigtail macaques (Tashian *et al.*, 1971). Recently, the amino acid sequences of the CA Ia and CA Ib proteins were determined, and it was found they differed by two amino acid substitutions at adjoining positions 241 and 242, which are, respectively, Ser and Glu in CA Ia, and Ile and Gln in CA Ib (Bergenhem *et al.*, 1992a). When the thermostability and proteolysis of the purified CA Ia and CA Ib proteins were compared, CA Ia was found to be more thermolabile and degraded more rapidly than CA Ib (Osborne and Tashian, 1984). Thus, it is possible that the lower levels of CA Ia can be attributed to these reduced stabilities.

The second deficiency is one in which the levels of red cell CA I have been reduced to trace levels ($\sim 1/5000$ of normal) in individuals homo-

TABLE 1

Genetic Variants within, or Associated with, the CA I, CA II, or CA III Genes of Mammals

Designation[a]	Mutation	Comments
Carbonic anhydrase I		
Human		
CA I Australia-1 (Aborigine)	Asp-8/Gly (GAT/GGT)	Polymorphic: allele frequency (0.025)
CA I Michigan-1 (Caucasian)	His-67/Arg (CAT/CGT)	Esterase activity toward naphthyl acetate activated by Zn^{2+}
CA I Nagasaki (Japanese)	Arg-76/Gln (CGA/CAA)	—
CA I Hiroshima-1 (Japanese)	86-Asp/Gly (GAC/GGC)	—
CA I Michigan-2 (Black)	Thr-100/Lys (ACA/AAA)	—
CA I London (English)	Glu-102/Lys (GAG/AAG)	—
CA I Hull (English)	Gln-225/Lys (CAA/AAA)	—
CA I Wisconsin (Caucasian)	Asp-236/Val (GCT/GTG)	—
CA I Guam [Micronesian (Marianas), Filipino, Malaysian, Indonesian (Java)]	Gly-235/Arg (GGT/CGT)	Limited polymorphism in Mindanao, Philippines (see text): allele frequency (0.256)
CA I deficiency [Greek (Icaria)]	Arg-246/His (CGC/CAC)	Allele may be polymorphic in Icaria
CA I Portsmouth (English)	Thr-255/Arg (ACA/AGA)	—

Pigtail Macaque (*Macaca nemestrina*) Electromorphs (polymorphic)		
CA Ia and CA Ib	His-96 (CAT/CAC)	CA Ia levels one-third those of CA Ib (see text)
	Glu-102 (GAA/GAG)	—
	Pro-238/(CCC/CCT)	—
	Ile-241/Ser (ATT/AGT)	Ca Ia (Ser). CA Ib (Ile)
	Gln-242/Glu (CAG/GAG)	CA Ia(Glu), CA Ib (Gln)
	Gln-249 (CAA/CAG)	
CA I deficiency	−85 C/G change forms AUG start codon 6 bp from 5' end of exon la	Creates upstream ORF terminating 5 bp from normal AUG (see text)
Horse (*Equus caballus*)[b] Electromorphs (polymorphic)		
CA I(A$_1$)	Ser-182/Arg	—
CA I(A$_2$)	Asp-81/Gly; Gly-82/Cys	—
	Pro-83/Phe	—
CA I(B)	Ser-182/Arg; Gln-222/Arg	—
CA I(D)	Ser-65/Gly; Asp-81/Gly	—
	Ser-115/His; Leu-157/Gly	—
	Cys-212/Tyr; Ser-224/Ala	—
	Ala-54/Asp	—
CA I(T)		
Carbonic Anhydrase II Human		
CA II Jogjakarta (Indonesian)	Lys-18/Glu (AAG/GAG)	—
CA II Melbourne (Caucasian)	Pro-237/His (CCC/CAC)	—
CA II deficiency [Belgian; Italian (U.S.)]	His-107/Tyr (CAT/TAT)	AccI site formed

TABLE 1 (*continued*)

Designation[a]	Mutation	Comments
Carbonic Anhydrase II		
Human		
CA II deficiency [Arabian (Algeria, Egypt, Kuwait, Saudi Arabia)]	Donor site intron 2 (GT/CT)	Impaired RNA processing
CA II deficiency [German (U.S.)]	Acceptor site intron 5 (AG/AC)	—
CA II deficiency [Hispanic (U.S.)]	Lys-228 (AAA/AA−) single bp deletion	Frameshift mutation *Mae*III site formed
CA II deficiency [Caucasian (U.S.)]	(15–20 kb) deletion that includes exon 1 or 2	—
CA II$_2$ (Black African)	Asn-253/Asp (AAC/GAC)	Polymorphic (see text)
RFLP (*Taq*I)	~1 kb upstream of exon 1	PCR typing developed
RFLP (*Bst*NI)	Leu-189 (TTG/CTG)	PCR typing developed
Mouse (*Mus musculus*)		
CA II deficiency	Gln-155/term (CAA/TAA)	Mutation induced by ENU (see text)
Strain difference (YBR, BALB/c)	Asp-31 (GAC/GAT)	YBR (GAC), BALB/c (GAT)
Electromorphs	Gln-38/His (CAG/CAC)	CA IIa (Gln), CA IIb (His)

RFLP (YBR)	Asp-178 (GAT/GAC)	GATC MboI site
RFLP (BALB/c)	3' noncoding region (A/G) at + 880	GCGC HhaI site
Horse (Equus caballus)[b]		
N-terminal polymorphism	Ser-2/Thr (TCC/ACC)	—
Ox (Bos taurus)		
Electromorphs	Arg-57/Gln	May be polymorphic
Sheep (Ovis aries)		
Electromorphs	Lys-36/Thr (AAA/ACA)	May be polymorphic
Carbonic Anhydrase III		
Human		
Polymorphism	Val-31/Ile (GTT/ATT)	Allele frequency ~50% in all human populations tested
RFLP (TaqI)	Anonymous marker D8S8 closely linked to CA3 at 8q13–21.1	Two separate sites: TaqI-A, TaqI-B
Mouse (Mus musculus/spretus)		
RFLP (PstI)	Interspecies hybrid difference	PstI alleles: Car-3[a] (M. musculus) Car-3[b] (M. spretus)

[a]Sources: human CA I and CA II (Tashian et al., 1980; Deutsch, 1984; Jones and Shaw, 1982, 1984; Tashian et al., 1983; Mohrenweiser et al., 1989; Wagner et al., 1991; Venta et al., 1991; Venta and Tashian, 1990; 1991; Roth et al., 1992; Hu et al., 1992; pigtail macaque CA I (Bergenhem et al., 1991, 1992a,b); horse CA I and CA II (Jabusch et al., 1980; Jabusch and Deutsch, 1984); mouse CA II (Lewis et al., 1988; Venta et al., 1985; W. S. Sly, unpublished); ox CA II (Gulian et al., 1974); sheep CA II (Mallet et al., 1979); human CA III (Hewett-Emmett et al., 1983; Edwards et al., 1990); mouse CA III (Beechey et al., 1989).

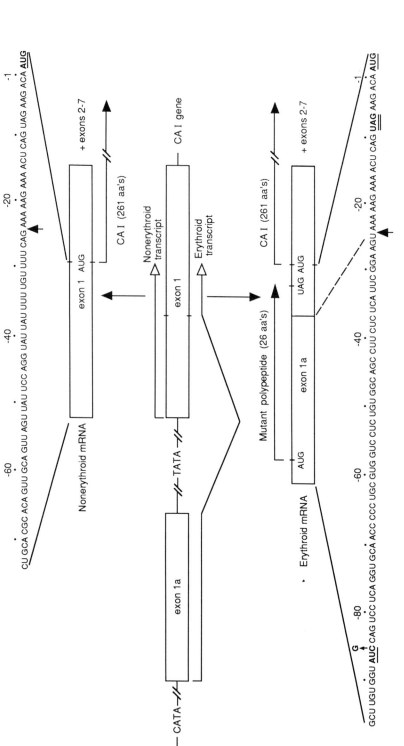

FIG. 8. Diagram of the 5′ end of the pigtail macaque CA I gene and the two mRNAs (erythroid and nonerythroid) produced by the two promoters. The proposed upstream open reading frame formed by the C → G change at −85 is shown along with the normal red cell CA I mRNA. Sequences of the 5′ ends (cap sites to AUG start site) of the erythroid and nonerythroid mRNAs are shown above and below their respective mRNAs. From Bergenhem *et al.* (1992b).

zygous for this CA I deficiency (Tashian *et al.*, 1971; DeSimone *et al.*, 1973a,b). The red cell CA I deficiency occurs as a widespread polymorphism in southeast Asia, where the frequency of the variant allele ranges from 0.55 in Thailand and 0.57 in Malaysia to 0.70 in Sumatra (Tashian *et al.*, 1971; Darga *et al.*, 1975). The fact that the levels of CA I appear to be normal in nonerythroid tissue (e.g., colon) indicates that the mutation is located in the erythroid-specific exon (exon 1a) or its promoter region (Tashian *et al.*, 1990). Sequence analysis of exon 1a revealed to C to G change 12 nucleotides downstream of the cap site in the variant red cell CA I mRNA (Bergenhem *et al.*, 1992b). As shown in Fig. 8, this mutation forms a new AUG start site and an open reading frame (ORF) encoding 26 amino acids that terminates six nucleotides prior to the normal AUG initiation codon for CA I.

This mutation is of unusual interest for several reasons. First, although upstream ORFs are known to inhibit translation of downstream initiation AUG codons, no example of a mutation of this type has been reported for a mammalian gene. Second, the similar, high frequencies for the deficiency allele in both mainland (Thailand, Malaysia) and island (Sumatra) populations indicate that the mutation originated early in a founder population that predated Pleistocene glaciation (cf. Fooden, 1975). In this respect, it will be of interest to examine pigtail macaques from the island of Borneo (known to be inhabited by this species) for this deficiency. Third, there is a cis-acting, 30% reduction in the product of the CA II gene that is linked to the mutant CA I deficiency gene (DeSimone *et al.*, 1973a,b). Thus, the levels of CA II in red cells of animals homozygous for the CA I deficiency gene are decreased by about 60%. It is difficult to conceptualize how the mutation in exon 1a might effect the expression of the CA II gene, whose promoter is separated from the erythroid CA I promoter by about 100 kb (see Fig. 7).

2. Human CA I Deficiency

A CA I deficiency variant was described from three members of a family living in Montreal, Canada, who had emigrated from the Greek Island of Icaria (Kendall and Tashian, 1977). Except for the reduction to trace levels of red cell CA I, no hematological or other abnormalities were noted in the individuals homozygous for this deficiency. The point mutation that is the probable cause of the CA I deficiency is an Arg to His change at position 246 in exon 7 (Wagner *et al.*, 1991). An Arg residue is found at this position (Fig. 6) in all CA isozymes sequenced to date, including those from the green alga, *Chlamydomonas reinhardtii,* and the CA-related proteins: D8 from vaccinia virus and

CARP from mouse and human (Hewett-Emmett and Tashian, 1991; Kato, 1990; Fukuzawa *et al.*, 1990; R. E. Tashian, unpublished). It thus appears that Arg-246 plays a critical role in the intermolecular structures of all CA molecules, and that the substitution of His at this position results in an unstable molecule.

Unlike the red cell CA I deficiency described in the pigtail macaque, which is seemingly restricted to erythrocytes, the human CA I deficiency probably results in greatly reduced levels of CA I in all tissues where it is normally expressed (e.g., gastrointestinal mucosa, erythrocytes, vascular endothelium). However, the apparent lack of any dysfunction in CA I-deficient individuals indicates that other CA isozymes or alternative processes can substitute for the function of CA I, or that trace levels of the mutant CA I may be present in nonerythroid tissue.

C. CA II Deficiency Mutations

1. Human CA II Deficiencies

In contrast to CA I deficiencies, human CA II deficiencies produce a syndrome characterized by osteopetrosis, renal tubular acidosis, and cerebral calcification. Varying levels of mental retardation and impaired growth are also associated with the syndrome. Over 40 cases of this syndrome have now been reported from a variety of populations, including European Caucasians (Belgian, German, Italian), Arabians (Algeria, Kuwait, Saudi Arabia, Egypt), Hispanics (U.S.), and Afro-Americans. [For recent reviews, see Cochat *et al.* (1987), Sly (1989), and Strisciuglio *et al.* (1990).] Those CA II deficiency mutations that have been determined are listed in Table 1 and their locations in the CA II gene are depicted in Fig. 9. The pattern that emerges is that two different mutations are found in the European families, another

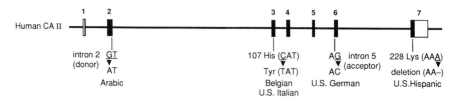

Fig. 9. Location of mutations on human CA II gene that are responsible for CA II deficiency syndrome in individuals of European (Belgian, Italian, German), Arabic (Kuwait, Saudi Arabia, Egypt, Algeria), and Hispanic origins. From Venta *et al.* (1991), Roth *et al.* (1992), and Hu *et al.* (1992).

is common to United States Hispanics, and one is unique to Arabic families.

a. Belgian Family. The patient in this family is the son of a consanguineous marriage and his CA II is characterized by an amino acid change (His-107 → Tyr) that results in an unstable CA II molecule (Venta *et al.*, 1991; Roth *et al.*, 1992). As seen in Fig. 6, His-107 is present in all CA isozymes sequenced to date, including the algal CA, the CA-related protein, and the vaccina transmembrane protein D8 (Hewett-Emmett and Tashian, 1991; Kato, 1990; Fukuzawa *et al.*, 1990). This indicates that His-107 may not only be involved in some aspect of the active site mechanism, but probably also plays a critical role in maintaining the structure the of CA II molecule. Furthermore, it provides *in vivo* evidence for the importance of a particular amino acid residue to the structure of all CA molecules. The fact that low levels of CO_2 hydrase activity could be detected in the mutant enzyme when expressed in bacteria suggests that if such residual activity is present in the patient, it may contribute to the apparent lack of mental impairment in the propositus (Roth *et al.*, 1992).

b. Double Heterozygous Patients (United States Family). This is the family from St. Louis in which the CA II deficiency was first described by Sly *et al.* (1983). The mutation found in the father (German ancestry) of the three affected daughters is characterized by a G to C substitution at the acceptor splice site of intron 5 and would be expected to interfere with the processing of the RNA transcript of the CA II gene (Sly *et al.*, 1983, 1985; Roth *et al.*, 1992). The mother in this family is of Italian extraction and carries the same His-107 → Tyr mutation described above in the Belgian patient (Venta *et al.*, 1991). The three affected children are therefore double heterozygotes for the amino acid substitution and acceptor site mutations.

c. Arabic Mutation. This mutation has been identified as the substitution of C for G at the invariant GT donor splice site of intron 3 in Arabic families with CA II deficiency syndrome from North Africa and the Middle East (Hu *et al.*, 1992). As with the "German" mutation in the St. Louis family described above, this splice site mutation must disrupt the normal processing of the RNA transcript. All affected individuals tested were the children of consanguineous parents and homozygous for this mutation.

d. Hispanic Mutation. This mutation is due to the deletion of one adenine nucleotide in the codon for Lys-228 (AAA) in exon 7 (Hu *et al.*,

1992). Translation continues for an additional 11 amino acids and terminates at a UAA stop codon. The truncated CA II molecule of 240 amino acids is expected to be unstable because it lacks the seemingly invariant Arg-246, the substitution of which to His was demonstrated to be the cause of the human CA I deficiency described in Section V,B,2.

2. Mouse CA II Deficiency

A CA II-deficient allele was produced in a mouse by ethylnitrosourea treatment, and homozgous CA II-deficient mice were obtained by appropriate matings (Lewis *et al.*, 1988). Surprisingly, even older CA II-deficient mice showed no evidence of the osteopetrosis and cerebral calcification that are pathognomonic for the human deficiency. They did, however, exhibit growth impairment, renal tubular acidosis, and vascular calcification (Lewis *et al.*, 1988; Spicer *et al.*, 1989). This mutation was found to be a C to T substitution in the codon for Gln-155 (CAA), which results in a termination codon (UAA) near the beginning of exon 5 (W. S. Sly, unpublished).

Because CA II is completely absent in all cells of the CA II-deficient mice, they are obviously ideally suited as controls for studies on the histochemical localization of CA II (Spicer *et al.*, 1989; Ghandour *et al.*, 1989; Ridderstrale *et al.*, 1992) as well as physiological studies relating to the functions of this widely distributed and most active of the CA isozymes (Brechue *et al.*, 1991). Furthermore, these mice would be expected to serve as useful animal models for gene therapy studies.

D. ALLELIC POLYMORPHISMS

A number of electrophoretic variants (electromorphs) of CA I and CA II occurring at polymorphic frequencies have been described in humans and other species of mammals (for reviews of early reports, see Tashian and Carter, 1976; Tashian *et al.*, 1980); however, only those in which the amino acid change or nucleotide substitution has been determined are listed in Table 1, and some of the more interesting alleles are briefly discussed in this section.

1. Carbonic Anhydrase I Locus

a. Horse (Electrophoretic Variants). Five electromorphs of CA I from erythrocytes have been described in Japanese farm horses (Jabusch *et al.*, 1980). A total of 11 amino acid substitutions were found, varying from one to five for each of the five electromorphs.

b. Pigtail Macaque (Nucleotide Variants). Sequence analyses of two (CA Ia and CA Ib) of the four electromorphs of CA I in pigtail macaques revealed five nucleotide changes in addition to the one responsible for the a and b electrophoretic types, i.e., 242-Gln/Glu (Bergenhem *et al.*, 1992a). The product of the CA Ia allele differs in that its levels are about one-third lower (in red cells) than the products of the other three alleles (DeSimone *et al.*, 1973a,b). It is of interest that biochemical analyses of the purified CA Ia molecule showed that it is less thermostable and more readily degradable by proteolysis than the CA Ib molecule (Osborne and Tashian, 1984). However, it is not clear how the substitution of Glu for Gln at position 242 and Ile for Ser at position 241 produces a less stable (proteolytically and thermally) molecule, and how this might contribute to the relatively lower levels of CA Ia. Certainly, the levels of their respective mRNAs will have to be quantitated to determine to what extent alterations in translation or transcription might be responsible.

c. CA I Australia-1. This variant (Asp-8 → Gly) is found among Australian Aborigines, in which the allelic frequency is 0.025 (Jones and Shaw, 1982). It is noteworthy that this position is Gly in mouse CA I, and Gly in all CA IIs sequenced with the exception of mouse, in which it is Asp (Hewett-Emmett and Tashian, 1991).

d. CA I Guam. This variant (Gly-235 → Arg) is found at subpolymorphic frequencies in Micronesians (Mariana Islands), Filipinos, Malaysians, and Indonesians (cf. Tashian *et al.*, 1980). However, the variant allele occurs at polymorphic frequencies (0.256) in the Negrito (Mamamwa) tribes of Northern Mindanao in the Philippines (Omoto, 1979). The distribution of the variant suggests that this mutation probably originated in Southeast Asia.

2. Carbonic Anhydrase II Locus

a. Human CA II$_2$. This variant (Asn-253 → Asp) appears to be widespread among black Africans. In East Africa, the allelic frequencies range from its absence in Ethiopia and Uganda, to 0.054 in Kenya; and in West Africa, the frequencies are 0.158 in Zambia and Nigeria, 0.103 in Gambia, and 0.044 in the Cameroons (Carter, 1972a; Welch, 1975; Kendall, 1979). In blacks not living in Africa, the frequencies range from 0.088 in Jamaica to 0.101 and 0.118 in the United States (Moore *et al.*, 1977; and R. E. Tashian, unpublished). Subpolymorphic frequencies (0.01) of what is purported to be the CA II$_2$ allele have been reported in Italian (Caucasian) populations (Beretta *et al.*, 1980).

b. Human RFLPs. Two restriction site polymorphisms have been identified. One is a *Taq*I site located about 1 kb upstream of exon 1

(Venta and Tashian, 1990), and the other is a *Bst*NI site at Leu-189 (TTG → CTG) in exon 6 (Venta and Tashian, 1991). These RFLPs will be useful for linkage studies in CA II deficiency patents as well as linkage to other CA genes. Polymerase chain reaction (PCR) typing has been developed for both RFLPs.

   c. Mouse RFLPs. In laboratory mice, two RFLPs have been described. In one, a T → C change in the third codon position of Asp-178 creates a GATC *Mbo*I site; in the second, an A → G substitution in the 3' noncoding region (at +880) creates a GCGC *Hha*I site (Venta *et al.*, 1985).

   d. Horse N-Terminal Variant. A polymorphism involving the codon for the N-terminal residue of the CA II gene (Ser-2/Thr) has been reported from Japanese farm horses (Jabusch and Deutsch, 1984).

3. Carbonic Anhydrase III Locus

   a. Human. The two CA III alleles produced by an amino acid change at position 31 (Val/Ile) were found to occur at frequencies of about 50% in all human populations examined (Hewett-Emmett *et al.*, 1983). It is of interest that this position is either Val or Ile in all CA III molecules of other species sequenced to date (Hewett-Emmett and Tashian, 1991). To what extent polymorphisms exist at this position in other species has not been determined.
   Allelic variation was found at two separate *Taq*I sites (*Taq*I-A and *Taq*I-B) at the human *D8S8* locus, which is closely linked to the CA III gene. Allelic frequencies at the A site were 0.78 and 0.27, and 0.94 and 0.04 at the B site (Edwards *et al.*, 1990).

   b. Mouse. A *Pst*I RFLP, linked to the CA III locus, *Car-3*, was found in hybrids between *Mus musculus* and *Mus spretus* (Beechey *et al.*, 1989).

VI. Future Directions

   It is likely that additional members of the carbonic anhydrase gene family will continue to be discovered, and it is logical to assume that as we learn more about these CA or CA-related genes, we may soon be in a position to begin to unravel some of the elusive evolutionary events that have guided the remarkable diversity of this unusually versatile multigene family.

Research directions on the CA genes that should prove rewarding would appear to be (1) elucidation of the regulatory elements controlling their tissue-specific and developmental expression, as well as factors involved in their hormonal control, (2) determination of the evolutionary origins of their regulatory elements, especially the multiple-promoter regions that control the erythroid and nonerythroid expression of the CA I genes, (3) determination of their possible physiological functions, and (4) identification of hereditary disorders resulting from deficiencies in CA genes other than CA I and CA II.

Recent advances in the use of transgenes to study tissue-specific expression, and the use of gene-targeted inactivation to study the physiological function of a specific CA isozyme gene, are some of the powerful procedures that can be used for the analyses of some of the proposed research areas (cf. Joyner, 1991). Also, results from site-directed mutagenesis studies are proving very useful in establishing the importance of various residues that participate in the active site mechanisms of the carbonic anhydrases (cf. Silverman, 1991). However, the roles of residues not associated with the active site that may be critical for folding, structural stability, and binding functions remain to be defined by this technique.

ACKNOWLEDGMENTS

I am indebted to Drs. Patrick Venta, William Epperly, and David Hewett-Emmett for their welcome contributions and for the many fruitful discussions I had with them during the preparation of this review. I also thank Dr. William Sly and his colleagues, Drs. Donna Roth and Peiyl Hu, for promptly and generously providing me with their unpublished data, particularly those relating to the CA II deficiencies. I am especially grateful to Ms. Judy Worley for the indispensible care, cooperation, and patience that she exhibited while putting this manuscript together. The data presented from my laboratory were supported by NIH Grant GM24681.

REFERENCES

Aldred, P., Fu, P., Barrett, G., Penschow, J. D., Wright, D., Coghlan, J. P., and Fernley, R. T. (1991). Human secreted carbonic anhydrase: cDNA cloning, nucleotide sequence, and hybridization histochemistry. *Biochemistry* 30, 569–575.

Arlot-Bonnemains, Y., Fouchereau-Peron, M. B., Moukhtar, M. S., Benson, A. A., and Milhaud, G. (1985). Calcium-regulating hormones modulate carbonic anhydrase II in the human erythrocyte. *Proc. Natl. Acad. Sci. U.S.A.* 82, 8832–8834.

Beechey, C., Tweedie, S., Spurr, N., Ball, S., Peters, J., and Edwards, Y. (1989). Mapping of mouse carbonic anhydrase-3, *Car-3:* A further locus in the homologous region on mouse chromosome 3 and human chromosome 8. *Genomics* 4, 672–696.

Beretta, M., Mattiuz, P. L., and Barrai, I. (1980). Red cell carbonic anhydrase II: A

CA^2_{II} allele in a sample of the Italian population. *Ann. Dell Univ. Ferrara (Biol.)* **2**, 77–82.

Bergenhem, N. C. H., Venta, P. J., Hopkins, P. J., Kim, H., and Tashian, R. E. (1992a). Variation in coding exons of two electrophoretic alleles at the pigtail macaque carbonic anhydrase I locus as determined by direct double-stranded sequencing of PCR products. *Biochem. Genet.* **30**, 279–287.

Bergenhem, N. C. H., Venta, P. J., Hopkins, P. J., Kim, H. J., and Tashian, R. E. (1992b). Mutation creates an open reading frame within the 5' untranslated exon of macaque erythroid carbonic anhydrase I mRNA that suppresses CA I expression and supports the scanning model for translation *Proc. Natl. Acad. Sci. U.S.A.* (in press).

Botrè, F., Gros, G., and Storey, B. T., eds. (1991). "Carbonic Anhydrase: From Biochemistry and Genetics to Physiology and Clinical Medicine." VCH Publishers, New York.

Boyer, S. H., Siegel, S., and Noyes, A. N. (1983). Developmental changes in human erythrocyte carbonic anhydrase levels: Coordinate expression with adult hemoglobin. *Develop. Biol.* **97**, 250–253.

Brady, H. J. M., Sowden, J. C., Edwards, M., Lowe, N., and Butterworth, P. H. W. (1989). Multiple GF-1 binding sites flank the erythroid specific transcription unit of the human carbonic anhydrase I gene. *FEBS Lett.* **257**, 451–456.

Brady, H. J. M., Lowe, N., Sowden, J. C., Edwards, M., and Butterworth, P. H. W. (1991). The human carbonic anhydrase I gene has two promoters with different tissue specificities. *Biochem. J.* **277**, 903–905.

Brechue, W. F., Kinne-Saffran, E., Kinne, R. K. H., and Maren, T. H. (1991). Localization and activity of renal carbonic anhydrase (CA) in CA II deficient mice. *Biochim. Biophys. Acta* **1066**, 201–207.

Burnell, J. N., Gibbs, M. J., and Mason, J. G. (1989). Spinach chloroplastic carbonic anhydrase. *Plant Physiol.* **92**, 37–40.

Butterworth, P. H. W., Barlow, J. H., Brady, H. J. M., Edwards, M., Lowe, N., and Sowden, J. (1991). The structure and regulation of the human carbonic anhydrase I gene. *In* "The Carbonic Anhydrases: Cellular Physiology and Molecular Genetics" (S. J. Dodgson, R. E. Tashian, G. Gros, and N. D. Carter, eds.), pp. 197–207. Plenum Press, New York.

Carter, N. D. (1972a). Carbonic anhydrase II polymorphism in Africa. *Hum. Hered.* **22**, 539–541.

Carter, N. D. (1972b). Carbonic anhydrase isozymes in *Cavia porcellus, Cavia aperea* and their hybrids. *Comp. Biochem. Physiol.* **43B**, 743–747.

Carter, N., and Jeffery, S. (1985). Carbonic anhydrase: Update and new horizons. *Biochem. Soc. Trans.* **13**, 531–533.

Chegwidden, W. R., Wagner, L. E., Venta, P. J., Yu, Y.-S. L., and Tashian, R. E. (1992). Human carbonic anhydrase I variant (CA I Michigan-1) is due to a 67-His to Arg change at the active site that results in the zinc-ion activation of its esterase activity toward naphthyl acetate. *Isozyme Bull.* **25**, 30.

Chin, C. C. Q., Anderson, P. M., and Wold, F. (1983). The amino acid sequence of *Escherichia coli* cyanase. *J. Biol. Chem.* **258**, 276–282.

Cochat, P., Loras-Duclaux, I., Guibaud, P. (1987). Déficit en anhydrase carbonique II: Ostéopétrose, acidose rénale tubulaire et calcifications intrâcraniennes: Revue de las litérature à partir de trois observations. *Pediatrie* **42**, 121–128.

Darga, L. L., Goodman, M., Weiss, M. L., Moore, G. W., Prychodko, W., Dene, M., Tashian, R. E., and Keen, A. (1975). Molecular systematics and clinal variation in macaques. *In* "Isozymes: Genetics and Evolution" (C. L. Markert, ed.), Vol. 4, pp. 797–812. Wiley-Liss, New York.

Davis, M., West, L., Barlow, J., Butterworth, P. H. W., Lloyd, J., and Edwards, Y. H. (1987). Regional localization of the carbonic anhydrase genes *CA1* and *CA3* on human chromosome 8. *Somatic Cell Mol. Genet.* **13**, 173–178.

DeSimone, J., Linde, M., and Tashian, R. E. (1973a). Evidence for linkage of carbonic anhydrase isozyme genes in the pigtail macaque, *Macaca nemestrina. Nature (New Biol.)* **242**, 55–56.

DeSimone, J., Magid, E., and Tashian, R. E. (1973b). Genetic variation in the carbonic anhydrase isozymes of macaque monkeys. II. Inheritance of red cell carbonic anhydrase levels in different carbonic anhydrase I genotypes of the pig-tailed macaque, *Macaca nemestrina. Biochem. Genet.* **8**, 165–174.

Deutsch, H. F. (1984). Primary structures and genetic changes in mammalian carbonic anhydrase isozymes. *Ann. N.Y. Acad. Sci.* **429**, 183–194.

Deutsch, H. F. (1987). Carbonic anhydrases. *Int. J. Biochem.* **19**, 101–113.

Disela, C., Glineur, C., Bugge, T., Sap, J., Stengl, G., Dodgson, S., Stunnenberg, H., Beug, H., and Zenke, M. (1991). *v-erbA* overexpression is required to extinguish *c-erbA* function in erythroid cell differentiation and regulation of the *erbA* target gene CA II. *Genes Dev.* **5**, 2033–2047.

Dodgson, S. J., Tashian, R. E., Gros, G., and Carter, N. D., eds. (1991). "The Carbonic Anhydrases: Cellular Physiology and Molecular Genetics." Plenum Press, New York.

Donehower, L. A., Stagle, B. L., Wilde, M., Darlington, G., and Butel, J. S. (1989). Identification of a conserved sequence in the non-coding regions of many human genes. *Nucleic Acids Res.* **17**, 699–710.

Dynan, W. S. (1986). Promoters for housekeeping genes. *Trends Genet.* **2**, 196–197.

Edwards, Y. (1990). Structure and expression of mammalian carbonic anhydrases. *Biochem. Soc. Trans.* **18**, 171–175.

Edwards, Y. H. (1991). Structure and expression of the carbonic anhydrase III gene. *In* "The Carbonic Anhydrases: Cellular Physiology and Molecular Genetics" (S. J. Dodgson, R. E. Tashian, G. Gros, and N. D. Carter, eds.), pp. 215–224. Plenum Press, New York.

Edwards, Y., Williams, S., West, L., Lipowicz, S., Sheer, D., Attwood, J., Spurr, N., Sarkar, R., Saha, N., and Povey, S. (1990). The polymorphic human DNA sequence *D8S8* assigned to 8q13-21.1, close to the carbonic anhydrase gene cluster, by isotopic and non-isotopic *in situ* hybridization and linkage analysis. *Ann. Hum. Genet.* **54**, 131–139.

Eriksson, A. E., and Liljas, A. (1991). X-ray crystallographic studies of carbonic anhydrase isozymes I, II, and III. *In* "The Carbonic Anhydrases: Cellular Physiology and Molecular Genetics" (S. J. Dodgson, R. E. Tashian, G. Gros, and N. D. Carter, eds.), pp. 33–48. Plenum Press, New York.

Fawcett, T. W., Browse, J. A., Volokita, M., and Bartlett, S. G. (1990). Spinach carbonic anhydrase primary structure deduced from the sequence of a cDNA clone. *J. Biol. Chem.* **265**, 5414–5417.

Fernley, R. T. (1988). Non-cytoplasmic carbonic anhydrases. *Trends Biochem.* **13**, 356–359.

Fooden, J. (1975). Taxonomy and evolution of liontail and pigtail macaques (Primates: Cercopithecidae). *Fieldiana (Zool.)* **67**, 1–169.

Forster, R. E., ed. (1988). Respiratory physiology: Velocity of CO_2 exchange. *Annu. Rev. Physiol.* **50**, 623–717.

Fraser, P., Cummings, P., and Curtis, P. (1989). The mouse carbonic anhydrase I gene contains two tissue-specific promoters. *Mol. Cell. Biol.* **9**, 3308–3313.

Fukuzawa, H., Fujiwara, S., Yamamoto, Y., Dioniso-Sese, M., and Miyachi, S. (1990).

cDNA cloning, sequence, and expression of carbonic anhydrase in *Chlamydomonas reinhardtii:* Regulation by environmental CO₂ concentration. *Proc. Natl. Acad. Sci. U.S.A.* **87,** 4383–4387.

Ghandour, M. S., Skoff, R. P., Venta, P. J., and Tashian, R. E. (1989). Oligodendrocytes express a normal phenotype in carbonic anhydrase II-deficient mice. *J. Neurosci. Res.* **23,** 189–190.

Guilloton, M. B., Korte, J. J., Lamblin, A. F., Fuchs, J. A., and Anderson, P. M. (1992). Carbonic anhydrase in *Escherichia coli:* A product of the *cyn* operon. *J. Biol. Chem.* **267,** 3731–3734.

Gulian, J.-M., Limozin, N., Charrel, M., Laurent, G., and Darrien, Y. (1974). Les deux isozymes de l'anhydrase carbonique erythrocytaire bovine different par une substitution Arg → Gln en position 56. *C. R. Acad. Sci. (Paris)* **278,** 1123–1126.

Härkönen, P. L., and Väänänen, H. K. (1988). Androgen regulation of carbonic anhydrase II, a major soluble protein in rat lateral prostate tissue. *Biol. Reprod.* **38,** 377–384.

Härkönen, P. L., Mäkelä, S. I., Valve, E. M., Karhukorpi, E., and Väänänen, K. (1991). Differential regulation of carbonic anhydrase II by androgen and estrogen in dorsal and lateral prostate of the rat. *Endocrinology* **128,** 3219–3227.

Hewett-Emmett, D., and Tashian, R. E. (1984). Origins and molecular evolution of the carbonic anhydrase isozymes. *Ann. N.Y. Acad. Sci.* **429,** 338–358.

Hewett-Emmett, D., and Tashian, R. E. (1991). Structure and evolutionary origins of the carbonic anhydrase multigene family. *In* "The Carbonic Anhydrases: Cellular Physiology and Molecular Genetics" (S. J. Dodgson, R. E. Tashian, G. Gros, and N. D. Carter, eds.), pp. 15–32. Plenum Press, New York.

Hewett-Emmett, D., Welty, R. J., and Tashian, R. E. (1983). A widespread silent polymorphism of human carbonic anhydrase III: Implications for evolutionary genetics. *Genetics* **105,** 409–420.

Hodgen, G. D., and Falk, R. J. (1971). Estrogen and progesterone regulation of carbonic anhydrase isoenzyme in guinea pig and rabbit uterus. *Endocrinology* **89,** 859–864.

Hu, P. Y., Roth, D. E., Skaggs, L. A., Venta, P. J., Tashian, R. E., Guibaud, P., and Sly, W. S. (1992). A splice junction mutation in intron 2 of the carbonic anhydrase II gene of osteopetrosis patients from Arabic countries. *Human Mutation* (in press).

Jabusch, J. R., and Deutsch, H. F. (1984). Sequence of the high activity equine erythrocyte carbonic anhydrase: N-terminal polymorphism (acetyl-Ser/acetyl-Thr) and homologies to similar mammalian isozymes. *Biochem. Genet.* **22,** 357–367.

Jabusch, J. R., Bray, R. P., and Deutsche, H. F. (1980). Sequence of the low activity equine erythrocyte carbonic anhydrase and delineation of the amino acid substitutions in various polymorphic forms. *J. Biol. Chem.* **255,** 9196–9204.

Jeffery, S., Carter, N. D., and Wilson, C. A. (1984). Carbonic anhydrase II isoenzyme in rat liver is under hormonal control. *Biochem. J.* **221,** 927–929.

Jeffery, S., Edwards, Y., and Carter, N. (1980). Distribution of CA III in fetal and adult human tissue. *Biochem. Genet.* **18,** 843–849.

Jeffery, S., Wilson, C., Mode, A., Gustafsson, J. A., and Carter, N. (1986). Effects of hypohysectomy and growth hormone infusion on rat hepatic carbonic anhydrase. *J. Endocrin.* **110,** 123–126.

Jeffery, S., Carter, N. D., Clark, R. G., and Robinson, C. A. F. (1990). The episodic secretory pattern of growth hormone regulates liver carbonic anyhydrase III. *Biochem. J.* **266,** 69–74.

Jones, G. L., and Shaw, D. C. (1982). A polymorphic variant of human erythrocyte carbonic anhydrase I with a widespread distribution in Australian Aborigines, CA I

Australia-9 (8-Asp → Gly); purification, properties, amino acid substitution and possible physiological significance of the variant enzyme. *Biochem. Genet.* **20**, 943–977.

Jones, G. L., and Shaw, D. C. (1984). Human red cell carbonic anhydrase variants in Australia. *Ann. N.Y. Acad. Sci.* **429**, 249–261.

Joyner, A. L. (1991). Gene targeting and gene trap screens using embryonic stem cells: New approaches to mammalian development. *BioAssays* **13**, 649–656.

Karrasch, M., Bott, M., and Thauer, R. K. (1989). Carbonic anhydrase activity in acetate grown *Methanosarcina barkeri*. *Arch. Microbiol.* **151**, 137–142.

Kato, K. (1990). Sequence of a novel carbonic anhydrase-related polypeptide and its exclusive presence in Purkinje cells. *FEBS Lett.* **27**, 137–140.

Kendall, A. (1979). Human erythrocyte carbonic anhydrase polymorphism in Keyna. *Hum. Genet.* **52**, 259–261.

Kendall, A. G., and Tashian, R. E. (1977). Erythrocyte carbonic anhydrase I (CA I): Inherited deficiency in humans. *Science* **197**, 471–472.

Kondo, T., Sakai, M., Isobe, H., Taniguchi, N., Nishi, S., and Kawakami, Y. (1991). Induction of carbonic anhydrase I isozyme precedes the globin synthesis during erythropoiesis in K562 cells. *Am. J. Hematol.* **38**, 201–206.

Laurila, A. L., Parvinen, E.-K., Slot, J. W., and Vaananen, H. K. (1989). Consecutive expression of carbonic anhydrase isoenzymes during development of rat liver and skeletal muscle differentiation. *J. Histochem. Cytochem.* **37**, 1375–1382.

Lewis, S. E., Erickson, R. P., Barnett, L. B., Venta, P. J., and Tashian, R. E. (1988). *N*-Ethyl-*N*-nitrosourea-induced null mutation at the mouse *Car-2* locus: An animal model for human carbonic anhydrase II deficiency syndrome. *Proc. Natl. Acad. Sci. U.S.A.* **85**, 1962–1966.

Lloyd, J., Brownson, C., Tweedie, S., Charlton, J., and Edwards, Y. H. (1987). Human muscle carbonic anhydrase: Gene structure and DNA methylation patterns in fetal and adult tissues. *Genes Dev.* **1**, 594–602.

Lowe, N., Brady, H. J. M., Barlow, J. H., Sowden, J. C., Edwards, M., and Butterworth, P. H. W. (1990). Structure and methylation patterns of the gene encoding human carbonic anhydrase I. *Gene* **93**, 277–283.

Lowe, N., Edwards, Y. H., Edwards, M., and Butterworth, P. H. W. (1991). Physical mapping of the human carbonic anhydrase gene cluster on chromosome 8. *Genomics* **10**, 882–888.

Ma, X., Fraser, P., and Curtis, P. J. (1991). A differentiation stage-specific factor interacts with mouse carbonic anhydrase form I gene and a conserved sequence in mammalian β-globin genes. *Differentiation* **47**, 135–141.

Majeau, N., and Coleman, J. R. (1990). Isolation and characterization of a cDNA coding for pea chloroplastic carbonic anhydrase. *Plant Physiol.* **95**, 264–268.

Mallet, B., Gulian, J. M., Sciaky, M., Laurent, G., and Charrel, M. (1979). Formes moleculaires multiples de l'anhydrase carbonique erythrocytaire ovine. *Biochim. Biophys. Acta* **576**, 290–304.

McIntosh, J. E. A. (1970). Carbonic anhydrase isozymes in the erythrocytes and uterus of the rabbit. *Biochem. J.* **120**, 299–310.

Mohrenweiser, H., Neel, J. V., Mestriner, M. A., Salzano, F. M., Migliazza, E., Simoes, A. L., and Yoshihara, C. M. (1979). Electrophoretic variants in three Amerindian tribes: The Baniwa, Kanamari, and Central Pano of Western Brazil. *Am. J. Phys. Anthropol.* **50**, 237–246.

Mohrenweiser, H. W., Larsen, R. D., and Neel, J. V. (1989). Development of molecular approaches to estimating germinal mutation rates. I. Detection of insertion/deletion/rearrangements variants in the human genome. *Mutation Res.* **212**, 241–252.

Montgomery, J. C., Venta, P. J., Eddy, R. L., Fukushima, Y.-S., Shows, T. B., and Tashian, R. E. (1991). Characterization of the human gene for a newly discovered carbonic anhydrase, CA VII, and its localization to chromosome 16. *Genomics* **11,** 835–848.

Moore, M. J., Funakoshi, S., and Deutsch, H. F. (1971). Human carbonic anhydrases. VII. A new C-type isozyme in erythrocytes of American Negroes. *Biochem. Genet.* **5,** 497–504.

Myers, R. M., Tilly, K., and Maniatis, T. (1986). Fine structure analysis of a β-globin promoter. *Science* **232,** 613–618.

Nafi, B. M., Miles, R. J., Butler, L. O., Carter, N. D., Kelly, C., and Jeffery, S. (1990). Expression of carbonic anhydrase in neisseriae and other heterotrophic bacteria. *J. Med. Microbiol.* **32,** 1–7.

Nakai, H., Byers, M. G., Venta, P. J., Tashian, R. E., and Shows, T. B. (1987). The gene for human carbonic anhydrase II is located at chromosome 8q22. *Cytogenet. Cell Genet.* **44,** 234–235.

Nicewander, P. H. (1990). Sequence and organization of a *Macaca nemestrina* carbonic anhydrase I gene. Ph.D. Dissertation, University of Michigan.

Nicholls, R. A., and Board, J. A. (1967). Carbonic anhydrase concentration in endometrium after oral progestins. *Am. J. Obstet. Gynecol.* **99,** 829–832.

Notstrand, B., Vaara, I., and Kannan, K. K. (1974). Structural relationship of human erythrocyte carbonic anhydrases B and C. *In* "Isozymes: Molecular Structure" (C. L. Markert, ed.), Vol. 1, pp. 575–599. Academic Press, New York.

Okuyama, T., Sato, S., Zhu, X. L., Waheed, A., and Sly, W. S. (1992). Human carbonic anhydrase IV: cDNA cloning, sequence comparison, and expression in COS cell membranes. *Proc. Natl. Acad. Sci. U.S.A.* **89,** 1315–1319.

Omoto, K. (1979). Carbonic anhydrase-I polymorphism in a Philippine aboriginal population. *Am. J. Hum. Genet.* **31,** 747–750.

Osborne, W. R. A., and Tashian, R. E. (1984). Genetic variation in the carbonic anhydrase isozymes of macaque monkeys. IV. Degradation by heat and proteolysis of normal and variant carbonic anhydrase isozymes of *Macaca nemestrina*. *Arch. Biochem. Biophys.* **230,** 222–226.

Parmentier, M., Passage, E., Vassart, G., and Mattei, M.-G. (1991). The human calbindin D28 (CALB) and calretinin (CALB2) genes are located at 8q21.3–q22.1 and 16q22–q23, respectively, suggesting a common duplication with the carbonic anhydrase loci. *Cytogenet. Cell Genet.* **57,** 41–43.

Ridderstråle, Y., Wistrand, P. J., and Tashian, R. E. (1992). Membrane-associated carbonic anhydrase activity in the kidney of CA II-deficient mice. *J. Histochem. Cytochem.* (in press).

Roeske, C. A., and Ogren, W. L. (1990). Nucleotide sequence of pea cDNA encoding chloroplast carbonic anhydrase. *Nucleic Acids Res.* **18,** 3414.

Roth, D. E., Venta, P. J., Tashian, R. E., and Sly, W. S. (1992). Molecular basis of human carbonic anhydrase II deficiency. *Proc. Natl. Acad. Sci. U.S.A.* **89,** 1804–1808.

Shapiro, L. H., Venta, P. J., and Tashian, R. E. (1987). Molecular analysis of G + C-rich upstream sequences regulating transcription of the human carbonic anhydrase II gene. *Mol. Cell. Biol.* **7,** 4588–4593.

Shapiro, L. H., Venta, P. J., Yu, Y.-S. L., and Tashian, R. E. (1989). Carbonic anhydrase II is induced in HL-60 cells by 1, 25-Dihydroxyvitamin D_3: A model for osteoclast gene regulation. *FEBS Lett.* **249,** 307–310.

Shaw, C. R., Syner, P. N., and Tashian, R. E. (1962). New genetically determined molecular form of erythrocyte esterase in man. *Science* **138,** 31–32.

Silverman, D. N. (1991). The catalytic mechanism of carbonic anhydrase. *Can. J. Bot.* **69,** 1070–1078.

Sly, W. S. (1989). The carbonic anhydrase II deficiency syndrome: Osteopetrosis with renal tubular acidosis and cerebral calcification. *In* "The Metabolic Basis of Inherited Disease": 6th Ed. (C. R. Scriver, A. L. Beaudet, W. S. Sly, and D. Valle, eds.), Vol. 2, pp. 2857–2866. McGraw-Hill, New York.

Sly, W. S., Hewett-Emmett, D., Whyte, M. P., Yu, Y.-S. L., and Tashian, R. E. (1983). Carbonic anhydrase II deficiency identified as the primary defect in the autosomal recessive syndrome of osteopetrosis with renal tubular acidosis and cerebral calcification. *Proc. Natl. Acad. Sci. U.S.A.* **80,** 2752–2756.

Sly, W. S., Whyte, M. P., Krupin, T., Sundaram, V., Tashian, R. E., Hewett-Emmett, D., Guibaud, P., Vainsel, M., Balvarts, H. J., Gruskin, A., Al-Mosawi, M., Sakati, N., and Ohlsson, A. (1985). Carbonic anhydrase II deficiency in 12 families with the autosomal recessive syndrome of osteopetrosis with renal tubular acidosis and cerebral calcification. *New Eng. J. Med.* **313,** 139–145.

Spicer, S. S., Lewis, S. E., Tashian, R. E., and Schulte, B. A. (1989). Mice carrying a *Car-2* null allele lack carbonic anhydrase II immunohistochemically and show vascular calcification. *Am. J. Physiol.* **134,** 947–954.

Strisciuglio, P., Sartorio, R., Pecoraro, C., Litito, F., and Sly, W. S. (1990). Variable clinical presentation of carbonic anhydrase deficiency: Evidence for heterogeneity? *Eur. J. Pediatr.* **149,** 337–340.

Stanton, L. W., Ponte, P. A., Coleman, R. T., and Snyder, M. A. (1991). Expression of CA III in rodent models of obesity. *Mol. Endocrinol.* **5,** 860–866.

Sutherland, G. R., Baker, E., Fernandez, K. W., Callen, W., Aldred, P., Coghlan, J. P., Wright, R. D., and Fernley, R. T. (1989). The gene for human carbonic anhydrase VI (CA VI) is on the tip of the short arm of chromosome 1. *Cytogenet. Cell Genet.* **50,** 149–150.

Suzuki, E., Fukuzawa, H., and Miyachi, S. (1991). Identification of a genomic region that complements a temperature-sensitive, high CO_2 requiring mutant of the cyanobacterium, *Synechococcus* sp. PCC7942. *Mol. Gen. Genet.* **226,** 401–408.

Tashian, R. E. (1969). The esterases and carbonic anhydrases of human erythrocytes. *In* "Biochemical Methods in Red Cell Genetics" (J. J. Yunis, ed.), pp. 307–336. Academic Press, New York.

Tashian, R. E. (1977). Evolution and regulation of the carbonic anhydrases isozymes. *In* "Isozymes: Current Topics in Biological and Medical Research" (M. C. Rattazzi, J. G. Scandalios, and G. S. Whitt, eds.), Vol. 2, pp. 21–629. Plenum Press, New York.

Tashian, R. E. (1989). The carbonic anhydrases: Widening perspectives on their evolution, expression and function. *BioEssays* **10,** 186–192.

Tashian, R. E., and Carter, N. D. (1976). Biochemical genetics of carbonic anhydrase. *In* "Advances in Human Genetics" (H. Harris and K. Hirschhorn, eds.), pp. 1–56. Plenum Press, New York.

Tashian, R. E., and Hewett-Emmett, D., eds. (1981). Biology and chemistry of the carbonic anhydrases. *Ann. N.Y. Acad. Sci.* **429,** 1–640.

Tashian, R. E., Goodman, M., Headings, V. E., DeSimone, J., and Ward, R. E. (1971). Genetic variation and evolution in the red cell carbonic anhydrase isozymes of macaque monkeys. *Biochem. Genet.* **5,** 183–200.

Tashian, R. E., Kendall, A. G., and Carter, N. D. (1980). Inherited variants of human red cell carbonic anhydrases. *Hemoglobin* **4,** 635–651.

Tashian, R. E., Hewett-Emmett, D., and Goodman, M. (1983). On the evolution and genetics of carbonic anhydrases I, II, and III. *In* "Isozymes: Current Topics in Biological and Medical Research" (M. C. Rattazzi, J. G. Scandalios, and G. S. Whitt, eds.), Vol. 7, pp. 79–100. Alan R. Liss, Inc., New York.

Tashian, R. E., Venta, P. J., Nicewander, P. H., and Hewett-Emmett, D. (1990). Evolution, structure, and expression of the carbonic anhydrase multigene family. *Prog. Clin. Biol. Res.* **344**, 159–175.

Tashian, R. E., Hewett-Emmett, D., and Venta, P. J. (1991). Diversity and evolution in the carbonic anhydrase gene family. *In* "Carbonic Anhydrase: From Biochemistry and Genetics to Physiology and Clinical Medicine" (F. Botrè, G. Gros, and B. T. Storey, eds.), pp. 151–161. VCH Publishers, New York.

Thierfelder, W., Cummings, P., Fraser, P., and Curtis, P. J. (1991). Expression of carbonic anhydrases I and II in mouse erythrocytes. *In* "Carbonic Anhydrases: Cellular Physiology and Molecular Genetics" (S. J. Dodgson, R. E. Tashian, G. Gros, and N. D. Carter, eds.), pp. 209–214. Plenum Press, New York.

Threadgill, D. W., Fries, R., Faber, L. K., Vassart, G., Gunawardana, A., Stranzinger, G., and Womack, J. E. (1990). The thyroglobulin gene is syntenic with the *MYC* and *MOS* protooncogenes and carbonic anhydrase II and maps to chromosome 14 in cattle. *Cytogenet. Cell Genet.* **53**, 32–36.

Tweedie, S., and Edwards, Y. (1989). Mouse carbonic anhydrase III: Nucleotide sequence and expression studies. *Biochem. Genet.* **27**, 17–30.

Venta, P. J., and Tashian, R. E. (1990). PCR detection of the *Taq*I polymorphism at the *CA2* locus. *Nucleic Acids Res.* **18**, 5585.

Venta, P. J., and Tashian, R. E. (1991). PCR detection of a *Bst*NI RSP in exon 6 of the human carbonic anhydrase II locus, *CA2*. *Nucleic Acids Res.* **19**, 4795.

Venta, P. J., Montgomery, J. C., Hewett-Emmett, D., and Tashian, R. E. (1985). Comparison of the 5′ regions of human and mouse carbonic anhydrase II genes and identification of possible regulatory elements. *Biochim. Biophys. Acta* **826**, 195–201.

Venta, P. J., Montgomery, J. C., and Tashian, R. E. (1987). Molecular genetics of carbonic anhydrase isozymes. *In* "Isozymes: Current Topics in Biological and Medical Research" (M. C. Rattazzi, J. G. Scandalios, and G. S. Whitt, eds.), Vol. 14, pp. 59–72. Alan R. Liss, Inc., New York.

Venta, P. J., Welty, R. J., Johnson, R. M., Sly, W. S., and Tashian, R. E. (1991). Carbonic anhydrase II deficiency syndrome in a Belgian family is caused by a point mutation at an invariant histidine residue (107 His → Tyr): Complete structure of the normal human CA II gene. *Am. J. Hum. Genet.* **49**, 1082–1090.

Villeval, J. L., Testa, U., Vinci, G., Tonthat, H., Beettaieb, A., Titeux, M., Cramer, P., Edelman, L., Rochant, H., Breton-Gorius, J., and Vainchenker, W. (1985). Carbonic anhydrase I is an early specific marker of normal human erythroid differentiation. *Blood* **66**, 1162–1170.

Wagner, L. E., Venta, P. J., and Tashian, R. E. (1991). A human carbonic anhydrase I deficiency appears to be caused by a destabilizing amino acid substitution (246-Arg → His). *Isozyme Bull.* **24**, 35.

Weiher, H., and Botchan, M. R. (1984). An enhancer sequence from bovine papilloma virus DNA consists of two essential regions. *Nucleic Acids Res.* **12**, 2901–2916.

Weiher, H., Konig, M., and Gruss, P. (1983). Multiple point mutations affecting the simian virus 40 enchancer. *Science* **219**, 626–631.

Weil, S. C., Walloch, J., Frankel, S. R., and Hirata, R. K. (1984). Expression of carbonic anhydrase and globin during erythropoiesis *in vitro*. *Ann. N.Y. Acad. Sci.* **429**, 335–337.

Welch, S. (1975). Population and family studies on carbonic anhydrase II polymorphism in Gambia, West Africa. *Humangenetik* **27**, 163–166.

Yoshihara, C. M., Lee, J.-D., and Dodgson, J. B. (1987). The chicken carbonic anhydrase II gene: evidence for a recent shift in intron position. *Nucleic Acids Res.* **15**, 753–770.

INDEX

A

a1, Mu elements of *Zea mays* and, 109, 113

A1, Mu elements of *Zea mays* and, 92

a1-mum2, Mu elements of *Zea mays* and, 98

a1-mum3, Mu elements of *Zea mays* and, 81

Abscisic acid, nitrate reductase and, 20

Ac, Mu elements of *Zea mays* and, 81, 111, 115

  properties of transposition, 105, 108

  structure, 91, 96

ACE1, superoxide dismutases and, 304

ace1, superoxide dismutases and, 277

ACE1, superoxide dismutases and, 276–277

Acetate, superoxide dismutases and, 280, 285, 290, 296

Acetylcholinesterase, mosquitoes and, 149–150

Actin

  mosquitoes and, 144

  Mu elements of *Zea mays* and, 112

Activator, somaclonal variation and, 52–54, 56

Adh1

  Mu elements of *Zea mays* and, 79–80, 92, 95, 109–112

  somaclonal variation and, 58

ADR1, superoxide dismutases and, 280, 285

Aedes, see Mosquitoes

Aerobiosis

  nitrate reductase and, 3

  superoxide dismutases and, 270, 275–276

    physiology of mutants, 289, 293–295, 305

Agrobacterium, nitrate reductase and, 28

Alcohol dehydrogenase, *see Adh1*

Alder, nitrate reductase and, 5

Alfalfa, somaclonal variation and, 53, 55

Algae, carbonic anhydrases and, 323, 343

Alleles

  carbonic anhydrases and, 337–338, 343, 346–348

  mosquitoes and, 149, 157, 160

  Mu elements of *Zea mays* and, 78–79

    applications, 114–115

    assays for monitoring, 81, 83

    characteristics, 87–90

    gene expression, 108–111

    properties of transposition, 100

    regulation, 98

  nitrate reductase and, 17–18, 21, 23, 27–28

  somaclonal variation and, 47, 53, 55, 64–65

  superoxide dismutases and, 262, 297

    molecular genetics, 282–283

    physiology of mutants, 299, 301–302, 305–306

Alu

  carbonic anhydrases and, 334

  telomeres and, 195

Amino acids

  carbonic anhydrases and, 322–323, 332

    genetic variation, 337, 343, 345–346, 348

  mosquitoes and, 150, 153–154

  nitrate reductase and, 2, 7

  somaclonal variation and, 59

  superoxide dismutases and, 264, 267, 272, 289, 305

  telomeres and, 216

Ammonia, nitrate reductase and, 3, 26

Ammonium, nitrate reductase and, 2, 11, 15, 19, 28

Ammonium sulfate, nitrate reductase and, 14

357

ISBN 0-12-017630-0

9 780120 176304

90040

268503